Fluid Transients in Pipeline Systems

To Lin, Hal, Star and Bo

Fluid Transients in Pipeline Systems

A Guide to the Control and Suppression of Fluid Transients in Liquids in Closed Conduits

Second Edition

A R David Thorley
Professor Emeritus of Fluid Engineering,
City University,
London, UK

Professional Engineering Publishing

Professional Engineering Publishing Limited,
London and Bury St Edmunds, UK

First published 1991 by D & L George Ltd.

This edition published 2004 by Professional Engineering Publishing Limited, UK.

ISBN 1 86058 405 5

A CIP catalogue record for this book is available from the British Library.

The various charts, graphs, tables and techniques in this book are presented in good faith, and it is expected that system designers will use them prior to, and not instead of, a more rigorous investigation, supported by a computer-based analysis. Neither the author nor the publisher can accept any liability for the use to which the information provided in this book is put, since it is outside their control.

The publishers are not responsible for any statement made in this publication. Data, discussion, and conclusions developed by the author are for information only and are not intended for use without independent substantiating investigation on the part of the potential users. Opinions expressed are those of the author and are not necessarily those of the Institution of Mechanical Engineers or its publishers.

MIX Paper from responsible sources FSC www.fsc.org FSC® C013604

Typeset by Data Standards Ltd, Frome, Somerset, UK.

Related Titles

Title	*Author*	*ISBN*
An Introductory Guide to Flow Measurement – Second Edition	R. C. Baker	1 86058 348 2
An Introductory Guide to Industrial Flow	R. C. Baker	0 85298 983 0
An Introductory Guide to Pumps and Pumping Systems	R. K. Turton	0 85298 882 6
Process Pump Selection – A Systems Approach	J. Davidson and O. von Bertele	0 85298 180 3
Guide to European Pumps and Pumping – Second Edition	B. Nesbitt	1 86058 286 9
Handbook for Process Plant Project Engineers	P. Watermeyer	1 86058 370 9
An Engineer's Guide to Pipe Joints	G. Thompson	1 86058 081 5
Pressure Equipment Technology: Theory and Practice	W. M. Banks and D. H. Nash	1 86058 401 2
Ageing Pipelines – Optimizing the Management and Operation	IMechE Conference Transactions	1 86058 232 X

For the full range of titles published by Professional Engineering Publishing (publishers to the Institution of Mechanical Engineers) contact:

Marketing Department
Professional Engineering Publishing Limited
Northgate Avenue
Bury St Edmunds
Suffolk IP32 6BW
UK
Tel: +44 (0) 1284 763277; Fax: +44 (0) 1284 704006
E-mail: marketing@pepublishing.com
www.pepublishing.com

About the Author

Professor A. R. David Thorley BSc, MPhil, PhD, CEng, FIMechE, FICE, MemASME.
Professor Emeritus of Fluid Engineering – City University, London, UK.

Professor Thorley's engineering career began as an apprentice in the automobile industry, followed by a period as a graduate engineer in the electricity supply industry. Since joining the academic sector, he has maintained strong links with industry through extensive consultancy, research and development. This latter activity has been concerned with a variety of projects in the field of fluid engineering, predominantly concerned with the design of safe and reliable piping systems for the water, oil, nuclear and petrochemical industries, for which he has acquired an international reputation.

The principal themes in this work have been the need to apply engineering common sense and judgement to challenge underlying assumptions in, and designs of, real piping systems, since most such systems are unique and yet a single mistake can have catastrophic consequences.

Arising from his academic and R&D activities, Professor Thorley has published some 62 journal and conference papers, 50 industrial reports and briefing papers to project sponsors. He has also written, or been joint author of, books on fluid transients and the design of anchor blocks.

In 1983 Professor Thorley founded the Thermo-Fluids Engineering Research Centre at City University and assembled around him a number of colleagues active in applying their skills to solving and understanding problems in thermo-fluids engineering, which would directly benefit UK industry. Much of their work related to the energy sector and attracted funding from industry at home and abroad, as well as from Research Councils.

From 1993 to 1999 Professor Thorley was Head of the Department of Mechanical Engineering and Aeronautics, then Dean of the School of Engineering from 1999 to 2001, embracing activities in aeronautical, civil, electrical and electronic, and mechanical engineering.

Among his professional activities, Professor Thorley has contributed to the organization of several international conferences on unsteady and transient flows in pipeline systems and been Chairman of three of them. He has been an active member of ESDU working parties for Design Items, and Scientific Advisor and representative of the EEC Commissioners for a transient flow project at the Delft Hydraulics Laboratory in the Netherlands.

He has travelled and lectured overseas professionally on numerous occasions, visiting India, Indonesia, North and South Africa, and North and South America, often by invitation and sometimes at the instigation of the British Council. He is registered with the World Bank as a consultant and has been an expert witness in the High Court in London.

Professor Thorley is married with three grown-up children. They share many activities as a family, but specific interests include sailing (the Caribbean, Europe, the Mediterranean, and the Far East), and travel. He is a member of the Little Ship Club and Cruising Association, and a Governor of the RNLI. Since retiring from full-time employment he has extended his sailing including a trans-Atlantic passage to the Caribbean where he is currently cruising among the islands.

Contents

Part 2

Acknowledgements

I am greatly indebted to many people and organizations for helping make this book possible. It is virtually impossible to mention them all by name, however special thanks are due to the following: Hans Horlacher and the late Harald Graze kindly gave permission for the use of the design charts for air vessels in Subsection 3.3.1, and Jean Marie Matringe of Charlatte provided data on bladder tank and hybrid vessels. Adnan Chaudry, a research assistant, collated data for the new four-quadrant pump characteristics in Section 3.4, and Erik Faithfull, who was sponsored for an MSc programme by Scott Wilson Kirkpatrick, undertook much of the work on pump and motor inertias. Data were supplied by Dresser UK Ltd, Flygt Pumps Ltd, SPP Ltd, Weir Pumps Ltd, Worthington SpA in Italy and Dr Adrian Boldy of HydroSim Consultants. Brook Crompton Parkinson kindly provided data on motor inertias to add to previously published data. The valve charts, in Section 3.5, were provided by Don Wood of the University of Kentucky. Data and illustrations on the various valves for vacuum relief, air release, and ventilation were provided by the Apco Willamette Valve and Primer Corporation, Illinois, by Sam Gilbert of Glenfield Valves, Kilmarnock, by Joseph Lescovich of G.A. Industries in Mars, Pennsylvania, and by Allistair Balutto of DFC (Pty) Ltd for the Vent-O-Mat valves. Sources for performance data on valves include: Don S. Miller, from the second edition of his book *Internal Flow Systems*; Glenfield Valves, Kilmarnock; Goodwin International plc in Stoke-on-Trent; NoReVA Valves of Mönchengladbach; and Alsthom-Atlantique/Neyrtec in France. The BHR Group, Cranfield, Bedfordshire, have kindly given permission for me to utilize material from the Proceedings of the many International Conferences on Pressure Surges that they have sponsored since 1972. The graphical output from computer analyses of the various examples used was generated with the aid of SURGE software developed by Don Wood and Jim Funk, of the University of Kentucky, for the analysis of transient flows in complex pipe networks.

On a more general level, and over a period of many years, I have benefited from the experience and wisdom of many members of the international transient family. In particular, I would like to mention Adrian Boldy, Hanif Chaudhry, Keith Enever, John Fox, the late Harald Graze, C. Sam Martin, Steve Murray, Hemmat Safwat, Vic Streeter, John Swaffield, Alan Vardy, David Wiggert, Don Wood, E. Ben Wylie and Werner Zielke, not forgetting 'Jim' – James W.R. Twyman – who, for me, started it all off in the early 1960s, and the many students who have studied on my courses.

The other essential ingredient to produce this book has been time and patience on the part of colleagues and family. In this respect I would like to thank colleagues in the Thermo-Fluids Engineering Research Centre at City University. Thanks also go to Sheril Leich, Lou Attwood and their colleagues at Professional Engineering Publishing.

A special mention must go to my daughter Leila who did a thorough and conscientious job of checking proofs while I was galivanting around the Caribbean! Last, but not least, I wish to thank my wife Lin, and Hal, Star and Bo, for the long days, evenings and weekends that I have been closeted with my keyboard to produce the text that follows.

Preface to the First Edition

There are already some excellent modern books on fluid transients and pressure surge – so why another? The main reason is that this one is aimed at a different readership. The existing books, dealing in some detail with the mathematical and numerical treatment of the subject, appeal mainly to graduate students and researchers seeking to gain further insights into very complex time-varying fluid flows, and to engineers developing computer codes.

There are also a few undergraduate texts that introduce the subject, usually in relation to a rather idealized event such as a rapid valve closure in a simple pipe, but which rarely take the topic much further.

This text is intended to fit into the gap between the two. The principal readership will be engineers of graduate level involved with the planning, design and operation of pipelines for transporting liquids, and who need an insight into fluid transient phenomena. They will be seeking guidance on how transient flows arise, what the consequences may be, what the critical fault conditions are likely to be, how estimates of their seriousness can be made, what practical steps can be taken to alleviate the undesirable consequences, which events need to be analysed by computer models and what data will be needed to do so?

The order in which material is presented has been devised to try to meet the needs of readers having very diverse levels of prior knowledge. The overall aim has been to take readers from a state of (assumed) complete ignorance through to the point where they could, if required, present a well-defined problem to a specialist in computer-based analyses, complete with all necessary data. To this end, Part 1 covers the basic ideas of how fluid transients arise and concentrates on the fundamental physical concepts involved. Practical methods for the control and suppression of transients are reviewed, and only such theory as is deemed necessary is introduced at this stage.

Engineers with some experience of dealing with fluid transient problems will probably omit Part 1 and proceed directly to Part 2 which focuses on applications. After a short discussion on 'risk', an outline procedure that can be followed to help determine whether or not a system may be exposed to unacceptable conditions following a transient event is developed.

Eight demonstration examples are then discussed to illustrate these ideas, leading up to a short outline of computer modelling. This is not from the standpoint of writing computer programs, but is intended more for engineers wishing to use existing codes, either through a consultant or having purchased them. Part 2 is then concluded with a section on accidents and incidents. This helps to reinforce the point that transient events can occur at any time and, sometimes, from unexpected causes.

Part 3 is where numerous charts, tables and other data that are useful for transient assessments will be found, and it is hoped that this will prove useful as a source of reference and information, even to the regular analyst. The author's own experience is that, when faced with the question of the extent to which a system is at risk and what protective strategies need be devised, the complete picture is never to hand. It is always necessary to make some assumptions based on engineering judgement. Part 3 is intended to make this a little easier.

For readers who become captivated by the subject and wish to take their theoretical studies further, more advanced reading is suggested in a short Bibliography. The reference section contains sources directly relevant to the contents of this text. There are many more in the literature, and long lists are given in the books cited in the Bibliography.

A. R. D. Thorley
Hadley Wood
1991

Preface to the Second Edition

The first edition of this book was published in 1991. The customer base has primarily been engineers working on, or responsible for, the safe design of piping systems for a wide range of applications. These include water resources (including supply, distribution and effluent collection and disposal), nuclear and conventional power plant design, airport and aircraft fuel systems, the oil, petrochemical and offshore industries, fire protection systems, mine drainage systems, etc. Also, many copies have been bought by graduate students on MSc and Continuing Professional Development courses in the United States, South Africa and the United Kingdom.

Feedback from reviews, directly from users of the first edition and from authors of books adopting a more academically rigorous or 'scientific' approach, has been very positive. Encouragement to produce an update and revision has been received from several quarters. The overall structure of the book remains the same as before but a few sections have been repositioned, and some enhancements and additions have been made. These are summarized briefly below.

1. Section 3.1 has been revised, partially reordered and slightly extended to include an additional chart on wave speeds in plastically deforming tubes. Some additional equations for dealing with lined thin- and thick-walled pipes have been added.
2. The air vessel design charts in Section 3.3 have been extended with five new charts to improve the ease of use of those currently in the first edition. These will considerably improve the ability of users to size air vessels more accurately at the smaller end of the range.
3. Section 3.4 has been renamed Pump Data and expanded considerably by the addition of a significant amount of data on four-quadrant head and torque characteristics for 14 pumps – most of it being new.
4. Section 3.6, Air Relief and Vacuum-Breaking Valves, has had a major rewrite.
5. Other sections of Part 3 have benefited from relatively minor changes and updates of a more general nature.

The consequence of changes and improvements in Part 3 is that the case studies and worked examples in Part 2 have been revisited and updated, and four new ones have been added. Similarly, the section on accidents and incidents has been reviewed and a further incident presented. The section dealing with computer modelling has been moved from Part 3 into Part 2 and strengthened by including an introduction to the method of characteristics.

These should be seen as additional improvements for design engineers who may rely on this section to guide them through risk assessments for systems they are designing. This will apply particularly to engineers who undertake this activity perhaps only once or twice a year, or even less frequently, and who therefore welcome the availability of a 'guidebook' to assist them.

Finally, the Bibliography has been recast in a new Section 3.9 entitled 'Other Sources of Information'. This also provides guidance on current (i.e. in 2003) sources of information on the World Wide Web. It includes a list of some of the sites available and guidance for new users on how to explore the web for topics of interest to them. The list of references has also been updated.

A. R. D. Thorley
Hadley Wood
2004

PART 1

This section of the book is largely descriptive. It is intended for the reader, new to the topic, who is seeking an introduction to the subject of fluid transients, or water hammer, in pipelines and tunnels.

It begins with a discussion of unacceptable conditions that can occur following changes to an initially steady flow, how they can arise and what some of the consequences might be. Using a simple pipeline example, the concept of pressure waves being transmitted to and fro in a pipeline system is developed to emphasize that, when unsteady flow occurs, the pressure head and flow will vary with both position in the system and the time after a change has been initiated. Also introduced and explained here are some of the terms familiar to the transient analyst.

Having established that transient flows can produce undesirable conditions within pipelines and tunnels, the second half of Part 1 reviews practical methods for overcoming the problems. First, the physical reasoning behind the various strategies for control and suppression are introduced, then various methods are described and, finally, Part 1 concludes with a discussion on how the task of formulating a control and suppression strategy for a pipeline system might be approached.

After digesting Part 1, the reader should have acquired a qualitative understanding of how fluid transients arise and the problems they can cause, and the ability to propose suitable methods for dealing with them. In the rest of the book the reader will find guidance on how to develop these ideas further, including how to estimate the magnitude of possible problems, identify the critical design cases and size some components and devices that may be used for suppression purposes. Also included is a review of some accidents and incidents, with comments on lessons that may be learned from them.

SECTION 1.1

Introduction

The design of pipeline systems passes through several evolutionary stages – initial conception, feasibility studies, functional design, optimization and risk assessment. In the functional design stage the principal features are based on the primary function of the intended system – be it for water supply, sewage disposal, transportation of petrochemicals, etc. The optimization and risk assessment stages are concerned with the economics and hazards respectively of the construction and operation of the system. Both of these stages feed into the functional design prior to it being finalized.

Even during the functional design phase the prudent engineer will begin to speculate on the extent to which circumstances may arise that could lead to unacceptable operating conditions developing in the system. These unacceptable conditions, and specifically 'fluid transients', are a major part of the risk assessment. They should be one of the final, if not *the* final, stage in the design process before construction commences since any changes to the physical design of the system may invalidate the fluid transient analysis. It is the assessment and resolution of the risks posed by fluid transients that are the subject of this book.

1.1.1 Unacceptable Conditions

Even during normal operation, as the flows in pipelines vary with changing demands on the system, the pressures will also change. These are events over which the designer and plant operator have some control. Other events, such as power failures or self-excited resonance, will be unplanned and, possibly, unexpected, but the designer should still assess the risks of them occurring and make due allowance for any unacceptable conditions that may arise. Examples of these include the following:

(a) Pressures too high – leading to permanent deformation or rupture of the pipeline and components; damage to joints, seals and anchor blocks; leakage out of the pipeline, causing wastage, environmental contamination and fire hazard.

(b) Pressures too low – may cause collapse of the pipeline; leakage into the line at joints and seals under subatmospheric conditions; contamination of the fluid being pumped; fire hazard with some fluids if air is sucked in.

(c) Reverse flow – causing damage to pump seals and brush gear on motors; draining of storage tanks and reservoirs.

(d) Pipeline movement and vibration; overstressing and failure of supports, leading to failure of the pipe; mechanical damage to adjacent equipment and structures.

(e) Flow velocity too low – mainly a problem in slurry lines, causing settlement of entrained solids and line blockage.

Unacceptable conditions, or failures, such as these are hazards that can be created by unsteady or rapidly changing flows within a piping system. So what may cause them?

1.1.2 Causes of Unsteady and Transient Flows

Unsteady or transient flows may be initiated by the system operator, be imposed by an external event, be caused by a badly selected component or develop insidiously as a result of poor maintenance.

Probably the most common hazard situation is the uncontrolled pump trip, often resulting from a power failure. In this case, the rapidly falling pressure may go subatmospheric and even drop to vapour pressure, allowing a large vapour cavity to form. The very rapid, and usually large, pressure rise following the collapse of this cavity (assuming the pipe itself has not collapsed) is a serious problem.

Also associated with pump trips, and pump changeover operations, is the phenomenon of 'check valve slam'. This phenomenon is most likely to occur when one pump out of two or more running in parallel is tripped, especially on high-head installations, or where a system is protected by an air vessel. In the latter case the air vessel provides a high-energy source similar to that of other pumps continuing to run in parallel.

Analogous to the pump trip is the too rapid closure of pump delivery valves. This is less of a problem on pipelines of just a few hundred metres in length, but needs to be taken seriously on longer lines. This is an example of an event over which the system designer may have some influence by being able to specify a minimum time for the valve closure.

Valves and similar flow control devices elsewhere in the system can initiate unwelcome fluctuations in pressure and flow. Common examples include loading valves on oil tankers, emergency shutdown valves on boiler plant, guide vanes on hydroelectric plant and float-controlled valves on tanks receiving fluids.

Perhaps the most serious pump start problem is associated with borehole and submerged deep-well pumps having check valves mounted at ground level. Upon start-up, the pump may rapidly reach full speed against negligible resistance and tend to operate at the high-flow end of its characteristic. Water rushing up the riser suddenly encounters the closed check valve, and the rapid deceleration that the flow experiences can generate a very large pressure rise.

Pipeline supports are a matter of compromise. They need to allow sufficient freedom of movement to cater for the thermal effects of expansion and contraction, but incorporate adequate restraint to limit movement due to the passage of transient pressure waves which give rise to unbalanced forces on bends and junctions.

The potential for resonance to occur should also be considered. Resonance effects are due to the cyclic behaviour of system components – positive displacement pumps, and hunting of automatic control valves or between badly matched centrifugal pumps running in parallel, for example. Some system failures due to resonance can be extremely rapid and dramatic; others can be long delayed as cumulative fatigue builds up over weeks or months.

To provide a basis for good judgement in formulating suitable strategies for avoiding the potential calamities prophesied above, it will be helpful to look at what actually happens when a flow is initiated or interrupted.

SECTION 1.2

Unsteady flows in pipes and tunnels

1.2.1 Basic Ideas

For most purposes it is perfectly adequate to regard liquid flows in pipes and tunnels as incompressible and one-dimensional. However, when dealing with unsteady flows the assumption that liquids are incompressible becomes less valid and in the case of rapid transients is extremely misleading.

The most important consequence to note is that changes in the flow do not, in fact, occur everywhere in the system at the same instant in time, but after some short interval. The duration of this time interval depends upon two things: the distance between the point in the system where the change in flow is initiated and the point of interest, and a parameter known as the speed of propagation of transient pressure waves.

The latter is governed by the physical properties of both the fluid and of the pipe or tunnel in which it is contained. In the majority of cases of interest it will be of the same order of magnitude as the speed of sound in the fluid. For example, the speed of sound in water is approximately 1440 m/s, and the propagation speed of transient pressure waves through water contained within a steel pipe will normally be in the range 1000–1400 m/s.

1.2.2 A Simple Example

Take the case of a pump trip in a pipeline several kilometres long. Assume, for the moment, that the pump stops almost instantaneously and that it provides no further discharge. The liquid downstream still has forward momentum which will tend to keep it in motion. However, as it moves away from the pump discharge, the pressure at that point will fall since no more liquid can be supplied to take its place and the liquid next to the pump will be brought to rest.

Picture the liquid elements in the pipe as discs packed in a tube. These liquid discs are very slightly elastic and are attached to each other. The one next to the pump stops because it cannot leave the pump, even though the one on the downstream side is tending to pull it away. The first hangs onto the second one and brings it to a stop, the second one brings the third to a stop, and so on one after the other. This all takes a very small, though finite, amount of time. As the forward motion is reduced, so the pressure changes – in this case it falls since the liquid was initially moving away from the obstruction to the flow.

This moving interface, between the fluid that has been affected by the pump trip and the fluid that has not and across which there is a change in pressure, is known as a transient pressure wave. As the word 'transient' implies, it is this that travels at a speed close to that of sound in the fluid.

These effects are illustrated by reference to a simple rising main pipeline illustrated in Fig. 1.1. It comprises a pump delivering water against a static head of some 128 m to a downstream reservoir through a 300 mm diameter ductile iron pipeline 15 km long. The flowrate is 35 L/s and the frictional head is 11.6 m. A check valve having a fast response is fitted at the pump discharge.

Figure 1.2 shows what is usually known as a 'pressure–time history' at the pump discharge. Here, it commences 0.5 s before the pump is tripped, which is one strategy to check the program is functioning correctly. It shows the pressure falling quite rapidly immediately after the pump is tripped, appearing to reach a minimum, and then continuing to drop further, though much more slowly. Two effects are evident here, indicated by the initial relatively rapid change in pressure followed by the rather slower fall. The initial drop in pressure, which occurs

Fig. 1.1 *A simple rising main pipeline*

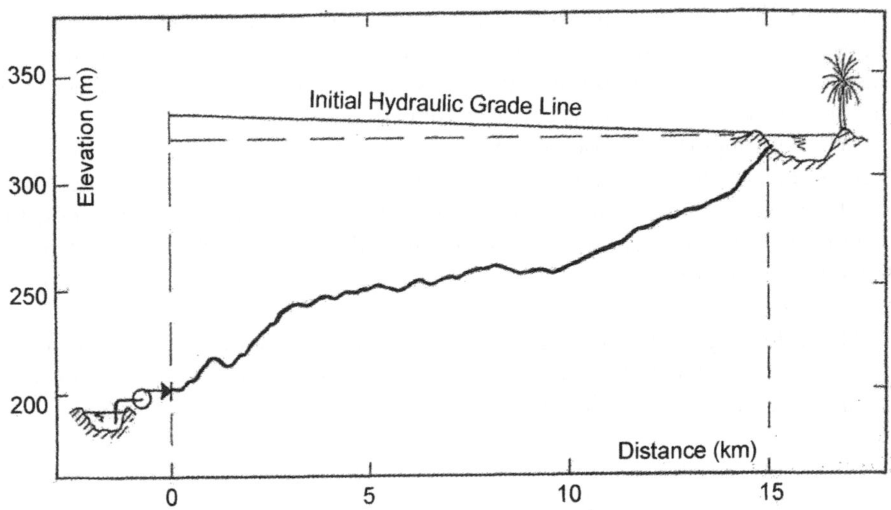

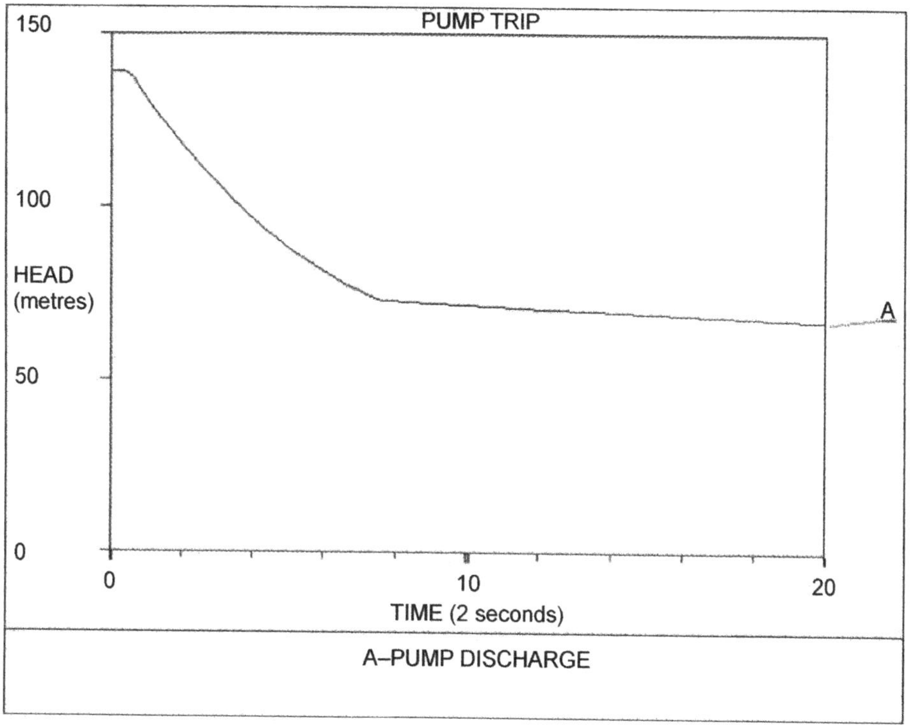

Fig. 1.2 *A pressure–time history at the delivery valve for the first few seconds after the pump is tripped*

simultaneously with a reduction to zero in the flow at the pump discharge, is the 'Joukowsky' head change, ΔH. This can be calculated with the *Joukowsky equation*

$$\Delta H = \frac{c}{g}\Delta v \qquad (1.1)$$

where

Δv = change in flow velocity
c = wave propagation speed
g = acceleration due to gravity

With wave propagation speeds likely to be in the region of 1000 m/s, it will be observed that, for each 1 m/s change in flow velocity, the corresponding change in the pressure head will be of the order of 100–120 m, or 10–12 bar.

The second effect, shown by the continuing slow fall in pressure, is due to the frictional head falling away. In many cases it is not unduly significant, but it should not be overlooked because in long-distance oil pipelines, for example, it can lead to total changes in pressure head that significantly exceed the Joukowsky head change.

Figure 1.3 is similar to Fig. 1.2, but a record of the pressure head at a point 8 km downstream of the pump has been added. Note how initially there is no change in the pressure at the latter point, even though it is changing at the pump. In this example, which will be used again later, the propagation speed for transients is 1250 m/s. It therefore takes 8000/1250 = 6.4 s for a change at the pump to appear at this downstream location.

Fig. 1.3 *Pressure–time histories at the pump delivery and at a distance of 8 km (just over halfway) down the line. Note the time delay between the pressure changes*

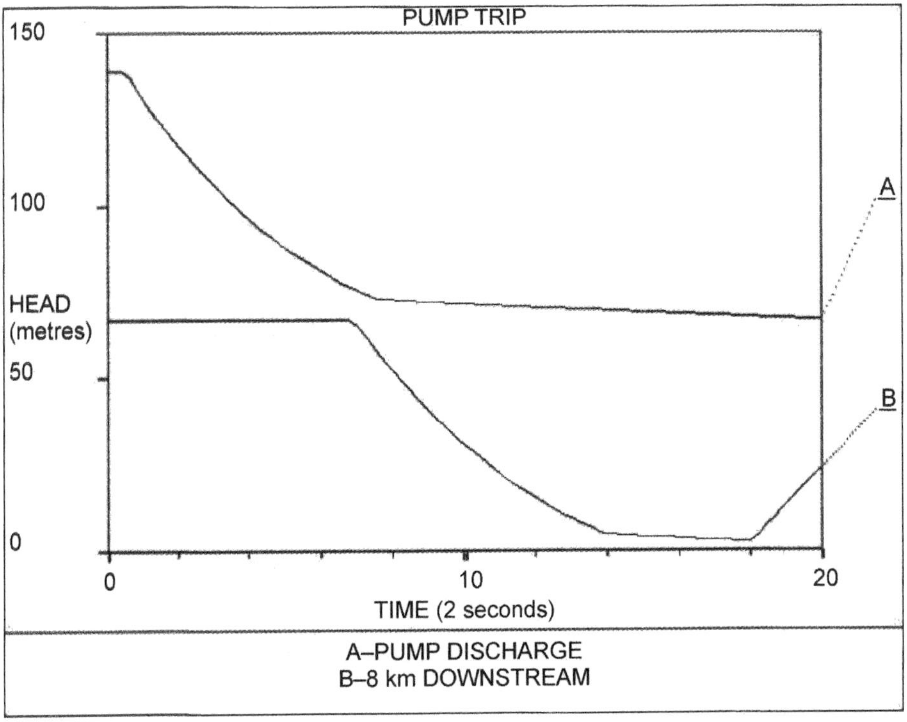

On the diagram, allowing for the initial period of 0.5 s of steady flow, the pressure at 8 km starts to fall at a time of 6.9 s. The rate of change is identical to that at the pump – it just occurs at a different time. The effect of the initial frictional gradient is again apparent.

1.2.3 Pressure Wave Reflections and Pipeline Period

The 'pressure wave' continues down the pipeline until it arrives at the reservoir at the end. If it were a step fronted wave, and there were no frictional effects, all the fluid in the pipeline would, at the precise instant it reaches the reservoir, be at rest, but the pressure in the pipe would be well below the local no-flow static head. This cannot continue, and so water will flow out of the reservoir and back into the pipe, raising the pressure and tending to bring the system back into a stable equilibrium condition. This 'reflected' pressure wave travels down towards the pump but reaches the 8 km point first, and hence its effect is noticed there first – see the increase in pressure commencing at about 18 s.

It continues on to the pump where it meets the, by now closed, check valve. A new reflected wave is created, this time travelling towards the downstream reservoir again. Each time the pressure wave reaches either the upstream or downstream end of the pipe it is reflected. The longer time-scales of Figs 1.4 and 1.5 serve to illustrate this. They also reinforce the point that, under unsteady and transient flow conditions, the pressures and flows in a piping system are dependent upon both time and position – the time after a transient event occurs, and the position of the point of interest in relation to where the transient started.

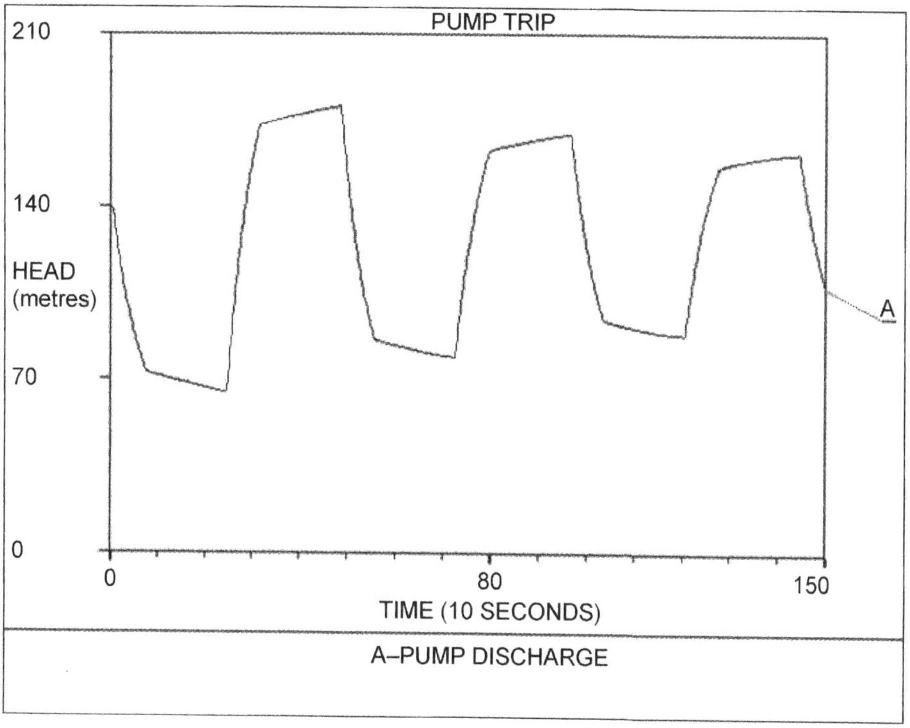

Fig. 1.4 *Extended pressure–time history at the closed check valve on the pump delivery*

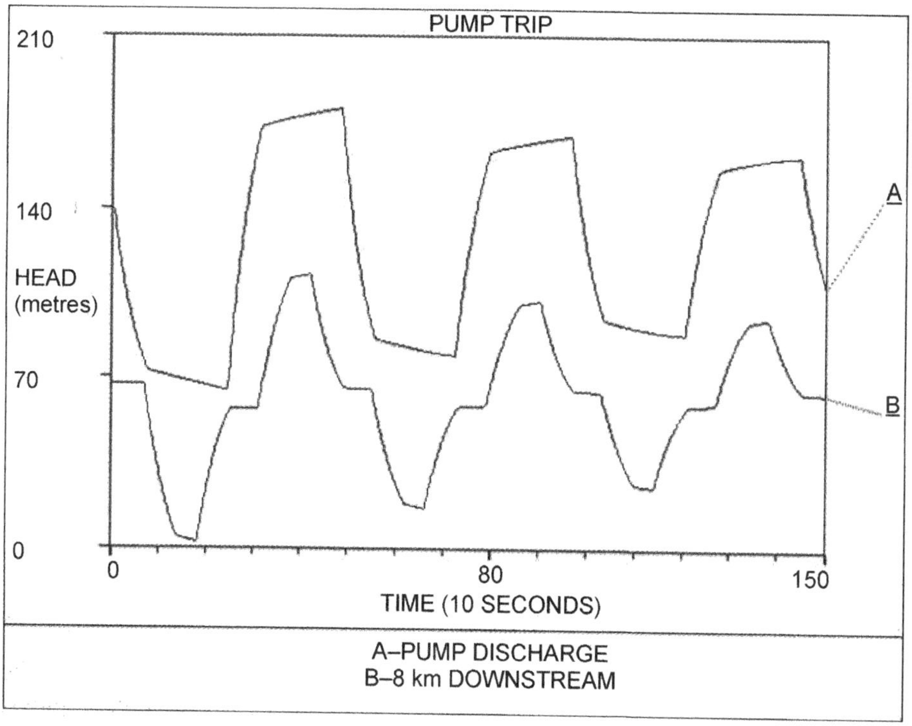

Fig. 1.5 *More extended pressure–time histories for the pump delivery and further down the line as in Fig. 1.3*

The time taken for a pressure wave to undertake a round trip of the pipeline is known as the *Pipeline Period*, or Periodic Time T. It is calculated by doubling the length of the pipe and dividing by the wave speed, i.e. in this case $T = 2L/c = 2 \times 15\,000/1250 = 24$ s.

1.2.4 A 'Rapid' Event

When introducing the illustrative example above it was said that the pump stopped almost instantaneously. The pressure–time histories show the pressure falling rapidly, as the flow stops, over a period of about 7 s. It remains at that low value, changing slightly owing to the frictional effect, until at a time of 24.5 s, i.e. one Pipeline Period after the pump trip commenced, the pressure suddenly starts to increase. This is caused by the pressure wave that was reflected at the downstream reservoir arriving back at the pump.

If the pipeline had been longer, the reflected wave would have taken longer to return to the pump. Conversely, if the pipeline had been shorter, the round trip travel time, i.e. the Pipeline Period, would have been shorter. If it had been very much shorter, say 4 km in this case, the returning pressure wave would have reached the pump before the flow had stopped.

The significance of this is that the event causing the transient would not have been completed before a reflected pressure wave arrived at the scene to modify the pressure changes being imposed on the system. In particular, the full Joukowsky pressure change would not be developed. Events, such as pump trips, valve closures, etc., that are completed within a Pipeline Period are known as *Rapid Events*.

1.2.5 Effects of Friction

In many systems where fluid transients occur, the principal component of the pressure fluctuations is the Joukowsky head. Nevertheless, friction does modify the process, and on very long pipelines it can add substantially to the magnitude of the pressure fluctuations. This is especially true for oil and slurry lines. Nevertheless, it is friction that causes the oscillations in pressure to be damped out eventually.

1.2.6 Max–Min Head Envelopes

Usually, it is the maximum and minimum pressures in a system that are of particular interest to designers. The most useful way to represent these is to plot the Max–Min Head Envelope directly onto the longitudinal section of the pipeline. Figure 1.6 is an example for the case discussed above. This provides a quick and easy way to identify critical design points in the system and is useful when reviewing potential surge control strategies.

1.2.7 Column Separation and Vapour Cavity Formation

In the simple example used above the pressures throughout the line, even though they rose and fell, were always such that the liquid column remained in that state. Suppose that in an alternative design the pipeline of Fig. 1.1 had followed the same route for the first 5 km and then taken a different, higher, route for the remaining 10 km (see Fig. 1.7).

Following a pump trip, the pressure head at the pump would fall and a pressure reduction wave would be transmitted along the pipe in exactly the same way as

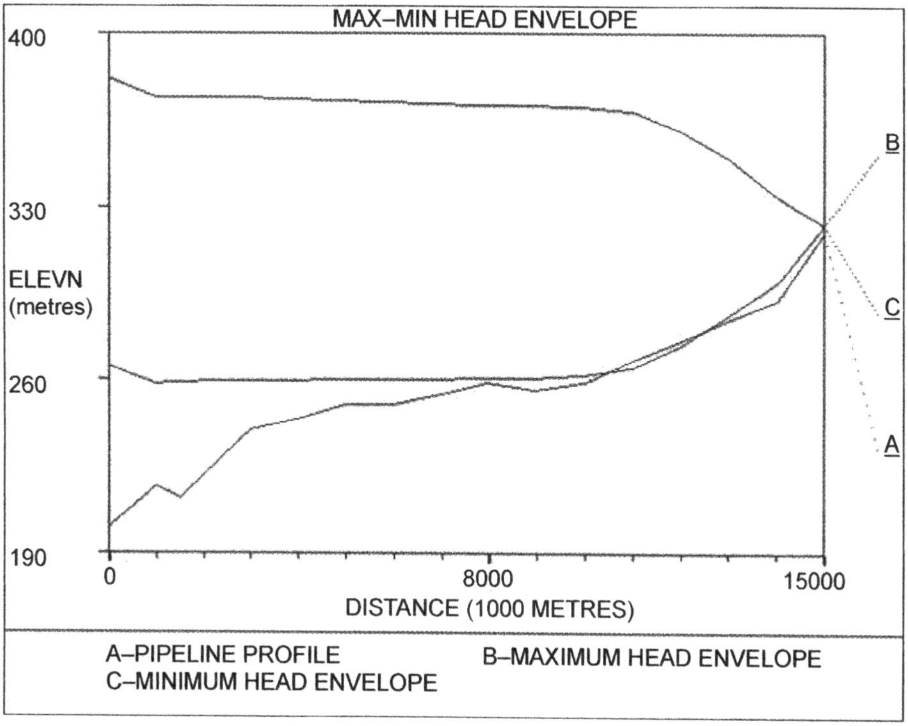

Fig. 1.6 *Profiles of the maximum and minimum head envelopes provide an effective summary of transient pressures*

before. However, approximately halfway along this new route, the pressure would not only fall below atmospheric, it would reach vapour pressure. The liquid column would part as the water, in effect, boiled, and the downstream section of the water column would continue to move with a slower deceleration. Eventually it would come to rest, reverse its direction of movement under the action of gravity and the static head at the downstream reservoir and recombine with the upstream part. At this instant the pressure at the point of recombination would increase suddenly, creating new pressure waves propagating in both the upstream and downstream directions. Compare Fig. 1.8 with Fig. 1.5.

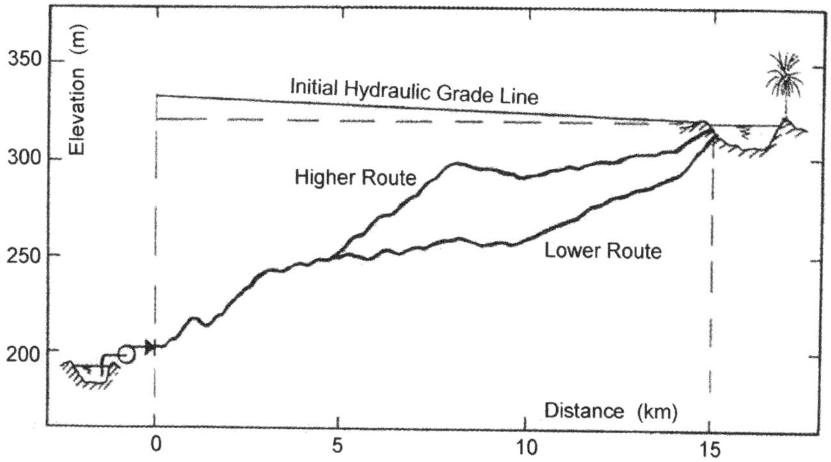

Fig. 1.7 *The pipeline could follow another route*

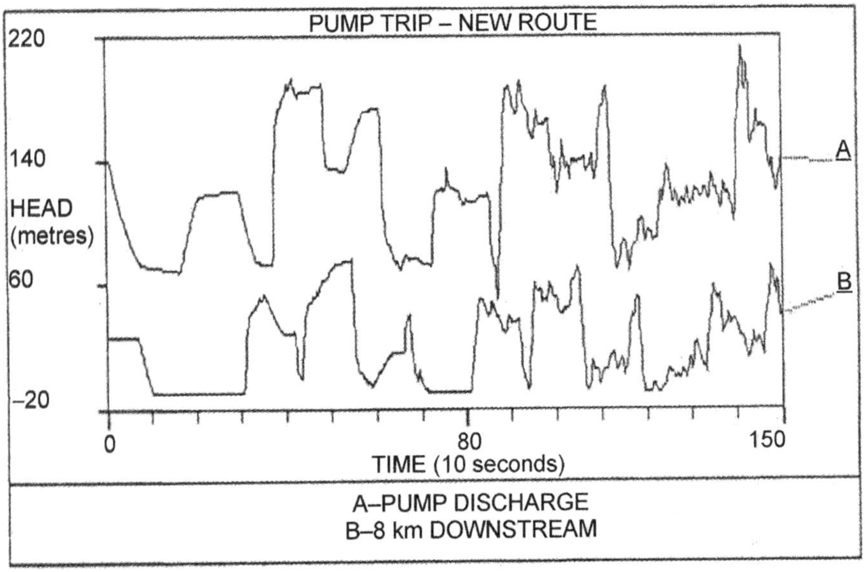

Fig. 1.8 *Pressure–time histories from the pipeline following the higher route to the downstream reservoir*

The pressure traces at the pump are initially identical, but in the second case a reflection arrives back at the pump much earlier. This is from the vapour cavity that is only some 7 km from the pump. The trace for the 8 km point clearly indicates the formation (twice) of a vapour cavity by the constant pressure line at about −10 m gauge. Except for a very short time immediately following the pump trip, i.e. before the cavity forms, the pressure traces are clearly very different.

Another aspect of transient flows that this helps to illustrate is that not only are the pressure changes in a system dependent upon position and time, but also no two systems are really the same. Simply changing one feature, e.g. the pipe wall material, or the size or, in this case, the topography, can significantly change the nature of the fluid transient response. Furthermore, it also means that what might be the most appropriate strategy for surge control in the one system could be quite unsuitable in another.

1.2.8 Air and Gas Entrainment

The presence of both dissolved and free air and gas can have a very marked effect on how a system behaves under fluid transient conditions. Dissolved air and gases will come out of solution when the pressure drops, but the rate at which they can be reabsorbed is so slow that it can be ignored.

The way in which systems respond to the presence of this free air and gas depends upon how it is distributed. In a stationary or slowly moving flow it will tend to collect in pockets. If these are large, they can behave like air cushions and become points of reflection for transient pressure waves. When dispersed throughout the fluid as bubbles, the effect of even very small quantities is to reduce the wave propagation speeds to as little as a quarter or less of that in the pure liquid.

Generally, the presence of air and similar gases in water and other liquid flow systems is regarded as highly undesirable. Quite apart from problems such as the risk of contamination of potable water or explosions of flammable vapours, extremely high shock loads can be generated when moving slugs of liquid following pockets of gas suddenly encounter valves, pipe bends and similar obstructions to the flow. For example, on dry riser fire sprinkler systems it is important that the spray nozzles are properly attached to the pipes supplying them. If not, forces created by the impact of the water as the last of the air is vented can blow them straight off.

1.2.9 Fluid–Structure Interaction

Although most analyses and investigations of fluid transients concentrate on the rapidly changing events within the fluid, these do interact with the pipes and supports. Stress waves propagate through the pipe walls in just the same way as through the fluid.

In many piping systems there are really two significant coupling mechanisms associated with fluid–structure interaction (FSI): Poisson coupling and junction coupling.

Increasing the pressure in a pipe causes the pipe diameter to increase slightly, and the Poisson effect will cause a corresponding axial strain. When a pressure wave is propagated down a line, the associated strain waves are also transmitted through the pipe wall. These 'precursor waves' travel at much higher velocities and also react with the, as yet, undisturbed flow ahead of the main transient. As well as causing movements of the pipeline, they also generate small fluctuations in the fluid pressure.

Junction coupling occurs at pipe bends, changes in cross-sectional area and blank ends. Dynamic pressures exert axial loads on pipes which can cause significant movement. This movement in turn generates changes in the pressure in the fluid.

The net effect of FSI is that the changes in pressure that a piping system experiences under transient flow conditions can exceed the Joukowsky pressure heads. However, as with the influence of friction, energy is transmitted out of the fluid system, and so FSI also contributes to the dispersion and attenuation of the transient pressure waves.

One of the implications for engineers designing pipeline supports is that the pipe must be physically restrained by its supports. It is not satisfactory simply to rest a pipe on a saddle – FSI can too easily cause the pipe to jump out, distort and possibly rupture.

1.2.10 Water Hammer in Steam Pipelines

A related phenomenon, also referred to specifically as 'water hammer', occurs in high-temperature steam pipelines. Two operating scenarios are the usual causes. In one it occurs when relatively cold water, or 'condensate', comes into contact with pipe walls or components that are at elevated temperatures, causing the water to vaporize almost instantaneously into superheated steam. The rapid and enormous expansion associated with this tends to propagate slugs of water at high speed through the system.

Conversely, if steam comes into contact with cold parts of the system, it may condense and create a vacuum. Should this vacuum be relieved, e.g. by opening a valve such that steam from a high-pressure, high-temperature zone is admitted into the evacuated space, slugs of the condensate can again be propelled at high speed through the system.

In both these situations the slugs of water travelling at high speed can cause violent motion of the pipes as they collide with bends and fittings. It is noisy and quite dangerous.

1.2.11 Mass Oscillation and Rigid Column Behaviour

There are a few transient flow situations where it may be possible to ignore the elastic effects without undue error. These are systems in which the time taken for valves to close or pumps to run down is so slow compared with the Pipeline Period that the mass of liquid involved may be assumed to behave as one unit along its entire length.

In a pipe of constant diameter, for example, the velocity, and changes in velocity, can be assumed to be the same at all points. Examples include some hydroelectric plants protected by surge shafts (where this behaviour is often known as 'mass oscillation'), and some small-scale systems incorporating 'slow' or 'two-stage' valve closures. Another example known to the author is water-ballasted floating cranes in which the maximum overall pipe length is about 100 m.

To help illustrate this phenomenon, suppose that in the simple pipeline following the higher route used earlier (see Fig. 1.7) a large-diameter vertical shaft (a surge shaft) is connected into the pipe at the peak 8 km from the pump station. This will be possible technically, though it is perhaps not financially viable since the height of the shaft must exceed the elevation of the hydraulic grade line in order to contain the oscillating water column when it flows into the surge shaft. Assume water can flow freely between this shaft and the pipe, and consider again what will happen in the system following a pump trip.

A transient pressure wave is initiated at the pump and propagated downstream. At the high point, where a vapour cavity was formed before, the local pressure is maintained by the elevation of the water in the surge shaft, and the pressure wave arriving from the pump will be reflected much as it would be from a constant-head reservoir, with the reflected wave heading back to the pump. The pressure–time histories will initially be similar to those of Fig. 1.5, but the Periodic Time for them will be that associated with the pipe between the pump station and the surge shaft, i.e. 12.8 s instead of 24 s for the whole pipe.

Meanwhile, on the downstream side of the surge shaft, the water is still flowing away from it towards the reservoir. Since no further water is coming from the pump station, a flow is initiated from the surge shaft as the pressure in the pipe begins to fall. This inflow from the surge shaft will tend to maintain the forward flow. The velocity in this section of the pipe will fall as the water level in the surge shaft drops, though at a far slower rate.

Basically, the whole water column in the pipeline between the surge shaft and the reservoir will tend to oscillate with simple harmonic motion as if it were in a giant U tube. If the pipe and surge shaft both had the same diameter, the time period for a complete mass oscillation cycle would be of the order of 120 s. With a larger-diameter surge shaft it would be considerably longer.

This description is rather oversimplified. Some elastic effects will be superimposed on the mass oscillation, but the extent to which this is so depends very much on the particular system of interest. The benefit to the analyst of being able to assume mass oscillation, or rigid column, behaviour is that the mathematical description, and subsequent numerical solution, is easier. However, this approach is not always suitable, and it is important that the analyst recognizes that the effects of transient pressure waves being reflected and transmitted throughout the system should be adequately taken into account.

1.2.12 Resonance and Auto-oscillation

Most fluid transient problems are associated with 'single event' changes in flow from one steady state condition to another – pump start/stop, valve operation to adjust the rate of flow, etc. A few situations arise where the disturbances to the flow are repetitive, with corresponding cyclic variations in the pressure in the system. This oscillatory behaviour may be generated by a component, such as a positive displacement pump, which acts as a periodic forcing function, or there may be some feature of the system that encourages the initiation of self-excited oscillations in the liquid column with associated changes in pressure.

As with other aspects of unsteady flow, the basic ideas are more easily described in terms of a simple pipe system. Consider a uniform pipeline supplied by a reservoir at the upstream end and terminated by a control valve at the downstream end. If the valve is closed rapidly, transient pressure waves (starting with an increase) are propagated to and fro, but die out after a while owing to friction and other losses.

Similarly, if the valve is opened rapidly, transient pressure waves will again be generated (now initially a decrease) and die away. However, if the valve is successively closed and opened, even only partially, at a frequency that is close to a natural frequency of the liquid column in the pipeline, pressure changes of a high order can result as the small increases and decreases in pressure are added together.

Figure 1.9 illustrates the pressure head envelopes in a simple pipeline for (a) the first harmonic and (b) the third harmonic. The critical period for the first harmonic is $4L/c$, where L is the pipe length and c is the transient velocity. Note that the critical period for the first harmonic is twice the Pipeline Period.

The maximum variations in pressure will occur at the valve and at certain other locations. The precise location of these other peaks will depend upon the frequency at which the pressure fluctuations are being excited and which harmonic of the system is reacting.

This phenomenon is not restricted to simple systems, but may be encountered in series pipe systems as well as branched and looped networks. The fundamental and higher critical frequencies are not so easily identified, but the systems will still possess the ability to resonate if a suitable form of excitation is present.

When a system is being 'forced', as with a positive displacement pump, the frequency of oscillation imposed on the system is dictated by the speed of the pump and the number of chambers from which fluid is injected into the system. If this matches a critical frequency of the system, resonance can occur. Figure 1.10 illustrates how very small pulsations in the discharge from a three-cylinder ram pump can be rapidly magnified. In this example, the pump is delivering a mean flow of 16 L/s to a 150 mm diameter, 800 m long pipe.

Self-excited oscillations can be set up when components of a system interact in

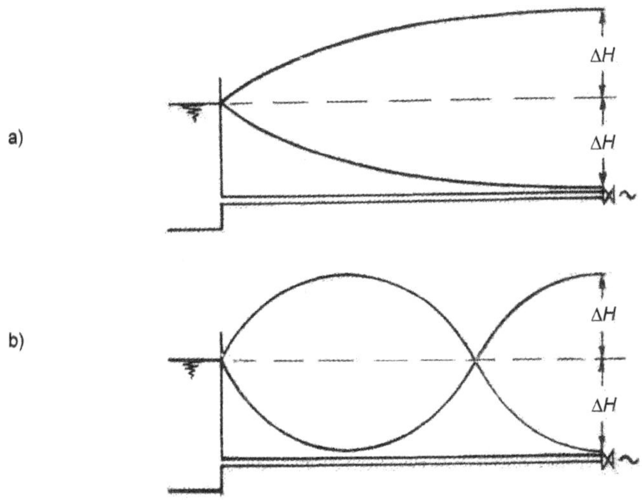

Fig. 1.9 *Maximum and minimum pressure head envelopes for the first and third harmonics in a simple pipeline*

such a way that there is a build-up of energy within the system. One example of this is the hunting of automatic flow control valves. For a normal valve, the flow increases as the upstream pressure increases. If the converse occurs, e.g. owing to the influence of fluid transients (i.e. the flow through the valve increases when the pressure drops and then decreases as the pressure rises), pressure fluctuations can increase.

Historically, most early instances of hydraulic resonance were associated with hydroelectric power plants. The causes have been attributed to faulty components, cavitation-induced flow vibrations, interactions between impeller blades passing guide vanes, leaking valve seals and hunting of the turbine governors.

More recent examples have occurred in oil hydraulic systems driven by positive displacement pumps such as those found in mining equipment, aircraft and ship control systems and diesel fuel systems. Fuel systems for aircraft and space rockets and oil and petrochemical pumping lines have also suffered from the effects of resonance. While it is not the principal general hazard from transient

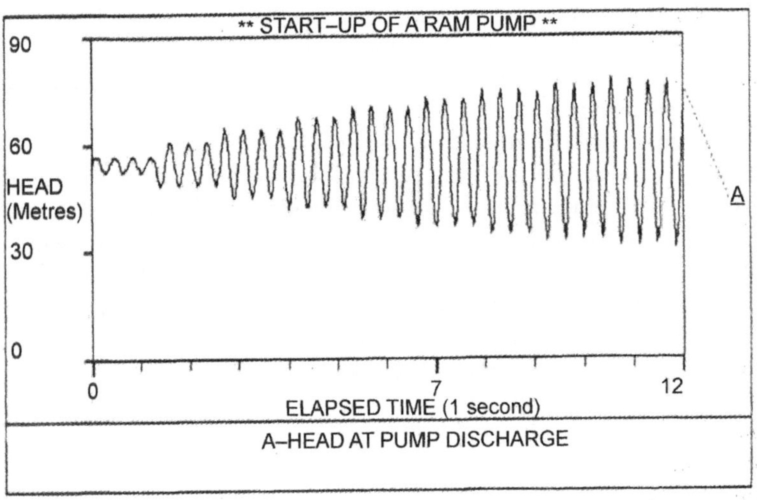

Fig. 1.10 *Rapid increase in pressure pulsations generated by a three-cylinder ram pump in a process plant*

flow behaviour the potential for it to occur should be recognized. A detailed treatment is outside the scope of this text but, in Part 2, the illustrative example given above is explored in greater detail and an incident due to resonance in the boiler feed water system of a coal-fired power station is described, including its cause and how it was resolved.

1.2.13 Key Points Developed in Sections 1.1 and 1.2

The preceding sections have sought to demonstrate that:

- changes to the steady flow in any pipe system lead to changes in pressure head;
- the changes, if too 'rapid' in the context of the system, can generate unacceptable conditions, e.g. pressure too high, too low, etc.;
- the changes are not experienced everywhere in the system at the same time – they are propagated through the system, albeit usually at high speed;
- the 'pressure waves' interact at 'boundaries' such as pumps, valves, reservoirs, etc.;
- these interactions produce reflections that can counteract the change in pressure (at junctions and open ends), while at closed boundaries such as dead-ends and closed valves they can be additive – i.e. they magnify the change in pressure.

Since these pressure changes can be unacceptable and therefore need to be controlled, ways of achieving this must be considered. Various practical methods of doing so have evolved over the years, and they are reviewed descriptively in the next section. Later, in Part 2, their application and, in many cases, procedures for determining their sizes and/or operating characteristics are developed.

SECTION 1.3

Suppression of fluid transients

The principal causes of the failures and unacceptable conditions that occur on account of fluid transients are the magnitude of the pressure changes, the speed with which they occur and the length of time that system components and pipelines, or often just parts of pipelines, are exposed to them.

The magnitude of the change in pressure across a single transient pressure wave is given by Joukowsky's well-known equation quoted earlier. In practice, this equation is usually only directly applicable to quite simple pipe systems and when rapid collapse of vapour cavities occurs. However, it does provide a useful parameter for assessing risks and for rendering a number of other parameters such as system data, or the performance characteristics of components such as valves and air vessels, non-dimensional.

1.3.1 Practical Methods of Surge Suppression

It is important to recognize at the outset that no two pipeline systems are quite the same, and there is no single, simple solution to transient problems that is universally applicable. Every project, every pipeline and even every operating scenario for a given pipeline has to be assessed individually and treated on its merits. It follows that any transient control devices or operating strategies must be chosen accordingly.

In pipe networks, pressure changes are the consequence of multiple interactions of pressure waves being reflected and transmitted from reservoirs, junctions, pumps, valves and other components. Quite often, the combination of these wave reflections is beneficial and can be exploited to advantage.

Several technical strategies are available to the designer to help mitigate the effects of excessive pressure and flow changes, and other unacceptable conditions. The final choice normally involves cost, and sometimes aesthetic or environmental considerations, but this text focuses only on the technical issues. The technical strategies themselves may be categorized broadly under two headings: direct action and diversionary tactics. The former involves making changes to specific components to influence how they perform, whereas the latter involves energy (and hence fluid) transfer into or out of the system. The two categories are described below and then explored in more detail in Part 2.

1.3.2 Direct Action

Under this strategy, action is taken to contain increases in pressure within the system and to influence the behaviour of the causes of the flow changes, such as valve or pump operations.

1.3.2.1 Stronger Pipes

There are still a few situations where there is no alternative but to increase the pressure rating of at least a part of the pipe or tunnel and associated components, such as valves.

Pipelines conveying fluids that, if they escaped, would cause a major incident and have the potential for serious injury and loss of life are prime examples of where containment within the basic system is essential. Examples include corrosive and toxic chemicals and high-temperature liquids that would vaporize on release from the system, e.g. some chemicals and very hot feed water in boiler plant and nuclear installations.

Hydroelectric plant is another example. Typically, the pipeline and tunnels in the high-pressure region immediately upstream of the power house are designed to withstand the full range of transient pressure heads that will occur when the turbine inlet valves and guide vanes are shut rapidly. This is necessary to prevent overspeeding of the turbine and alternator in the event of a complete loss of the electrical load.

1.3.2.2 Rerouting

In the simple example used earlier it was demonstrated that the response of the system to a pump trip was different when the route of the pipeline was changed. If the system designer has some freedom of choice over the route his pipes may follow, he may be able to reduce, even if he cannot eliminate, some of the hazards associated with fluid transients. For example, in the simple rising main, the pipeline following the lower route will be much less exposed to vapour cavity formation. This is because the route of the pipeline stays above the Minimum Pressure Head Profile for most of its length and only coincides with it in the second half – the profile may be termed 'convex downwards' – have another look at Fig. 1.6.

Changing the route can be achieved either by going around obstacles or through them. In one system known to the author it proved beneficial, economically and technically, to drive a tunnel through a mountain rather than to take a pipeline over it.

Increasing the pressure rating of pipelines or rerouting them both tend to be expensive methods of overcoming fluid transient problems, especially when used as the only solution. Hence, the general principle upon which most fluid transient control and suppression is based is to reduce the rate at which changes to the flow occur. This is to give time for reflected pressure waves to arrive back at the source of the disturbance and control or suppress any further change.

1.3.2.3 Changing Valve Movements

If the closure time, T_c, of a valve in a pipeline of length L (to a junction or other terminal point) is less than the pipeline period, $T = (2L/c)$, the full Joukowsky head change and/or column separation will occur over at least a part of the line. However, by extending the valve closure over a time much longer than the pipeline period, T, the amplitude of the pressure fluctuations in the pipeline will be reduced. Note that the duration of the closure must extend over several pipeline periods as virtually all valves have highly non-linear characteristics. In some long gravity-fed lines, valve closure times of an hour or more are used for this reason.

A difficulty is that, in many situations, a suitable valve closure time in relation to surge suppression would not be acceptable from some other standpoint. Too fast a closure can lead to column separation, while a slow closure can permit tanks to overfill or a chemical reactor not to shut down quickly enough for the processes taking place within it. This technique is often restricted, therefore, to systems containing very short pipelines.

In some instances it may be possible to use a two-stage valve closure. This requires a very rapid movement of the valve spindle for the first 75–85 per cent of closure, with the final stages of closure taking considerably longer. Similarly for opening, the initial movement of the main valve can be very gradual and, once flow has been established, further opening can be more rapid.

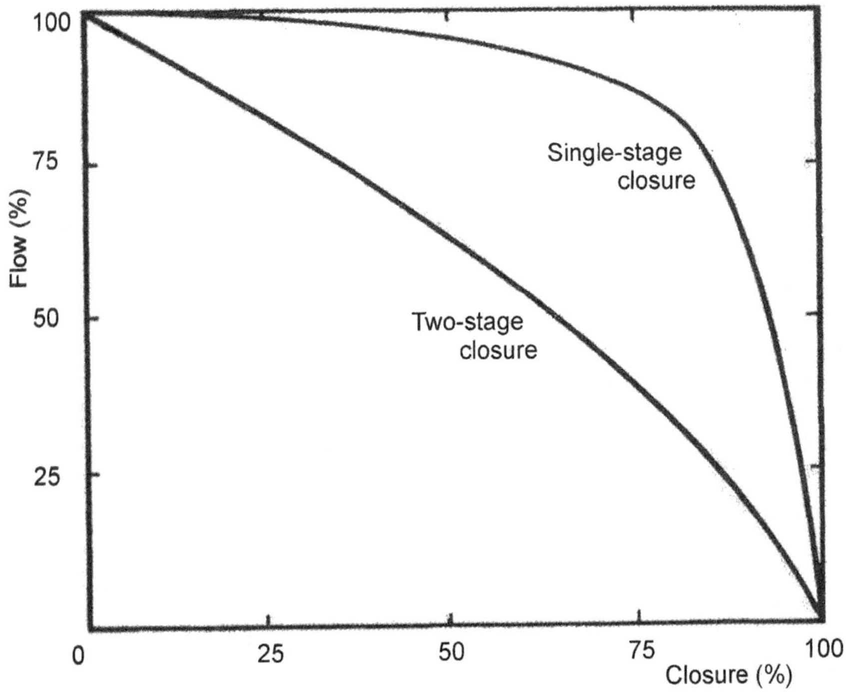

Fig. 1.11 *Comparison of the reduction in flow versus valve spindle movement for a constant closure rate and for one in two stages – initially very rapid, for about 80 per cent of travel, with the remainder taking considerably longer*

In high-pressure systems and lines of more than, say, 150 mm nominal diameter, bypass valves may be fitted. These are normally to enable pressures on both sides of the main valve to be equalized prior to opening to make the whole process physically much easier. The bypass valve can therefore be utilized in the context of controlling transient pressures – before closing the main valve, make sure the bypass is open so that it can become the final shut-off. Figure 1.11 compares the typical, highly non-linear, reduction in flow produced by many valves with what can be achieved by a two-stage closure.

The only satisfactory method for determining the optimum valve closure time is to use the relationship between the valve head loss coefficient and percentage of opening as data for an appropriate computer code. Charts have been produced to enable first estimates to be made and establish whether or not this is a viable option. A selection for various types of valve is presented and discussed in Parts 2 and 3.

1.3.2.4 Avoiding Check Valve Slam

Check valves are sometimes selected without proper thought to their response under transient flow conditions. The phenomenon of 'check valve slam' has been recognized for some time but its true causes were not fully understood until towards the end of the last century. Meanwhile, various designs of valve were modified, e.g. by fitting balance weights, dampers or springs to try to avoid the moving element 'slamming' onto its seat.

In reality, check valve slam is caused by valves that are not matched to the system of which they are an integral part. Systems that are most at risk are those where a high-energy source continues to exist downstream of a pump that has tripped. Two examples are:

- One pump trips out of two or more running in parallel.
- A pump trip in a rising main protected by an air vessel. High-head installations are especially vulnerable.

Figure 1.12 illustrates a 'before and after' scenario. The data are taken from a real power station in Europe that experienced check valve slam in the boiler feed water

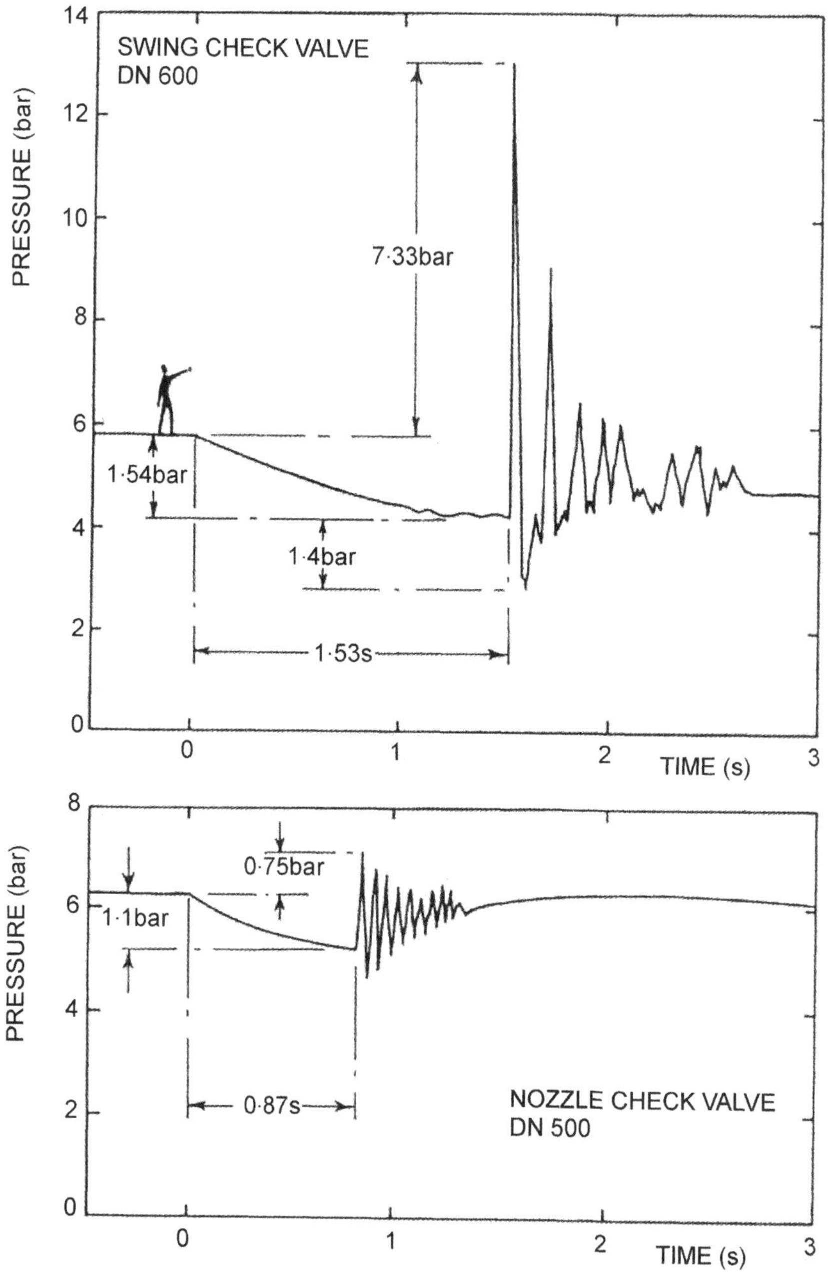

Fig. 1.12 *Replacing an unsuitable check valve can significantly reduce pressure fluctuations and check valve slam following a pump trip*

lines when one pump, out of three running in parallel, was tripped out. The original check valves were conventional swing check valves. These were replaced with nozzle-type check valves and the slamming was eliminated. The two diagrams give clues as to how this was achieved.

The principal clue is the significant time difference between completion of the closure motion of the two valves following the pump trip. The swing check valve takes about twice as long.

The ideal check valve will close at, or very close to, the time when the flow through the valve itself reaches zero. To achieve this the valve will be characterized by three features:

- the moving components will be of low mass;
- they will not have far to travel to reach the closed position;
- their closing motion will be assisted by a spring or springs.

Simple clapper valves tend to have a very poor response, spring-assisted split disc valves, especially with stronger springs, are a little better, whereas nozzle-type valves generally have an excellent response. Some comparisons are given in Fig. 1.13; more details about the various valve types and their Dynamic Performance Characteristics are provided in Subsection 3.8.2.

Although there are many situations where the simple swing valve will be quite acceptable, e.g. low-head, single-pump operation, general services and small-scale systems, etc., it is unwise to regard this type as the automatic choice. Equally, nozzle check valves, while essential for some systems (though usually much more expensive), need not be specified for every system. Split disc valves may represent a compromise in terms of both cost and performance. Recent designs, e.g. the Anti Pressure Surge (APS) split disc valve with pressure relief, could provide another solution.

There are a multitude of valve types and designs available from which the plant designer can select, but their manufacturers should be encouraged to provide Dynamic Performance Characteristics, just as pump manufacturers do for their products, to enable informed and satisfactory choices to be made and justified.

1.3.2.5 Increasing the Inertia of Pumps and their Motors

This direct-action strategy applies to the pump trip situation and is again restricted to relatively short systems. If the pump stops instantaneously, the pressure downstream will fall by the Joukowsky head or to vapour (and gas release) pressure, whichever it reaches first. In the former case, friction effects may also accentuate the drop in pressure. If the rundown time exceeds the Pipeline Period adequately, these limiting pressures are not reached. For a given pump the rundown time is governed mainly by the back pressure and the inertia of the rotating parts. Since the former is fixed, the question that arises is: can the inertia of the pump–motor unit be increased?

The addition of a flywheel will achieve this, but at the cost of the pump driving motor requiring a greater starting current and perhaps more sophisticated switchgear. A simple rule of thumb proposed by Stephenson (1981) is that

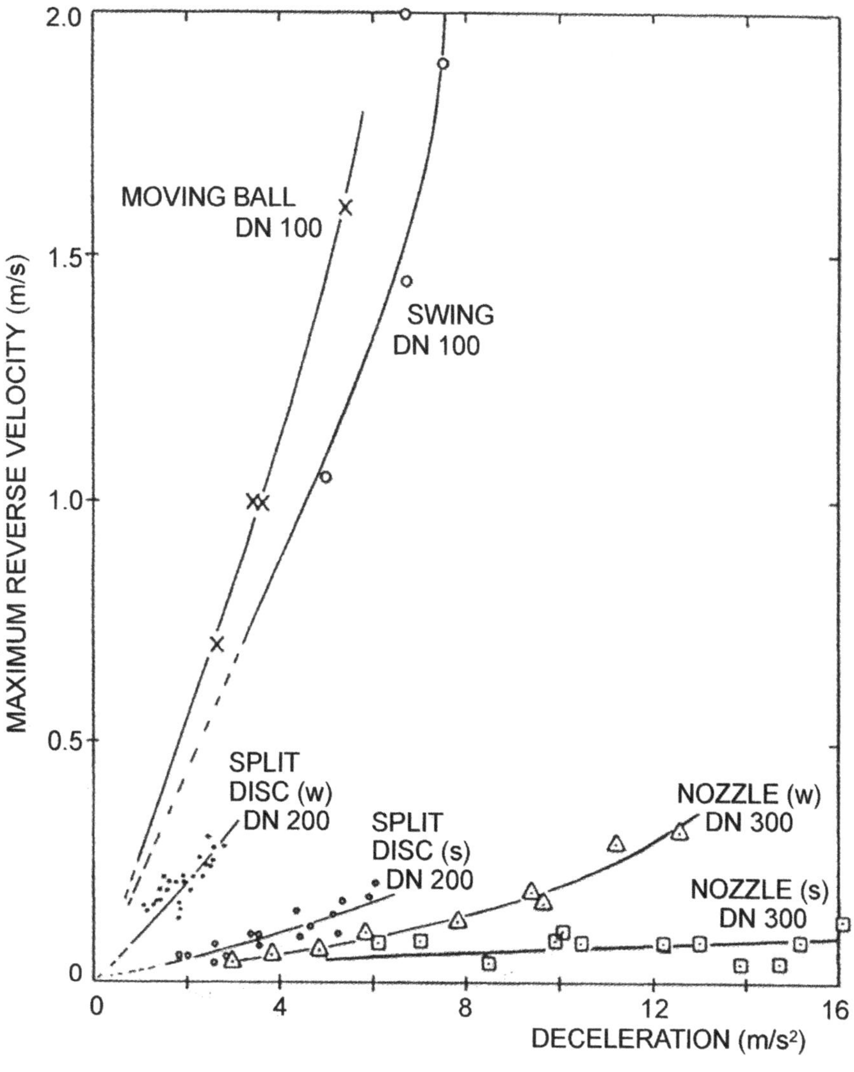

Fig. 1.13 *Typical dynamic performance characteristics for various types of check valve [Thorley (1989)]*

increased pump inertia, I, may be beneficial in reducing fluid transient effects when

$$\frac{IN^2}{\rho ALH_o^2} \geq 0.01 \ (\text{r/min})^2 \tag{1.2}$$

where N is the rotational speed (r/min), ρ is the fluid density (kg/m^3), A is the pipeline cross-sectional area (m^2), L is its length (m) and H_o is the head rise across the pump (m). The inertia, I, should be in kg m^2.

1.3.2.6 Minimizing Resonance Hazards

Resonance is associated with systems in which the flows can have an oscillatory or pulsatile component. If the cyclic element of the flow is imposed on the system by, say, a positive displacement pump, the basic mechanisms for control are to ensure that:

(a) the frequency of the disturbances does not coincide with any of the critical frequencies of the system;
(b) the amplitude of the pulsations is kept as low as possible.

The forcing frequency from the pump can be modified by changing the speed or the number of cylinders. Increasing the latter also reduces the amplitude of the pulsations, and modern high-speed multicylinder swash plate pumps do give a fairly steady output.

Small-scale systems can benefit from the relatively high attenuation of transient pressure waves afforded by the use of flexible hoses. On larger systems, small accumulators can be fitted in the pump delivery line. These are small vessels that are mounted on top of the pipeline and contain a flexible bag or membrane. The bag, or the space above the membrane, is filled with an inert compressed gas which acts as a cushion to smooth out the pressure fluctuations.

Flow control valves, used for such functions as boiler feed water control and in process plants, run the risk of hunting either by interacting with other similar components or as a result of slackness in the mechanisms as moving parts become worn. The latter can be prevented by regular inspection and maintenance. The former ought to be avoided by careful design of the control systems.

Float valves on discharge tanks can cause problems as the float bobs up and down on waves in the tank or reservoir. If problems do occur, or are anticipated, several simple strategies can be considered. If the discharge into the tank cannot be submerged, a guard or baffle can be mounted around the float to deflect waves before they reach it. Alternatively, its mass moment of inertia (and hence its frequency response) can be changed either by adjusting the length of the supporting arm or by hanging an additional mass onto it.

On long stretches of straight pipe above ground, the supports should be fixed at irregular intervals to ensure that the natural frequencies of transverse vibration of adjoining sections do not coincide. Failure to do so can lead to standing waves being set up which, although spectacular to watch, can lead to rupture of the line.

1.3.3 Diversionary Tactics

This is by far the most common strategy and is based on the concept of releasing surplus energy from the system or drawing needed energy into it. To achieve this, it makes use of various devices and methods by which fluid is expelled from, or drawn into, the system at a key point, or points, in order to reduce the rate at which the fluid decelerations in the overall system are changed. Examples include air vessels, surge tanks, feed tanks (sometimes called one-way surge tanks), relief valves, vacuum-breaking/air release valves, bypass lines, etc.

To be most effective, many of these devices should be installed in the system at, or near to, the point where the transient event is initiated, e.g. at the pump discharge or by the closing valve. Exceptions to this general rule include vacuum-breaking/air release valves and feed tanks.

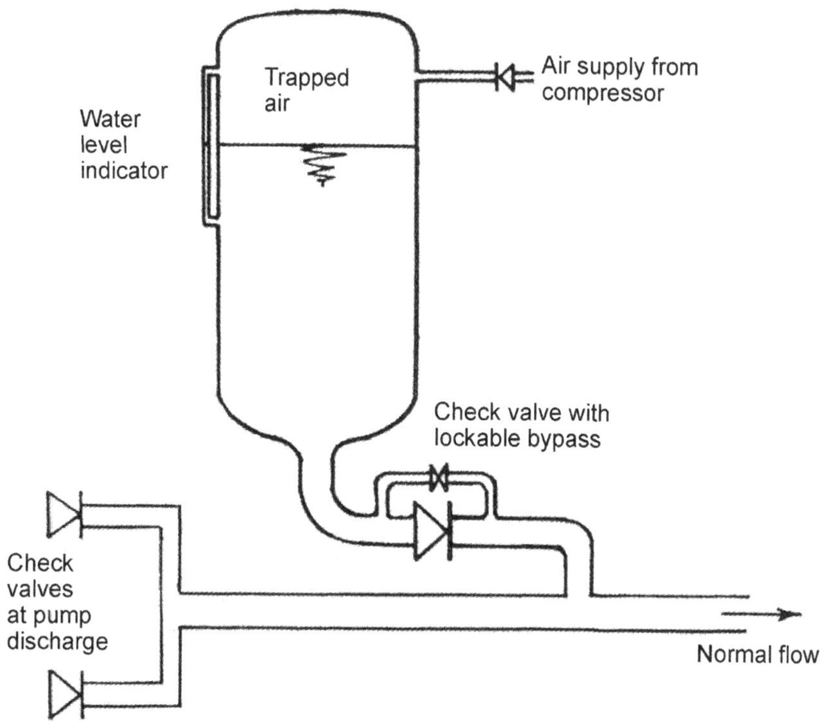

Fig. 1.14 *Typical arrangement for an air vessel protecting a pipeline from the consequences of a pump trip. Ancillary equipment is not shown*

1.3.3.1 Air Vessels and Air Cushion Surge Chambers

One of the most common devices used in the water industry for the suppression of fluid transients and water hammer, particularly to guard against the adverse effects of a complete pump stoppage, is the air vessel.

Following the loss of power to the pump driving motors, the air vessel takes over as the energy source tending to maintain the forward motion of the flow. As the air in the vessel expands and the driving pressure falls, the flow is allowed to decelerate in a controlled manner. This will be at a rate much slower than in the absence of an air vessel. In effect, an attempt is being made to convert a rapid fluid transient event into a controlled mass oscillation.

To protect a system against a pump trip, the most desirable location for the air vessel is at the upstream end of the pipeline and as near to the pump(s) as is practicable. A suitable check valve should be fitted between the pump discharge flange and the vessel to prevent reverse flow through the pump. It is important that the check valve should have an adequate response to minimize the risk of check valve slam.

The volume of air that is required in the vessel is governed by several factors: the length of the pipeline, the cross-sectional area, the initial flow velocity and the pipeline profile. For example, the pipeline following the higher route in the case discussed earlier would require a much larger initial air (and vessel) volume than the line following the lower route to achieve the same measure of protection.

The total volume of the vessel itself is determined by the extent to which the air expands as the pressure falls, plus an allowance to ensure that the vessel does not

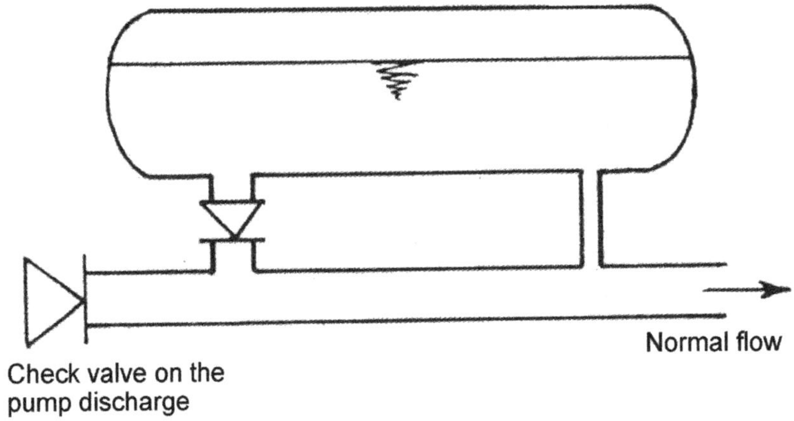

Fig. 1.15 *Large vessels may be mounted on their sides*

drain down completely. Preliminary estimates of the initial air volume required under normal flow conditions, and the volume to which it expands following a pump trip, can be made with the assistance of design charts [e.g. Graze and Forrest (1974) and Graze and Horlacher (1986)].

The rate at which water enters and leaves vessels may be controlled by throttles. Usually, the flow out should be unrestricted, and a smooth rounded exit from the vessel into the pipe connecting it to the main pipeline is often provided. The size of this pipe should be of the same order as the main pipe to minimize losses during outflow.

For the return flow into the vessel, some throttling is desirable to help damp out the pressure fluctuations and control the maximum pressures generated when the reversing water column is brought to rest. The determination of the optimum loss coefficient for flow into the vessel is a matter of trial and error and is best achieved with the aid of computer analyses, though the use of charts can be quite useful to obtain an initial figure.

In practical terms, the differential loss can be achieved by having a non-return valve in the connecting line from the vessel to the main pipe, with a bypass around the non-return valve. The bypass line should be of a much smaller diameter and may contain a throttle valve for fine-tuning when the system is commissioned. The throttle valve should then be locked to prevent inadvertent or unauthorized adjustment in the future. An alternative to the bypass line is to have two separate connecting pipes. This is sometimes preferred when the vessel is mounted on its side.

Mounting vessels on their side, especially large ones with separate inflow and outflow connections, has attractions from an environmental standpoint as well as requiring slightly less substantial foundations. When very large air volumes are required, it is often cheaper and more practical to mount two or more medium-sized vessels side by side and connected into a manifold.

If a convenient hill lies alongside the pumping station, it may be feasible (and cheaper) to construct the air vessels from several lengths of pressure pipe laid in parallel and buried within the hill. The upper ends would be blanked off and the lower ends connected together to form a manifold which is then connected into the discharge line from the pumping station.

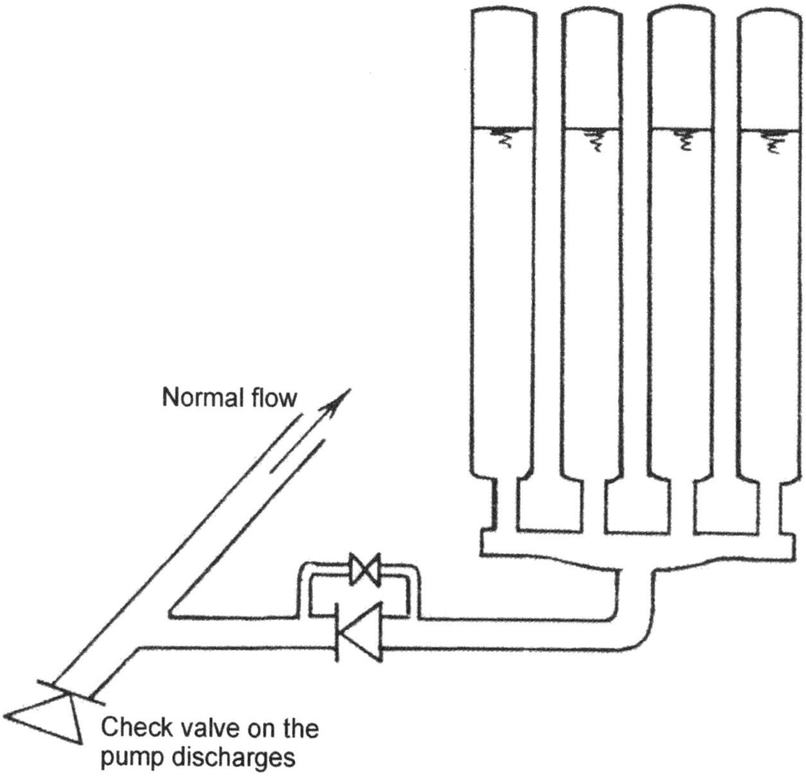

Fig. 1.16 *Multiple air vessels arranged in parallel. This is also a useful strategy when schemes develop in phases over several years as additional vessels can be added later*

The largest air vessels of all, sometimes referred to as Air Cushion Surge Chambers (ACSCs), are found in hydroelectric installations. Their total volumes [see Goodall *et al.* (1988)] have been as high as 110 000 m^3, though more often they are in the range of 5000–15 000 m^3. The corresponding initial air volumes are 75 000 m^3 for the largest chamber and 3250–10 000 m^3 for the more typical cases, with air pressures between 20 and 40 bar.

Air chambers of this type are really caverns excavated out of rock. Their use on this scale is relatively recent, but some successes have been achieved, notably in Norway. One of the major difficulties is the problem of air loss, especially leakage through cavern walls and the roof. Suitable locations tend to be those where the local subterranean strata are solid rock, devoid of fissures and with a low risk of

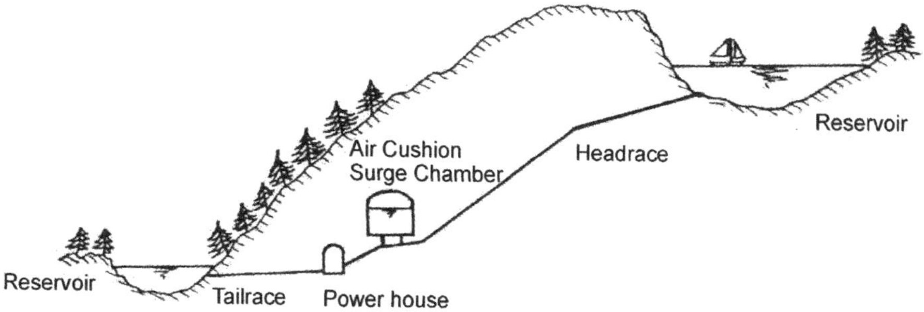

Fig. 1.17 *A very large air vessel – an air cushion surge chamber – excavated within a mountain to protect the hydroelectric plant from rapid flow changes*

ground movement. The cost of excavating and sealing the rock surfaces of the cavern may be offset by savings on the high-pressure penstocks and tunnels which would otherwise have to be designed to withstand higher pressures following emergency load reductions.

As a general comment on the use of traditional air vessels, the need for air compressors and associated instrumentation to monitor the internal water level and control the compressor introduces the constraint that a local power source should be available.

This requirement has tended to limit the use of air vessels and air chambers to locations close to pump stations and power plants. Their applications also tend to be restricted to systems conveying water, as, with other fluids, direct contact with air is often undesirable. A further traditional constraint has been that air vessels should not drain down and allow air to enter the system. However, new and innovative designs have broadened the field of potential applications.

One design variation is the *bladder tank* in which the bladder is precharged with a specific volume of air (or it can be an inert gas for systems conveying flammable liquids). For many years this strategy was only used with small accumulators (see below), but new materials and manufacturing techniques now enable vessels with internal bladders to be of virtually any size. Furthermore, they can also be used in sewage and effluent systems.

Another design variation that is particularly suitable for enabling air vessels to be used in systems with low, zero or even slightly negative static head is to fit a vacuum-breaking, or air admission, valve on top of the vessel. This has found very successful application on ash slurry lines to settling ponds. The idea behind their use is that, when the system is started up, air is compressed within the vessel to act as a stored energy source on pump shutdown. When this occurs, the liquid column in the system slows down gradually, without any tendency for vaporous cavitation. Naturally, there is column separation but there is no flow reversal and no high pressure rise when there is no static head to drive the liquid column backwards. Vessels utilizing this approach are sometimes called *hybrid tanks*.

A refinement to the *hybrid tank* when used in low-head systems, i.e. when there could be some return flow into the vessel, is to incorporate a *dipping tube* below the air admission valve. This helps control the time during refilling when the air valve closes, and hence the amount of air stored in the vessel for normal operation. For extended periods of continuous operation hybrid tanks may need to be fitted with an air compressor as well to maintain the internal air volume.

1.3.3.2 Accumulators

On relatively small-scale systems, e.g. oil hydraulic systems as used in a wide variety of applications, from transportation to the marine, petrochemical and process industries, etc., an accumulator may be suitable for suppressing transient pressures.

In design they are very small versions of the bladder tanks described above, and the gas will normally be nitrogen or another inert fluid. The flexible bladder prevents absorption of the gas by the liquid, and only very infrequent topping up from a portable gas cylinder should be required.

Being relatively small, typically 1–20 L, though they can be up to 60 L in volume, they are usually available 'off the shelf', and can easily be removed or exchanged for servicing.

The preferred location in the piping system where they should be fitted is adjacent to whatever device will initiate transient pressure waves. As pulsation dampers for positive displacement pumps, they would normally be fitted close to the discharge flange. If there is a long suction line, it may be necessary to include one by the suction flange as well. Fast-acting solenoid valves can be a source of high transient pressures, and it may be appropriate to locate an accumulator in their vicinity.

1.3.3.3 Surge Shafts

In large low-pressure applications, or where the hydraulic grade line is close to the pipeline profile, an open-topped air vessel, i.e. surge shaft, may be suitable. The most common use of surge shafts has, traditionally, been in the field of hydroelectric systems, where the pipelines or tunnels from the source reservoir could be arranged as a low-pressure tunnel followed by a high-pressure penstock. The surge shaft is at the interconnection between the two (see Fig. 1.18), where its main function is to protect the low-pressure tunnel.

The design of surge shafts for hydroelectric systems has evolved into a specialist subject in its own right. It is a principal feature of several books on hydraulic transients [e.g. Chaudhry (1987), Jaeger (1977), Pickford (1969) and Rich (1963)], and has been a focus of Technical Sessions at many pressure surge conferences. This being the case, surge shafts will not be covered in detail here except to provide an awareness of the principal features of their design since, in their simpler forms, they are sometimes used in water supply schemes.

Some of the various designs that have been used are illustrated in Fig. 1.19. There are many variations on these, especially the gallery variety where the galleries may be curved in a spiral, perhaps making use of access tunnels from the construction phase. The move away from the simpler shapes has been the result of trying to deal with two problems associated with surge shaft design: the need to damp out the oscillations of the water column between the shaft and the upstream reservoir in a reasonable period of time so that electrical loads can be picked up again, and to overcome problems of stability. In very early installations using simple surge shafts, unwanted resonance tended to occur [Jaeger (1960)] owing to interactions between the surge shaft and the turbine governor while the machine was on load.

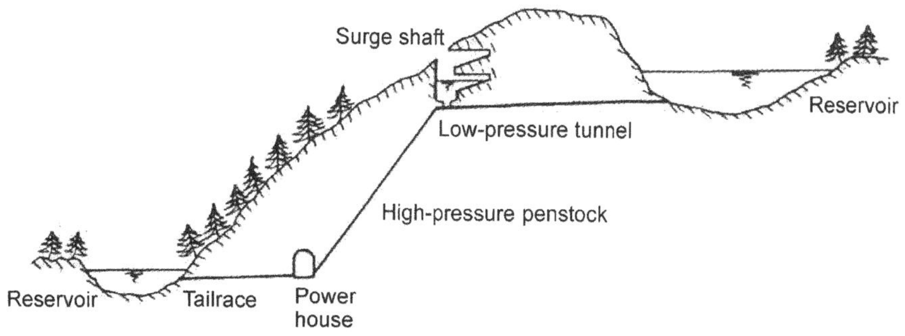

Fig. 1.18 *Surge shafts can be used to protect part of a system when the pipelines or tunnels are close to the hydraulic grade line*

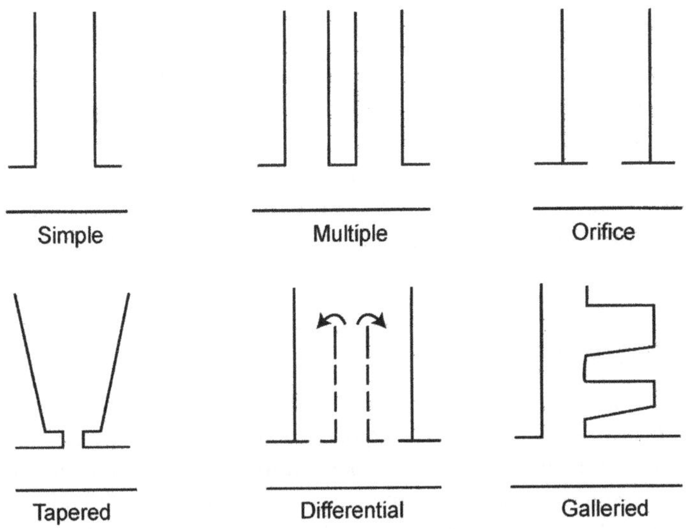

Fig. 1.19 *Some of the many variations in surge shaft design*

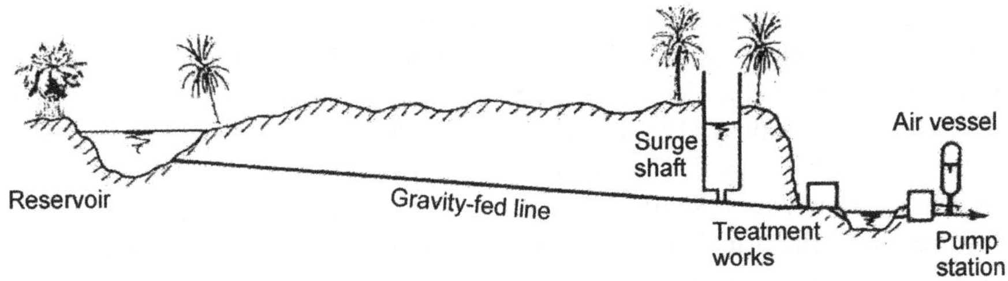

Fig. 1.20 *A surge shaft on a gravity-fed line to a treatment works. A pump station draws its supply from an intermediate reservoir and discharges it to a line protected by an air vessel*

This is not so much of a problem on low-head, gravity-fed lines between, say, a storage reservoir and a water treatment plant (see Fig. 1.20) – where a relatively simple surge shaft may be quite acceptable to protect the line when the valves controlling the flow to the treatment plant are closed.

1.3.3.4 One-Way Surge Tanks (Feed Tanks)

When the principal hazards to a part of a rising main are subatmospheric pressures following a pump trip, a feed tank, sometimes known as a one-way surge tank, can make a useful and economical contribution to the surge suppression strategy devised. The feed tank is a suitably sized vessel connected to the main pipeline by a pipe of nearly the same diameter in which a non-return valve is included (see Figs 1.21 and 1.22). The water level in the feed tank is maintained through a small bypass line and float valve.

Feed tanks only function when the local hydraulic grade line falls below the water level in the tank. Under transient flow conditions the places in a line where this is most likely to occur will be at significant reductions in an upward slope, and in the vicinity of peaks in the pipeline profile.

For very long, undulating pipelines there can be advantages in installing more than one feed tank. For systems where a significant proportion of the downstream end of the line is relatively high and close to the hydraulic grade line, the

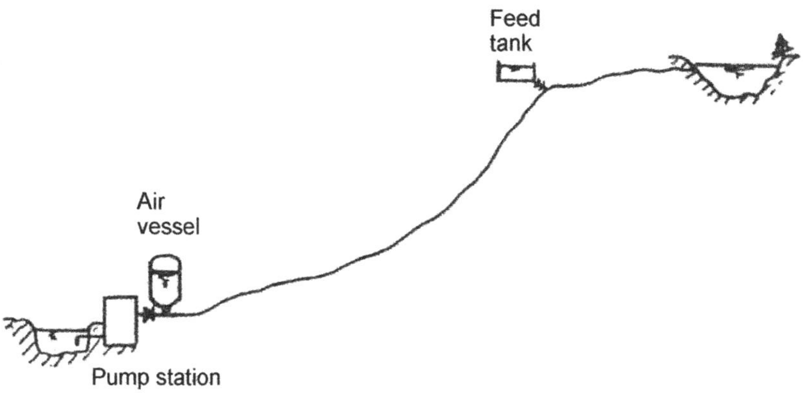

Fig. 1.21 *A typical location for a one-way surge tank*

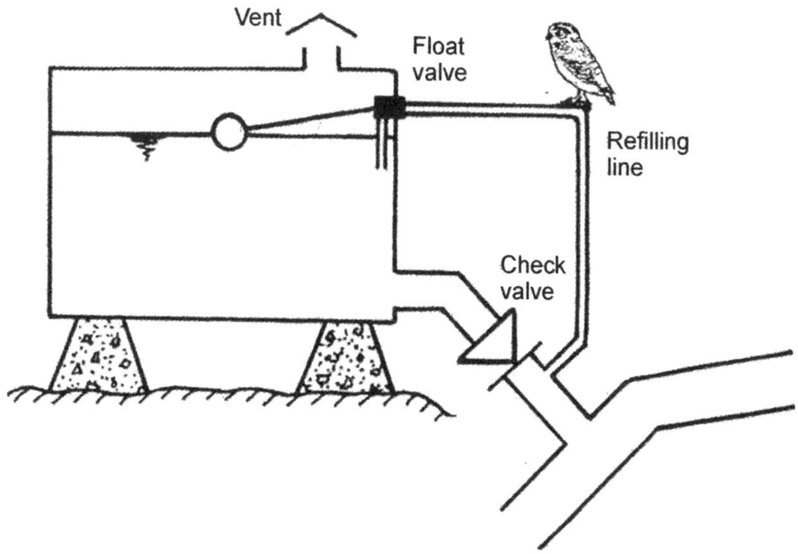

Fig. 1.22 *A closer view of the one-way surge tank in Fig. 1.21*

incorporation of a feed tank can lead to a useful reduction in the size (and cost) of an air vessel on the pump discharge (see Fig. 1.21).

Note that, after the feed tank has discharged to the line, it is necessary to allow adequate time for it to refill before restarting the pumps again. Also, feed tanks only directly provide protection against subatmospheric pressures in a system. They provide no protection against pressure rises other than indirectly through, for example, avoiding vapour cavity formation and the ensuing high pressures when the cavities collapse.

1.3.3.5 Vacuum-Breaking and Air Release Valves

Another device that can be used in certain circumstances to limit the development of subatmospheric pressures in a system is the vacuum-breaking valve. The aim in using this type of valve is to admit air into the line to form a cushion. This helps to damp out the pressure fluctuations over a period of time, after which it is important that the air is released very slowly. It is also important that the air pocket remains in a particular location rather than travelling about.

As with one-way surge tanks, these devices only function when the transient response of the system produces a reduction in line pressure, bringing the local hydraulic grade line down to the same elevation as the pipeline.

When considering the installation of 'air valves', using the term in its widest sense, it is essential to distinguish between the two main reasons for using them, i.e. for ventilation and for the control of fluid transients. These two roles are discussed in some detail in Section 3.6 so will not be repeated here.

When used for surge control they are mounted on peaks in the line and where an upward slope is reduced. As these are also the locations where air release valves would normally be installed, to release air during filling and any air that comes out of solution during normal operation, one of the various combination valves may be used if appropriate.

The operation of most air valves is controlled by one or more floats. There is evidence in the literature of floats in certain designs of valve becoming stuck and partially extruded through the large in-flow orifice. The risk of this happening will be highest in those valves using floats that seat directly into the in-flow orifice and where the liquid columns have been driven back into the valves at high velocities owing to high back pressures developing in the system.

It is also known that, in some systems, extremely high pressures (and temperatures) develop in the air cavity as the high kinetic energy acquired by the liquid column is converted into strain energy of pressure. These pressures can greatly exceed the Joukowsky pressure head and have led to ruptures of the pipe and components.

It is recommended that the use of air admission for vacuum relief should be restricted to low-pressure systems, e.g. where the local hydraulic grade line is normally no more than, say, 5 m head and where the initial flow velocities are modest.

A second cautionary note is that their use in potable water systems is often frowned upon owing to the risk of leakage into the pipeline and contamination of the water, although this can be resolved by arranging for the air valves to be protected in a covered pit.

Examples of where vacuum-breaking valves have been used successfully on their own are when they are mounted downstream of condensers on power station cooling water pipelines and in ash slurry disposal lines, including being mounted on top of air vessels. The first example is one where the static head on the system is quite small. In the second example, although the pumps can nevertheless still be delivering at a substantial pressure, they are only working against friction. By freely admitting air downstream of the condensers, or near the slurry pump discharge, the liquid column downstream gradually decelerates and stops with negligible reverse flow.

A useful strategy for protecting the risers on deep-well pumps on start-up combines an air release valve and a pressure regulating valve. Since under shutdown conditions the riser would be empty, or at a partial vacuum, the pump will run up to speed rapidly and tend to operate at the zero head/maximum flow end of its characteristic.

By venting the air at a pressure less than that holding the check valve closed, and then gradually closing the initially open pressure regulating valve, the flow in the main discharge line can be established in a controlled fashion.

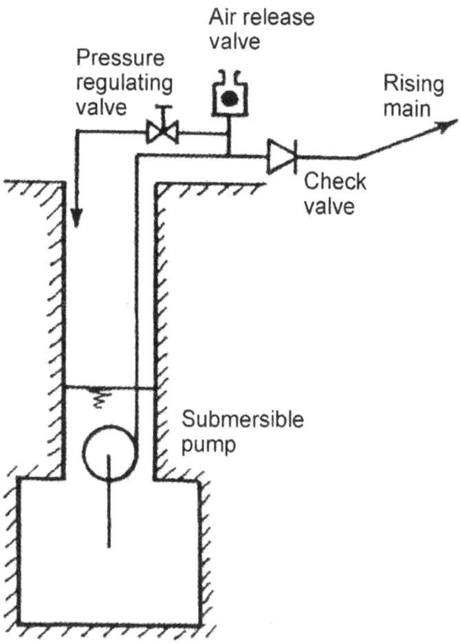

Fig. 1.23 *The combination of air release and pressure regulating valves may be used to suppress high pressures on start-up of deep-well pumps*

1.3.3.6 Pressure Relief Valves and Bursting Discs

When the flow in a line is interrupted by a control valve or similar device at the end, or at some distance from the source, the first transient effect on the upstream side of the valve will be a pressure rise. Typical situations include the rapid closure of oil tanker loading valves and control valves on process plants.

Since many of these pipelines are quite long, extending the valve closure time to many times the pipeline period is not feasible, and so the suppression strategy has to be based on the principle of releasing surplus energy from the system. Among the options open to the designer to achieve this are relief valves that open, or bursting discs that blow, to enable the product to be diverted, thereby maintaining the line pressure within acceptable limits.

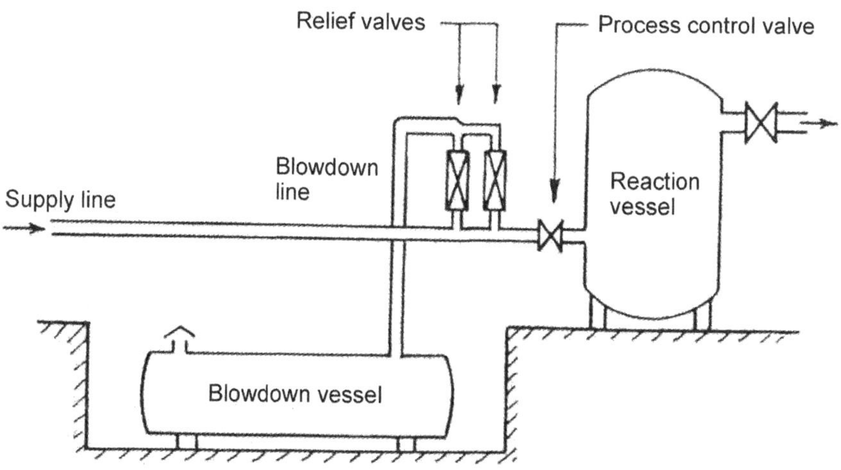

Fig. 1.24 *A pressure relief system protecting a reaction vessel*

Several varieties of pressure relief valves and ancillary equipment are available. They range from the simple spring-loaded safety valves to highly sophisticated surge anticipation valves that are triggered by remote sensors so that they open before the transient pressure wave actually arrives at the valve. Systems such as these are used in the oil industry. In one type, the movable element in the valve is a flexible sleeve surrounding a perforated tube, and is held in place by compressed nitrogen, as a gas spring.

It is good practice to use a minimum of two valves to allow for maintenance and servicing. They should be capable of responding rapidly to the increase in pressure and be set to open at different pressures. The closing pressures should be set lower than the opening pressures.

The liquid that passes through the relief valves should be collected in a suitable storage facility or blowdown tank. Whether liquid is returned directly to the system later, reprocessed or disposed of will depend on the fluid and other factors.

The pressure relief system should be capable of limiting the rise in pressure to within acceptable limits. This may sometimes mean taking the full line flow but more usually it would only be necessary to divert a proportion of the normal flow and to increase the back pressure to help decelerate the moving liquid column.

When sizing the relief valves, the anticipated frictional pressure drop in the line from the valves to the receiving vessel must be taken into account. Likewise, the capacity of the receiving vessel should be adequate to accept the flow until such time as the system can be shut down.

The use of pressure relief valves is not particularly common in water systems unless the water is at elevated temperatures such as in power stations and some process plants. If transients there are caused by pump trips leading initially to a drop in pressure, a relief system offers no benefit. However, if a reactor must be shut down quickly, Emergency Shut Down (ESD) valves may have to close, leading to excess pressures. It may be desirable for the high-temperature water, which may flash off to steam if relieved into the atmosphere, to be diverted to blowdown tanks and recovered later.

Other exceptions are fire hydrant systems if it becomes necessary to protect them from high pressure rises when the nozzle valves at the ends of the hoses are shut off quickly.

Another example is an interesting case [Griffiths (1972)] where pressure relief valves have been used successfully to limit the rise in pressure in the penstocks of a hydroelectric plant, following closure of the turbine guide vanes.

The *bursting disc* is another form of pressure relief that has the advantage of permitting a high flow capacity very quickly. Discharge would again usually be to a receiving tank, and in some systems this may be precharged with an inert gas and pressurized to offer an increasing resistance to flow as the vessel fills. By appropriate sizing of the vessel, choice of precharge pressure and even the use of orifice plates in the discharge line, the reduction in relief flow from the main system can be accurately controlled – a very useful attribute where space may be limited as on oil tankers loading and unloading offshore.

The main disadvantage of this approach to surge control is that, before the system can be restarted, the receiving vessel must be emptied and the bursting disc replaced. Related disadvantages are that human exposure to the products in the system may be undesirable, making disc replacement hazardous. Also, metal discs have been known to suffer from fatigue, although this can be overcome by using non-metallic discs provided they are compatible with the product.

Many of the systems where pressure relief valves and/or bursting discs are normally used come under specialized codes of practice, regulations and standards. These should be consulted during the design stage [see, for example, AICE (1998), Parry (1991) and Inst. Pet. (2001)].

1.3.3.7 Bypass Lines

The protection of pipelines against low pressures following a pump trip by making use of air vessels can prove expensive. For some types of installation, a bypass line around the pumps may be a cheaper and satisfactory alternative.

One such installation is the rising main where the pump is discharging against a low static head. Following a trip, if the discharge pressure falls below the available suction head, it could be possible for water to be fed around the pump to limit the reduction. Figure 1.25 is a situation where this can work. However, to be successful, the bypass line should cause a negligible friction loss.

A second typical application is the in-line booster pump (see Fig. 1.26). Note that the bypass line is the same size as the main line and contains a non-return valve to prevent flow recirculation. If this pump trips, it represents an obstruction

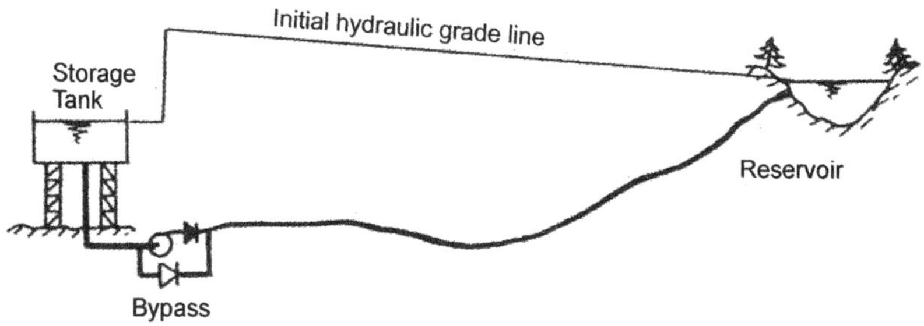

Fig. 1.25 *A bypass line around the pumps in low-head pumping mains having a positive suction head*

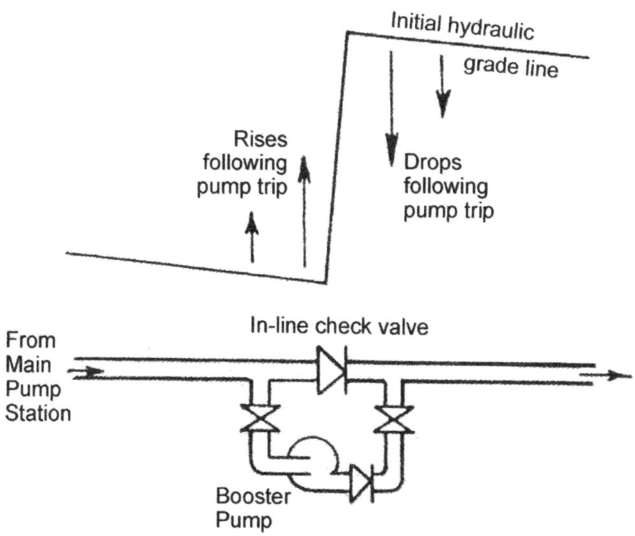

Fig. 1.26 *A typical bypass arrangement for a booster pump. Note how the hydraulic grade lines move following a pump trip*

to the normal flow. A positive pressure builds up on the suction side, while a negative pressure change develops downstream.

The precise manner in which the pressure changes develop depends on the lengths of the suction and discharge lines, but in many cases the pressure changes can be considerable. In the event that the pressure on the pump suction exceeds that downstream, it can be relieved by feeding liquid through to the downstream side, thus helping to minimize the pressure drop there.

1.3.3.8 Avoiding Water Hammer in Steam Pipelines

In Subsection 1.2.10 the phenomenon of water hammer in steam pipelines was described. The underlying causes are associated with the occurrence of liquid water together with superheated steam in the same location and their interaction causing slugs of the water to be propelled at high speed through the system – with potentially catastrophic effects.

The problems are most acute when steam plant is being brought back into commission, even after a temporary shutdown of short duration, but can also occur during cooling off periods if adequate drainage of the condensate is not provided.

Inevitably, condensate will form as a line is being charged from cold with hot steam. The best avoidance strategies will therefore be to charge the line slowly, ensure that adequate water traps and drainage points are provided, that the pipes will slope downwards slightly in the direction of steam flow and that there will be no dead-legs or low points where water could collect.

Charging the line will often be done via a bypass valve round the main steam valves, partly to help balance the pressures on both sides to make opening the main valve easier as the plant comes on load. It is important to generate a steam flow through the system to warm it through properly, with the steam and condensate usually running through to a blowdown tank. The drainage system should be left open, even as the system picks up load, until it is clear that the pipework is sufficiently heated through for no further condensate to be created.

During shutdown periods, the water traps and drains in the steam lines should be open to allow any pockets of water that form to drain away to a suitable receiving point. During periods when the plant is shut down, drainage from the system should be complete – no condensate should be allowed to remain within the system.

1.3.4 Choice of Protection Strategy

It is best wherever possible to avoid rapid changes in the flow. Unfortunately, this is not always possible in most pipeline systems, and the main functional design of the system must be modified to ensure that unacceptable conditions do not arise.

No two systems are completely identical, and hence the preferred transient protection strategies will usually differ. However, the approach to assessing and alleviating potential problems will follow the same general pattern.

The first step is to list all the possible causes of changes to a steady state flow in the system in question. Include all normal operational changes such as start-up, shutdown and changes that would be expected, together with any unplanned events and fault conditions to which the system may be exposed. Then prioritize them to identify the critical design cases.

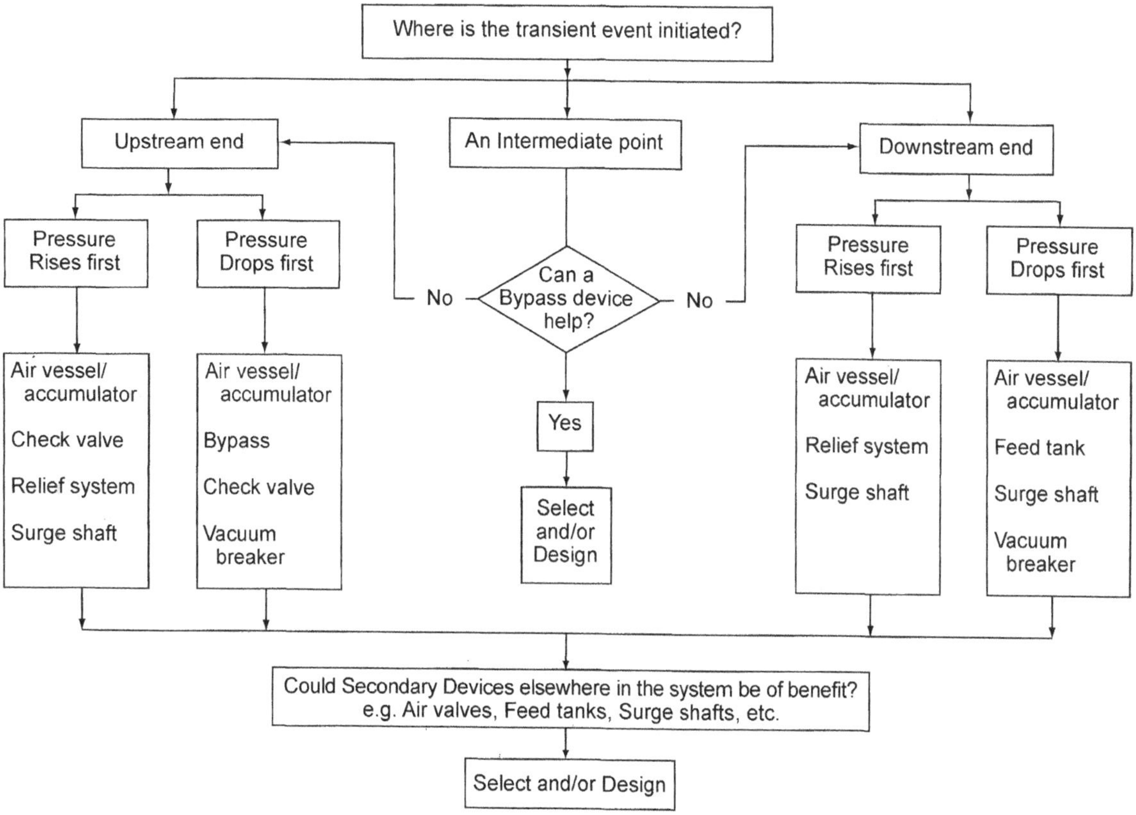

Fig. 1.27 *A guide to the selection of components for the control and suppression of transients in pipeline and tunnel systems*

Note the locations where the events may be initiated and check whether or not any form of direct action (i.e. changing the rate of valve operation, changing the route of the line, or strengthening the pipe, etc.) might be suitable to overcome, or at least reduce, the problem without introducing others. More often, however, it is necessary to incorporate specific control and suppression devices in the system.

Figure 1.27 suggests how a list of the more suitable options for a particular situation may be compiled. The basic philosophy is to look for the source of the transient pressure waves and at whether the initial change is a rise or fall. If the rate and magnitude of this initial change can be reduced, the later fluctuations due to reflected pressure waves returning from elsewhere will also be reduced.

Some of the preliminary options (listed alphabetically in Fig. 1.27) may be eliminated immediately for some systems on the grounds that, technically, they are inadmissible, e.g. air admission to oil or petrochemical lines. The choice will tend to narrow down to a principal control device to be located at the source of the disturbance.

In a number of situations secondary control devices may be introduced elsewhere to alleviate problems at critical points in the system and perhaps reduce the scale, and cost, of the primary device.

Two examples of the possible locations of various control devices are shown in Figs 1.28 and 1.29. The first example is more appropriate to the water resources field, whereas the second is taken from the oil, petrochemical and process

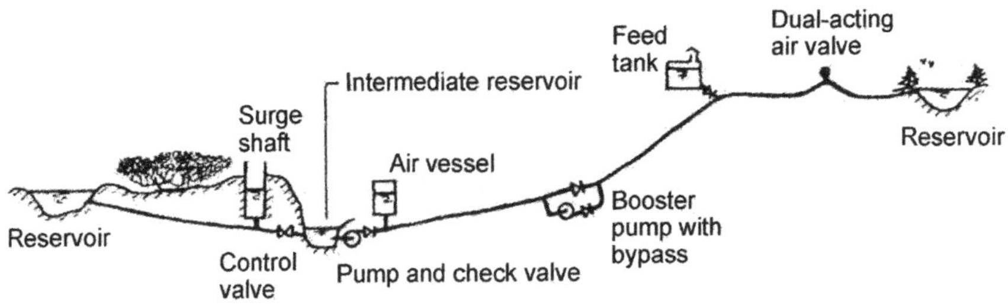

Fig. 1.28 *Examples of the more usual locations for various transient protection devices in pipelines associated with water resources projects*

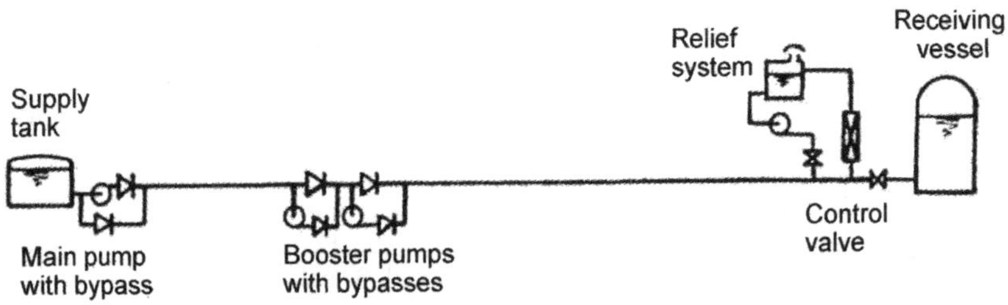

Fig. 1.29 *Examples of the location of transient protection devices in oil, petrochemical and process lines*

industries. The 'receiving vessel' could represent many things, from a chemical reactor vessel to an oil tanker at an off-shore terminal.

In some situations, especially in petrochemical plants, interlock systems may feature in protection strategies. For example, in Fig. 1.29, if, say, the valve at the downstream end is closed rapidly, a radio or telemetry signal would be sent to the upstream end to trip the pump. Depending on the precise application, there may be codes of practice or regulations that will influence the final choice of surge control strategy and system operating instructions.

In addition to the technical suitability and relative costs of the available options, other factors will also influence the final choice. These include reliability, space and power requirements, the amount of maintenance and supervision needed and the availability of suitably skilled labour and possibly planned enhancements to, and uprating of, the system. Environmental factors, such as the appearance of large air vessels, have also been known to influence decisions.

Having decided upon a preferred strategy, the initial design and sizing of the components may be undertaken, for which some guidance follows in Part 2. Checking of the adequacy of the proposed solution should generally be effected with the aid of a computer-aided analysis, with final confirmation being obtained from a test programme during the commissioning of the main project.

1.3.5 Summary of Part 1

The intention in Part 1 has been to give the reader new to the topic an introduction to the subject of fluid transients, the problems they can cause, an understanding of the basic concepts involved and an appreciation of the various means that can be employed for their alleviation. In particular, it should now be appreciated that:

- changes to the steady flow in any pipe system lead to changes in pressure head;
- these changes, if too 'rapid', in the context of the system, can generate unacceptable conditions, e.g. pressure too high, too low, etc.;

- the changes are not experienced everywhere in the system at the same time – they are propagated through the system, albeit usually at high speed;
- the 'pressure waves' interact at 'boundaries' such as pumps, valves, reservoirs, etc.;
- these interactions produce 'reflections' which can counteract the change in pressure (at junctions and open ends), while at closed boundaries such as dead-ends and closed valves they can be additive – i.e. they magnify the change in pressure;
- several strategies are available that can be used to contain or reduce excessive pressure changes and other unacceptable conditions;
- no two systems are the same – they must all be treated on their merits and appropriate protective measures identified on a case by case basis.

This summary represents the 'prior knowledge' that it is assumed the reader will have before embarking on Part 2, in which the application of the various protective strategies is explored in greater detail, making use of the concepts of wave propagation and reflection, the Joukowsky pressure change [$\Delta H = (c/g) \times \Delta v$], etc.

PART 2

This section of the book is intended for those seeking guidance on how to approach the task of identifying and managing risks associated with fluid transients in pipe systems. It is assumed, by way of prior knowledge, that the reader has an appreciation of:

- the common causes of unsteady flow;
- the initiation, transmission and reflection of transient pressure waves;
- the consequences, i.e. the unacceptable conditions that can arise;
- practical methods and strategies for the control and suppression of fluid transients.

In addition, the reader should understand enough basic theory to determine the speed at which transient pressure waves are propagated through a system, and the Joukowsky pressure change across a step wave.

After an introduction to the topic of Risk Assessment, a series of typical situations is described in order to:

- develop and illustrate procedures for identifying and resolving problems;
- indicate how to achieve preliminary designs for control strategies;
- produce a suggested specification for computer-based analyses to refine the designs;
- help identify appropriate computer codes (and analysts) to undertake these analyses;
- assist the designer to discuss the results from a position of knowledge and understanding.

The early examples are discussed in some detail. The later, more complex examples extend the range of applications covered and embrace a variety of transient flow situations. Also, a number of incidents that have been reported in the literature are reviewed as there are a few lessons that may be drawn from them.

SECTION 2.1

Assessment and management of risk

2.1.1 Introduction

Risk assessment has become an established management tool that can be applied to the whole enterprise. It is not restricted just to Health and Safety issues but can be applied, for example, to the financial as well as the technical viability of the business as a whole. Its attraction is that it brings a disciplined approach to the process of identifying and evaluating *all* threats to the enterprise and how to respond to them. The following diagram illustrates the elements involved.

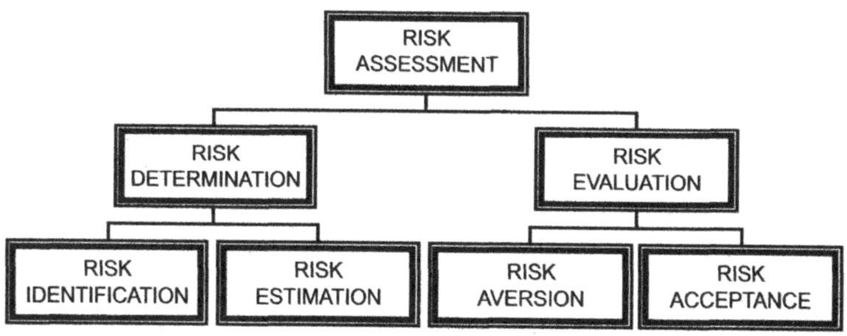

The evolution of Health and Safety legislation and the threat of legal action following accidents and incidents [e.g. Anderson (2000), Baker and Ramos (2000) and Sykes (1995)] have made it more incumbent on plant designers, operators and owners to be able to demonstrate that they have undertaken full risk assessments and followed through with appropriate action to discharge their responsibilities to safeguard their employees and all third parties against accident and injury.

Part of this must embrace the risks associated with changes to the steady flows within piping systems. It involves:

- understanding the nature and extent of the risks;
- the extent and category of risks that are regarded as acceptable;
- the likelihood of the risks concerned materializing;
- the options available to reduce the incidence and impact of the risks;
- the periodic review of the effectiveness of the control systems designed to minimize the effects of the risks.

Fluid transient risk assessments usually arise from two situations:

1. Feasibility studies have been completed, the functional design is in progress and the question is posed – what are the risks from fluid transients?
2. A system has been in operation for some years, demand is growing and there is a decision to uprate and/or extend the original system. Then the question is again asked – what are the risks from fluid transients?

There are two other scenarios where the same question *ought* to be asked but sometimes is not. They are:

1. The design has been finalized, risk assessments including transient analyses have been completed and then, at the last minute or during construction, there is a change.
2. Commissioning and start-up tests will often be at part load and, perhaps, with sections of the system shut down or, at least, in a non-standard configuration.

If the risk analyses are not repeated, at least insofar as they need to take account of the changes, there can be unfortunate consequences – as examples later in Part 2 illustrate.

A useful and relevant definition of 'risk' [Rowe (1979)] that embraces these ideas is that it is 'the potential for the realization of unwanted consequences from impending events'.

The unwanted consequences can range from the mildly inconvenient to the disastrous. The former may be a temporary drop in pressure or interruption of a water supply, while the latter can be the destruction of a power plant, a major environmental disaster, an explosion and/or a fire.

Among the more well-known disasters to affect power plants are the Lac Blanc–Lac Noir pumped storage scheme which was destroyed during commissioning tests and six engineers were killed [Roccard (1937)], and the rupture of the high-pressure penstocks at the Oigawa Power Plant in Japan [Bonin (1960) and Chaudhry (1987)]. Further examples, from the water, power, nuclear and petrochemical industries, are documented in the literature [Arastu *et al.* (1999), Jaeger (1963), Kletz (1998), Serkiz (1983), Thomas (2003), Thorley (1976), Thorley and Spurrett (1990) and Trenke (1979)].

Some of these undesirable consequences are not only less spectacular but may even develop insidiously to confront the unwary plant operator with problems. Examples include pollution – pollution of the fluid in the pipe owing to leakage into it, or a gradual pollution of the environment owing to leakage from the system.

A summary of these unacceptable conditions, i.e. unwanted consequences in, and from, pipeline systems, and expressed in terms familiar to engineers, was listed in Subsection 1.1.1. It is repeated here for convenience:

(a) Pressures too high – leading to permanent deformation or rupture of the pipeline and components; damage to joints, seals and anchor blocks; leakage out of the pipeline, causing wastage, environmental contamination and fire hazard.
(b) Pressures too low – may cause collapse of the pipeline; leakage into the line at joints and seals under subatmospheric conditions; contamination of the fluid being pumped; fire hazard with some fluids if air is drawn in.
(c) Reverse flow – causing damage to pump seals and brush gear on motors; draining of storage tanks and downstream reservoirs, overfilling of upstream sumps.
(d) Pipeline movement and vibration; overstressing and failure of supports, leading to failure of the pipe; mechanical damage to adjacent equipment and structures.
(e) Flow velocity too low – mainly a problem in slurry lines, causing settlement of entrained solids and line blockage.

The risk assessment process must therefore address the following questions:

- What 'impending events' could give rise to these situations?
- What is the probability that they will occur?
- Does it matter?
- If so, what can be done about them?

In the present context, 'impending events' are mainly those events that lead to changes to an initially steady flow, including one that may initially be zero. Table 2.1 gives examples of what might appear in a formal risk assessment report. It is offered here as an initial checklist to help generate ideas. It is essentially a brainstorming exercise based on 'constructive pessimism'.

The various events may be classified as either 'controlled' or 'external' (i.e. uncontrolled) events. The design engineer and system operator should have some influence over controlled events, e.g. shutdown of pumps, so that, although there may be a high likelihood that they will occur, their impact should be quite low provided adequate surge protection measures are taken.

Conversely, they will have much less, if any, direct influence at all over the occurrence of external events such as power failures. Unfortunately, this will not necessarily absolve them from the responsibility of anticipating such events, or of ensuring that the system is adequately protected, particularly given the trend in safety regulations that place the onus on the designer, operator and owners to ensure their plants are safe [Baker and Ramos (2000) and Sykes (1995)].

Once all the potential threats have been identified, they should be reviewed to establish those that constitute the critical design cases. In many pipe systems there could be quite a number of potential situations involving changes to the flow, but in most cases there will be, at most, three critical design cases. The intention then should be that strategies devised to protect against these cases would automatically cover the other less hazardous events.

Considering all the things that may occur to a system, what is the likelihood that any of them will actually happen, what is 'the potential for their realization'?

Table 2.1 Specimen List of Potential Causes of Fluid Transients in a Pipe System

Item	'Impending events', e.g. possible causes of flow changes
1	The system is started up, possibly with parts of it shut down and 'isolated'
2	The system is shut down
3	The power to the motors driving the pumps fails
4	A pump is tripped owing to thermal overload
5	A pump delivery valve is shut in a time of t seconds
6	One pump trips but others keep running
7	A routine pump changeover operation takes place
8	A pump is restarted t seconds after being tripped
9	A control or Emergency Shut Down valve closes rapidly
10	Tanker loading valves on the end of a long pipeline are shut in, say, 15 s
11	An operator opens/shuts a process control valve y quickly
12	Component x (e.g. an automatic flow control valve, pressure relief valve, etc.) malfunctions
13	The route of a pipeline is changed
14	The demand on the system increases/decreases
15	Basic design data (e.g. heads, flows, component operating characteristics, materials specifications, fluid properties and quality, etc.) are unreliable by $\pm x$ per cent?
16	Changes are made to the original design of the system
17	A turbine trips owing, say, to a fault on the electrical system
18	The surge suppression strategy/control devices fail
19	A third party causes corrosion or mechanical damage to the system
20	Any other potential causes that a safety review or audit identifies

Furthermore, if they do occur, does it matter and, if so, how adequately is the system protected against them?

These are subjective issues to which statistical evidence can sometimes be brought to bear. Also, in some industries, e.g. nuclear power, transport and disposal of toxic substances, etc., the process of risk assessment is institutionalized, with regulatory bodies laying down guidelines and standards [see, for example, AICE (1998), Inst. Pet. (2001), Merilo *et al.* (1990) and Peterson and Kim (1993)]. In the field of water resources the same is true with regard to supply pressure and water quality. Frequently, judgements have to be made in order to find an acceptable balance between the probability that a particular risk will actually occur and the cost of reducing or eliminating it.

The sections that follow are intended to provide guidelines on how to approach the problem of estimating whether or not there could be a risk in some typical systems, of assessing the nature and scale of the potential problems, identifying the probable critical design cases and formulating strategies for their control and suppression. A good understanding of the physical nature of pressure wave propagation and reflection, as described in Part 1, is assumed, and use will be made of the charts and other design aids presented in Part 3.

It is important to recognize that these assessments are just that, assessments or estimates. They are not substitutes for a detailed computer-assisted study, although they are a necessary precursor since computer codes for transient flow analysis do just that – analyse – they do not 'design' protection strategies. The purpose of this text is to help the designer produce candidate protection strategies that can then be refined with the aid of the more accurate calculations that can be undertaken by a good computer code. Often, the cost of these studies, if properly focused, can be recouped by improvements in the design.

2.1.2 A Procedure for Fluid Transient Risk Assessments

The following is a series of steps that can be followed to form an assessment of the hazards that may face typical pipeline systems as a result of transient flows. They are set out briefly here, and then illustrated with some demonstration examples. The reason for this is that, in the realm of fluid transients, no two systems are really quite the same. Even apparently small changes in a component, the flowrate or pipeline profile can significantly change the nature of the problem and hence the control and suppression strategy that may need to be adopted. The suggested steps are:

1. Define the system – the pipelines, components and devices, etc. At this stage include all pumps and valves, etc.; idealizations will be considered later.
2. Define the safe and/or specified operating limits and any conditions or circumstances that would be unacceptable.
3. Develop longitudinal sections of the pipeline(s), showing all principal features. Include the hydraulic grade lines for the initial steady conditions and the maximum and minimum permissible pressure head envelopes. Idealizations of the system may now be introduced, provided they have negligible influence on the transient events. For example, a uniform wave propagation speed may be assumed in a composite system if the speeds in individual sections vary about the mean by no more than, say, 8–10 per cent; two or three pumps running in parallel may be combined into an equivalent pump when the transient event is a power failure to all pumps; and a number of valves in series may also be aggregated into a single unit by summing the loss coefficients.
4. List all potential causes for changes to steady flows in the system – normal and abnormal, 'controlled' and 'external' events, etc. (see Table 2.1). Review what can go wrong – where, how, why and in what order may the flow change? Both the magnitude and the rate of change are important.
5. Review the list created in step 4 above to identify the probable worst cases, in terms of the consequences, e.g. those that give rise to the highest and lowest pressures, contravene any statutory regulations or exceed the specification of the system. These cases will often, though not always, be the uncontrolled events outside the direct control of the system operator, such as power failures and closure of loading valves on oil tankers or of Emergency Shut Down Valves (ESDVs). The purpose here is to identify the critical cases to be used for assessing the magnitude of the problem that will provide a basis for formulating whatever control strategies may be required.
6. Assemble data on basic parameters such as the wave propagation speeds, the pipeline period, the estimated pump rundown and/or valve operating times, the Joukowsky pressure head, etc. Then, build up a picture of how the pressures and flows in the system will change following the initiation of the transient event with a view to identifying if, and where, unacceptable conditions may arise.
7. In the light of the results from step 6, identify and review options for control and suppression strategies. Develop preliminary designs using, if appropriate,

relevant charts, etc. from Part 3. On technical grounds there may be two or three possibilities, but other considerations such as cost, space and maintenance will influence the final choice.

8. If the project is part of a phased development, consider the implications of future transient risk assessments. Usually, the flows and pressures will increase, and there may be modifications to the system itself. Insofar as the future plans are known, review the consequences for the transient control and suppression strategy. If, for example, there is likely to be a need for greater air vessel capacity, or more pressure relief or air and vacuum valves, the provision of additional tee-offs with valves and blank flanges can reduce later interruptions to system operation. Space may also be an issue for locating air vessels.
9. Prepare a specification for the computer study to confirm the assessments of the problems and their solutions and to refine the control and suppression strategies as appropriate.
10. With respect to fluid transient problems, finalize the system design and ensure that operational constraints and guidelines incorporate recommendations in accordance with the validated control and suppression strategies.
11. Devise a test programme, for incorporation into the commissioning procedures, to confirm the response of the suppression and control strategy under selected transient flow conditions. These need not necessarily be the 'worst-case' scenarios. For example, a pump trip or valve closure from, say, 25 per cent flow could be used to compare with corresponding data from one of the computer simulations. It is also essential that this is included in the computer-assisted assessments as testing of sections of a complex system at part load can bring its own problems.

Several demonstration examples are presented in the next section. They are all derived from real systems and situations, though much of the physical data, such as pipeline dimensions and flows, etc., have been changed so that they are now 'typical' rather than 'actual' situations.

The early examples are quite simple systems that are discussed at some length, closely following the procedure outlined above. The later examples increase in complexity and are discussed in more general terms. They help to broaden the scope of the applications to different industries as well as enabling a variety of transient problems and control strategies to be reviewed.

SECTION 2.2

Demonstration Examples

The examples that follow are typical cases that have been derived from real situations. They have been selected to help illustrate the outline procedure for risk assessments presented in the previous section, and to reinforce one or two important aspects of fluid transient work.

To repeat a point made earlier, no two systems are really quite the same, and the problems faced by apparently similar systems may well vary. Furthermore, there is seldom a unique solution to these problems, at least on technical grounds, and different engineers may prefer different suppression and control strategies for various good reasons based on their own previous experiences. However, by developing a structured approach to the assessment of fluid transient events, it is hoped that some of the mystery and uncertainty surrounding them can be dispersed.

2.2.1 Rising Main Example 1

This first example, which was used in Part 1 to illustrate how transient events develop, is a rising main for conveying untreated water from a river to a storage reservoir. Figure 2.1 shows the longitudinal section of the system, including the option that two possible routes exist for part of the line. In this first example, only the lower route will be considered.

The overall system is to be developed in two stages. The initial design specification is to provide a flowrate of 35 L/s, but it is anticipated that the system will need to be uprated to provide a flowrate of 55 L/s in approximately 10 years time.

An assessment is required of the hazards to which the system may be exposed from fluid transients, and recommendations are needed for transient control to ensure its safety for phase 1 operation. A preliminary indication is also required of the consequences, with regard to fluid transients, of uprating the system to meet the anticipated phase 2 demand.

Following the procedure proposed in Section 2.1, the first step is to collate the

Fig. 2.1 *General layout for a 15 km ductile iron rising main pipeline for conveying untreated water from a river to a storage reservoir*

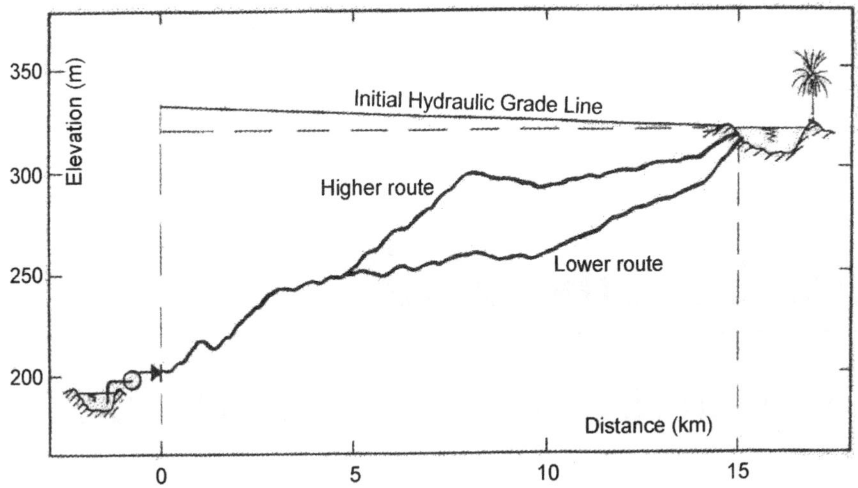

data describing the system. Figure 2.1 gives the general layout for the ductile iron pipeline and can be used as the longitudinal section.

At the pumping station, three identical pumps are to be installed in parallel, two of which will be in use normally, with one on standby. Two pumps will give a combined delivery of 35 L/s. Each pump will be fitted with suction and delivery valves, and a check valve. At this stage, firm orders have not been placed for either the pumps or the valves, and hence engineering judgement will be required with respect to selecting suitable performance characteristics. However, the following data are available:

Pipeline diameter	300 mm
Pipeline wall thickness	12 mm for 7 km
	10 mm for 5 km
	8 mm for 3 km
Elevation above river level	192 m above datum
Elevation of storage reservoir	320 m above datum

The operating limits for the system, step 2, are that:

(a) Vapour cavities and column separation should not occur, though subatmospheric pressures are acceptable down to 6 m water gauge vacuum.
(b) The pressures should not exceed the hydraulic test pressures of the pipeline or the rated maximum pressure of any of the components, such as valve casings and anchor blocks, whichever is the lower.
(c) Shock loads to the system and its mountings should be avoided.

Step 3 is to prepare the 'longitudinal section' of the system and collect additional relevant data. Figure 2.2, shown here on a very reduced scale, is a longitudinal section, similar to Fig. 2.1, to which has been added the static head line, the hydraulic grade line for the initial steady flow and the maximum permissible head line based on the hydraulic test pressures to be used as the system is built. These tests will be on the three line segments of different wall thickness given in the data list above. The steps in the maximum permissible head line correspond to where the wall thickness changes.

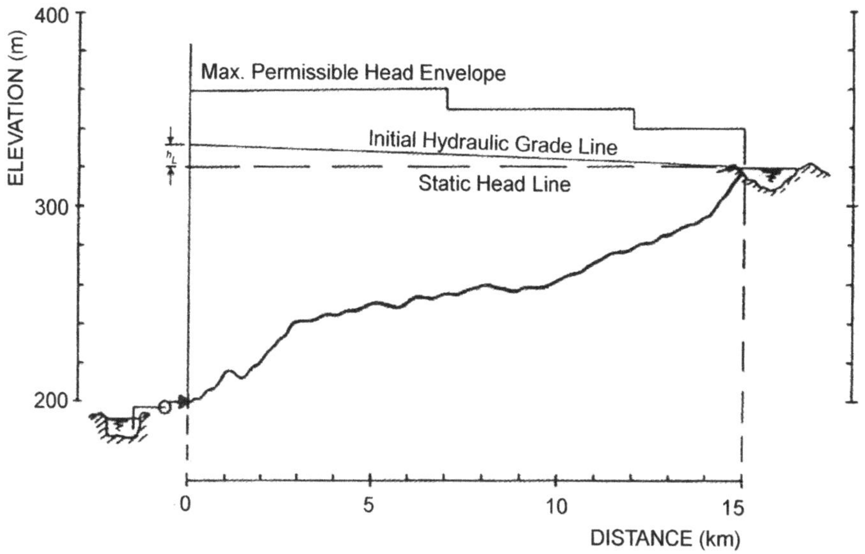

Fig. 2.2 *Longitudinal section to be used to show pictorially the pressure changes developing in the system*

The next step, step 4, is to list all the likely causes of changes to initially steady flows in the system (see Table 2.1). These include pump starts, pump shutdown preceded by closure of the discharge valves, loss of power to all pumps simultaneously and the loss of one pump out of two running, e.g. it trips out owing to a thermal overload or a high bearing temperature.

Of these potential threats to the system, only the first generates an initial rise in pressure, while the others cause the pressure to drop from the location where the transients are initiated. Provided the start-up of the pumps follows good practice, i.e. the pumps are started against closed discharge valves (one at a time), and the discharge valves are opened slowly, the pressure changes in the system will be quite modest and not give rise to any problems.

In terms of their probable significance, step 5, the flow reductions may be listed in the following order of descending importance: full pump trip (e.g. loss of power), tripping of one pump out of two and closure of the discharge valves followed by tripping out the pumps. The critical case, therefore, to be studied first is the power failure, which is an 'external' or uncontrolled event.

Moving on to step 6, the frictional head loss, h_L, was calculated from the Darcy–Weisbach equation using a friction factor of 0.0185. This diagram, i.e. the longitudinal section, on a larger scale, will be used to sketch out how the changes to the pressure heads in the system develop.

The speed at which the transients are propagated through the pipelines can be estimated from Fig. 3.2. From the pipe data given above, the D/e ratios for the three sections of pipe are 25, 30 and 37.5. These give wave speeds of about 1200–1250 m/s. A value of 1250 m/s will be used to err on the safe side, as it will imply slightly higher pressure changes.

The Joukowsky pressure change, h_j, for a complete stoppage is now calculated from equation (1.1). Noting that the flow velocity corresponding to 35 L/s through a 300 mm pipe is 0.495 m/s, $h_j = 1250 \times 0.495/9.81 = 63.07$ m head.

One final piece of information that is needed is an estimate of the time it takes, following the pump trip, for the flow at the pump discharge to fall to zero. At this stage this can only be sheer guesswork, supported by engineering judgement and experience.

To put some limits on this, in general, modern pumps tend to run down in perhaps 1–10 s when tripped on load, depending on the back pressure. Small pumps run down relatively quickly, while larger pumps, with their greater inertia, take longer to run down although, again, the magnitude of the back pressure will also have some influence on this. The flow usually stops and tends to reverse before the pump speed is zero. For the present example, it will be assumed that the flow drops to zero at the pump discharge 7 s after the power is lost to the pumps.

The next stage, still within step 6, is to transfer this information onto the longitudinal section in a way that provides a reasonable indication of how the transients start to propagate through the system. Refer now to Fig. 2.3.

As soon as the pump is tripped, the pressure immediately downstream starts to fall and the hydraulic grade line (HGL) will start to change. At the precise moment the flow reaches zero at the pump, i.e. 7 s later, when, ideally, the check valve just closes, the pressure has dropped by the full Joukowsky head, h_j. Also, the leading edge of the wavefront has arrived at point A in Fig. 2.3. The distance to this point is 8.75 km, i.e. the distance travelled in 7 s by the leading edge of the wavefront moving at 1250 m/s.

For this particular instant, a curved line may be sketched in for that part of the hydraulic grade line between the pump and point A. Strictly speaking, only two points are known on this curve, one is at the pump discharge where the pressure head has dropped by h_j, and the second is on the initial hydraulic grade line

Fig. 2.3 *Longitudinal section of the pipeline showing the changing hydraulic grade line (HGL) as transient pressure waves propagate through the system following failure of power to the pumps*

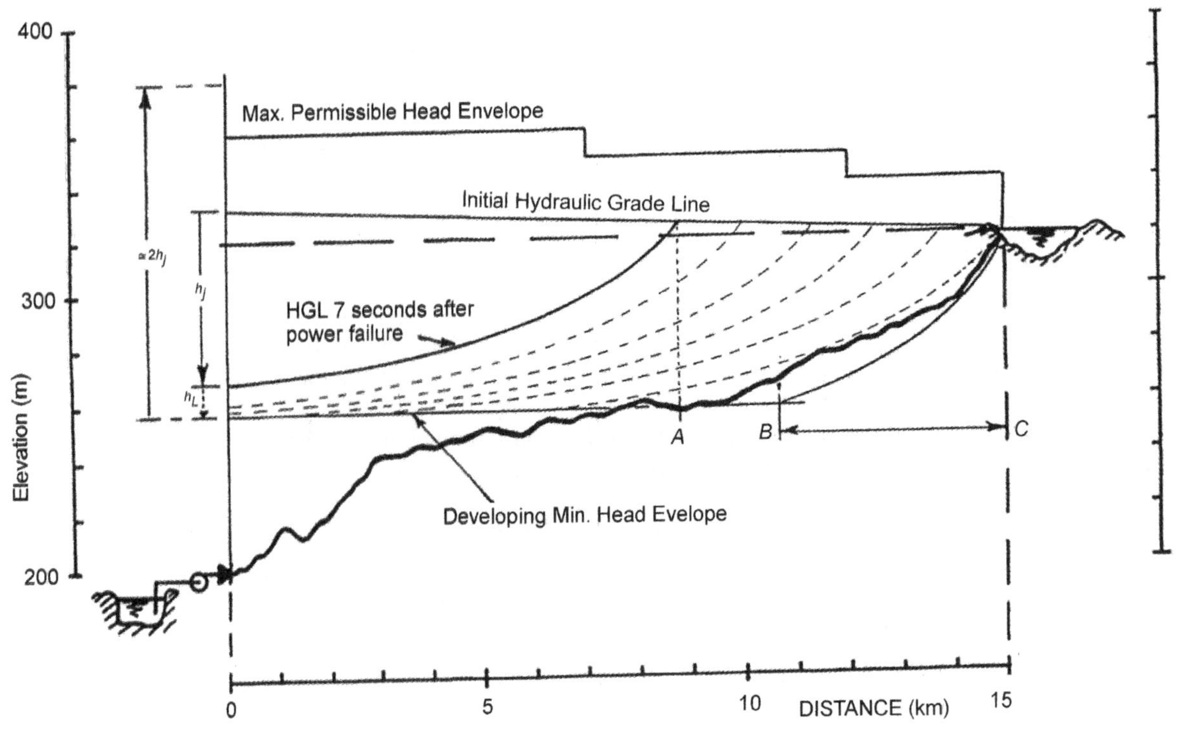

above A. However, it is known from experience that, when a pump is tripped, the pressure falls rapidly at first, and then more slowly, and hence the curve should be convex downwards. It is impossible to say precisely how curved it will be, but the general shape will be similar to the solid line shown in Fig. 2.3.

As the pressure wave, which extends over 8750 m, continues to move towards the reservoir, the hydraulic grade line changes continuously. The dashed lines shown are at 1 s intervals. The pressure at the pump discharge continues to fall very slightly owing to the friction gradient levelling out – see Figs 1.2 to 1.5 in Part 1 – and the Minimum Head Envelope, i.e. the lower limit to which the pressure along the line falls, can be seen developing.

Eventually, the leading edge of the wavefront reaches the reservoir, though the trailing edge is 8.75 km behind it. The leading edge is reflected back towards the pump station, tending to maintain the pressure at the reservoir head. The returning leading edge and the trailing edge still moving towards the reservoir pass each other at point B on the diagram, where the distance BC is equal to half the length, i.e. 4.375 km, of the pressure wave.

Point B represents the approximate limit along the pipeline from the pump station that experiences the full Joukowsky pressure change. The outlet pressure to the reservoir will always be at atmospheric pressure, so point B can be joined up to the end of the pipe as shown. This provides a reasonably good estimate for this system of the Minimum Head Envelope.

The indications from this sketch are that about half the line remains at a positive pressure, but, from about the 10 km point, subatmospheric pressures can be expected if no transient control is adopted. Figure 2.3 suggests the pressure falls almost to a full vacuum. A computer analysis of the same problem predicts pressures of 3–4 m below atmospheric.

The pressure wave, meanwhile, returns towards the pump station and is reflected from the closed check valve. Because it is a closed end, the consequence is that the pressure change will be additive and will increase by approximately twice the Joukowsky head from the level to which it had fallen. As can be seen from Fig. 2.3, this would take it above the Maximum Permissible Head Envelope.

To summarize progress thus far, as a result of sketching out the changes in pressure head in Fig. 2.3, the following conclusions may be drawn:

1. Part of the system will experience subatmospheric pressures.
2. The maximum permissible pressure head will be exceeded.
3. A further deduction may also be made, namely that the discharge valve on the pumps must be closed over a time period that is rather longer than the pump rundown time, in order to avoid negative pressures. With a view to the specification to be formulated later for the computer analysis, and bearing in mind the non-linear nature of valve characteristics, the preliminary suggestion might be a valve closure of, say, 60 s, i.e. about 8 or 9 times the pump rundown time for this particular system.
4. Surge protection will be required in order to reduce the peak pressures in the system to within acceptable limits.

This leads to step 7, which is to review possible options for dealing with the problem of the excess pressures. These include strengthening the pipe, pressure relief through suitable valves and the use of an air vessel.

To specify pipes having a higher pressure rating is quite straightforward, and the main feature to address will be the increased cost compared with the other options. Pressure relief valves, which release water back into the river and which close slowly, will also alleviate the problem of high pressures and be relatively economical.

However, the most common approach to the problem is the use of an air vessel. The reason for this is that it addresses the change that occurs *first*, namely the initial reduction in flow and drop in pressure. By replacing the pump as a source of energy, it serves to reduce the *rate of change* and, because there will then be time for beneficial wave reflections to return from the open end the subsequent increases in pressure will be lower. Also, looking to the future, an air vessel will probably be required anyway for the higher flowrate in phase 2 of the project.

To explore the air vessel option further, use will be made of the design charts described in Subsection 3.3.1. The Joukowsky head for complete stoppage, h_j, has been found to be 63.07 m, and the friction parameter, from equation (3.37), is $11.55/63.07 = 0.18$. The most appropriate chart is therefore Fig. 3.17a, reproduced below as Fig. 2.4.

From equation (3.39)

$$\bar{h}_o = \frac{(320 - 200) + 11.55 + 10}{63.07} = 2.24$$

From Fig. 2.3 the maximum permitted head is 160 m at the pump discharge, and using this in equation (3.40) gives a maximum head ratio

$$\mathrm{HR}_{\max} = \frac{160 - 120}{63.07} = 0.634$$

For completeness, the Minimum Head Ratio at the pump, $\mathrm{HR}_{\min}$, will also be calculated. The minimum pressure that occurs there is seen in Fig. 2.3 to be about 60 m, and hence

$$\mathrm{HR}_{\min} = \frac{60 - 120}{63.07} = -0.95$$

As this is off the bottom of Fig. 2.4, the inference may be drawn that negative pressures are not a major problem for the low flowrate in this system.

Now interpolate a line for $\bar{h}_o = 2.24$ on the top half of Fig. 2.4, which refers to the maximum head ratio, and then draw in a line for the value of 0.634 calculated above. From the intersection of these two lines, drop another line down to the x axis, which it meets at a value for K in the region of 2.8 (see Fig. 2.4); K is the composite air vessel and pipeline parameter, given by equation (3.42), i.e.

$$K = 2.8 = \frac{cC_o}{nAv_oL} = \frac{1250}{(\pi/4) \times 0.3^2 \times 0.495 \times 15\,000}\,\frac{C_o}{n} = 2.38 \times \frac{C_o}{n}$$

Assigning a compromise value of 1.2 to n, the volume C_o of air required in the air vessel under normal operating conditions of a flow of 35 L/s is 1.41 m^3.

When the pumps are tripped this air volume expands and the pressure falls to a minimum, for which the Minimum Head Ratio is about –0.645, as shown in

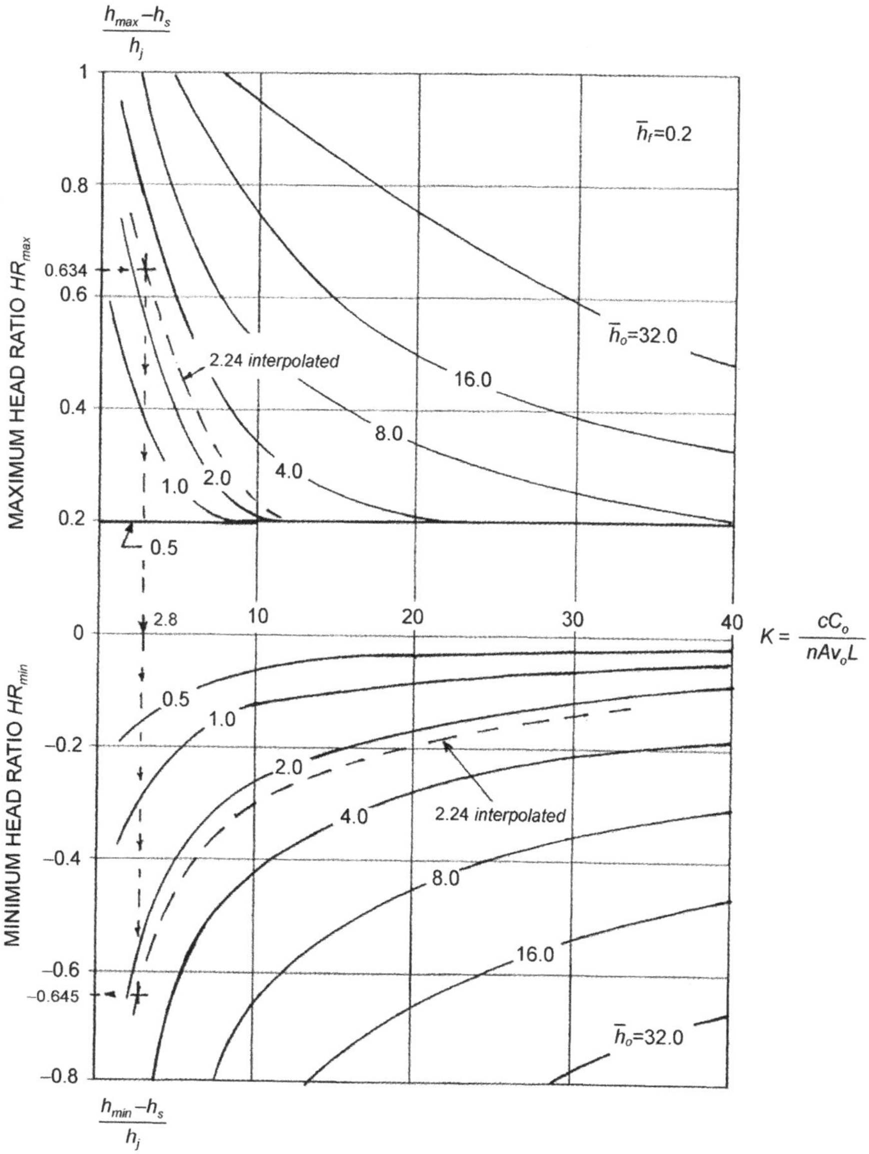

Fig. 2.4 *The air vessel design chart, Fig. 3.17a, that is appropriate for the first demonstration example*

Fig. 2.4. Substitute this value back into equation (3.41) to obtain the Minimum Head that occurs in the air vessel

$$h_{\min} = -0.645 \times 63.07 + 120 = 79.32\,\text{m}$$

This is the pressure to which the 1.41 m^3 volume of air in the vessel expands as the flow comes to rest. The corresponding air volume, from equation (3.43), is

$$C' = C_o\left(\frac{h_o^*}{h_{\min}^*}\right)^{1/n} = 1.41 \times \left(\frac{120 + 11.55 + 10}{79.32 + 10}\right)^{0.833} = 2.07\,\text{m}^3$$

Allowing an initial Factor of Safety of about 20 per cent, given that these preliminary design figures are to be checked with a computer-based analysis, the estimated total volume for the air vessel may be taken as $2.5\,m^3$.

The loss of two pumps will be more severe than the loss of a single pump running on its own, and such an air vessel would also provide adequate protection for this eventuality.

When one pump out of two running in parallel is tripped, the pump that continues to run will maintain pressure in the discharge manifold. Flow reversal in the discharge branch from the tripped pump will therefore occur rapidly. It is important that the check valve closes as near as possible to the instant this flow becomes zero to avoid the risk of 'check valve slam'. Since the delivery pressure for this system is quite high, the rate at which the flow decelerates should be investigated in the computer study, and it may be found necessary to make recommendations on the type of check valves to be used.

To summarize progress again for Phase 1 – surge protection is required to contain the pressures in the system within acceptable limits and, on technical grounds, may take the form of stronger pipes, a pressure relief system or an air vessel of about $2.5\,m^3$ capacity. The closure time for the pump discharge valves should extend over perhaps 60 s or more, and care should be exercised over the choice of check valves to avoid shock loads to the pump and pipelines.

Before moving to Step 9, the preparation of a specification for a computer analysis, the consequences of uprating the system for Phase 2 operation should be assessed. This can begin by sketching out the changing Hydraulic Grade Lines as before, and is shown here as Fig. 2.5.

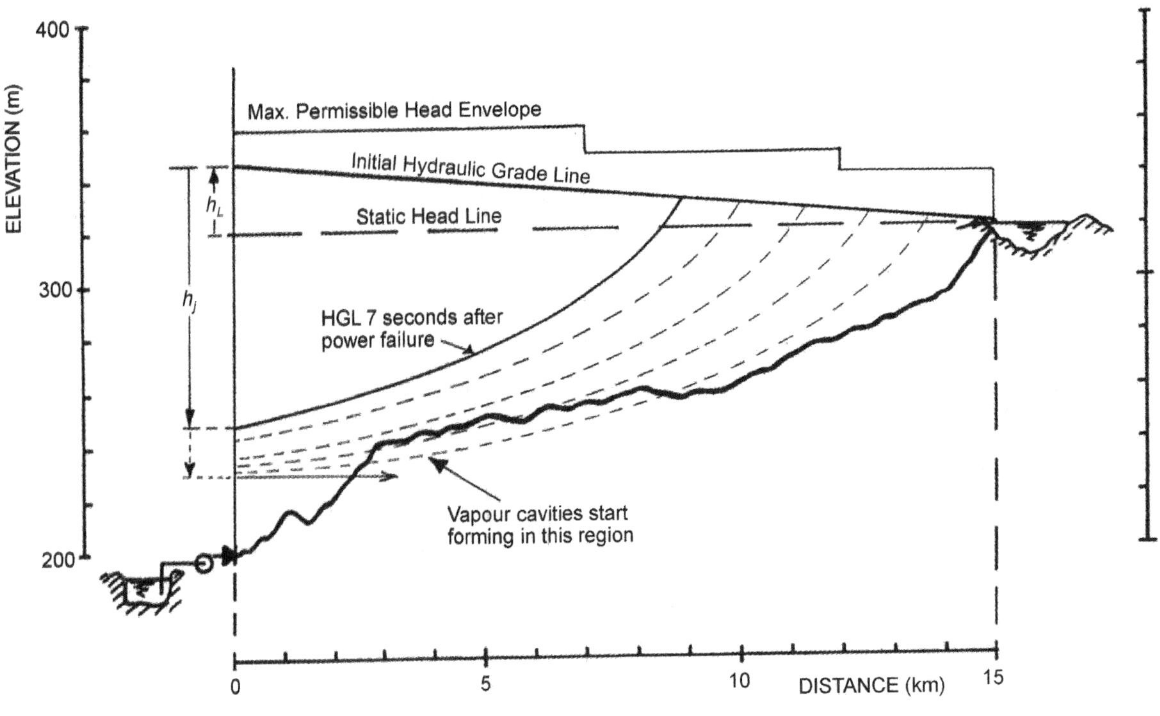

Fig. 2.5 *Longitudinal section, similar to Fig. 2.3, for Phase 2. Note the higher friction head and how the minimum head envelope drops below the pipeline indicating the development of vapour cavities and liquid column separation*

A higher pump discharge head is required, since the frictional head increases to 26 m at the new flowrate of 55 L/s (the Darcy–Weisbach friction factor is now 0.017), and the revised Joukowsky head following a power failure or full pump trip is 99 m.

The HGL profile at the moment the flow at the pump reached zero, just before the end of the pump rundown, is sketched in by hand and stepped along at 1 s intervals as for phase 1. It rapidly becomes apparent that subatmospheric pressures and vapour cavities will occur 3–4 km downstream of the pumps.

To continue to sketch out the HGL in this fashion has no physical significance, but this is unimportant since the objective has been achieved, namely that of assessing whether or not the pressure was likely to fall to unacceptably low levels. It does, and hence some strategy must be devised to maintain it at an acceptable level.

A bypass arrangement around the pumps will not work since the pressure head at the pump discharge remains above the suction pressure. It is also highly likely, especially in view of the assessment for phase 1, that the pressures following the collapse of the vapour cavities will be too high. Therefore, the most suitable option is to investigate the use of a larger air vessel.

The friction parameter

$$\bar{h}_{\mathrm{f}} = \frac{26}{99} = 0.2626$$

and

$$\bar{h}_{\mathrm{o}} = \frac{h_{\mathrm{s}} + h_L + h_{\mathrm{atm}}}{h_{\mathrm{j}}} = \frac{120 + 26 + 10}{99} = 1.58$$

This friction factor again points to the use of Fig. 3.17a for the estimation of an air vessel size and curves for $\bar{h}_{\mathrm{o}} = 1.58$ interpolated.

Before determining the Maximum and Minimum Head Ratios, $\mathrm{HR}_{\mathrm{max}}$ and $\mathrm{HR}_{\mathrm{min}}$ respectively, curves for the *Practical* Maximum and Minimum Head Envelopes need to be estimated. They must fall within the Theoretical Max/Min profiles but are unlikely to coincide with them. Their estimation is a matter of engineering judgement, but, from experience, they will take the general shape of the curves given in Figs 3.19 to 3.21 in Part 3.

The essential features are that the curve for the Practical Maximum Head Envelope should remain below the Maximum Permissible Head Envelope which was based on the hydraulic test pressures, and that, for the Practical Minimum Head Envelope to avoid subatmospheric pressures, it should be above the pipeline profile for its whole length. The suggested curves for this example are shown in Fig. 2.6. From this it is possible to obtain the Maximum and Minimum Heads that can be permitted at the pump discharge – 160 and 50 m head respectively. Hence

$$\mathrm{HR}_{\mathrm{max}} = \frac{160 - 120}{99} = 0.404$$

$$\mathrm{HR}_{\mathrm{min}} = \frac{50 - 120}{99} = -0.707$$

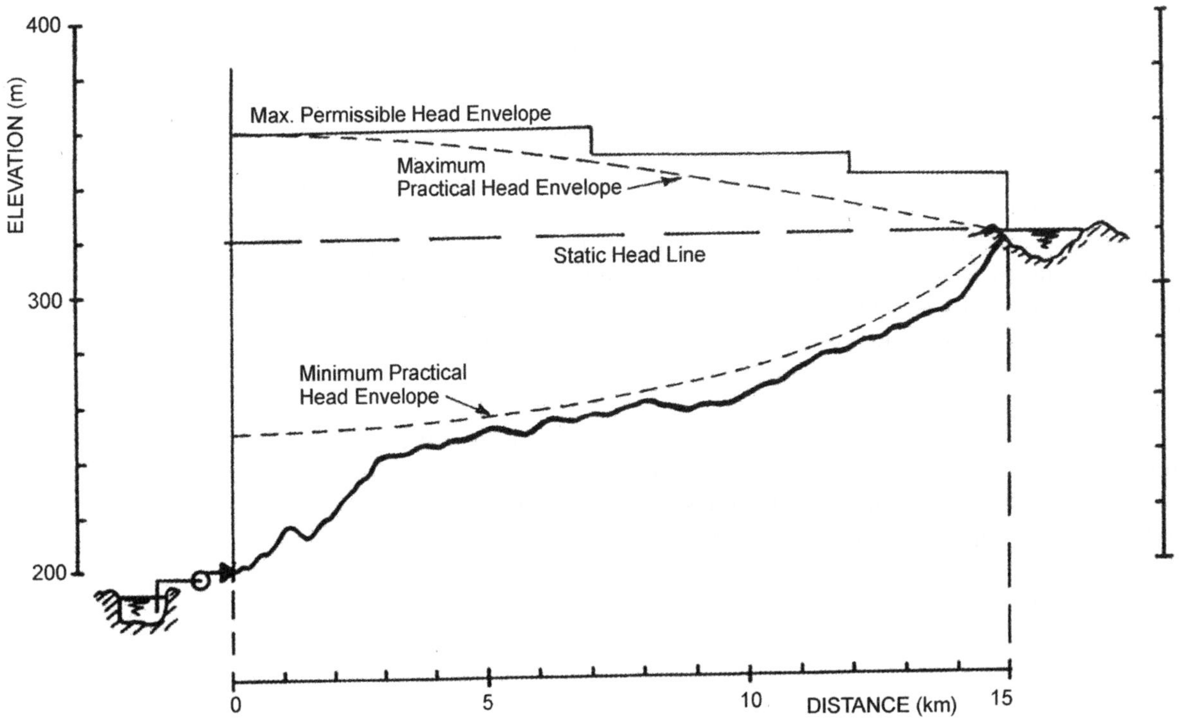

Fig. 2.6 *The longitudinal section of the pipeline showing the Practical Maximum and Minimum Head Envelopes which are intended to be those that occur when an air vessel is in service. They are sketched in by eye, using engineering judgement*

Each of these values will give an air vessel capacity from the design chart, Fig. 3.17a, repeated here as Fig. 2.7, and the larger of the two values is the one to be used. The maximum pressure is again the critical design case (the minimum pressure being almost off the bottom of the chart and hence not shown). A value of $K = 3.8$ may be read off the x axis, yielding a value of $C_o = 3.015\,\text{m}^3$ for the volume of air required in the vessel under initial steady flow conditions.

Reading down to the interpolated curve for $\bar{h}_o = 1.58$, as before, and across to the vertical axis gives a value of around –0.36 for the minimum head when the water column just comes to rest and the air in the vessel has expanded to its maximum volume, C'. Therefore

$$h_{\min} = -0.36 \times 99 + 120 = 84.36\,\text{m}^3$$

and

$$C' = 3.015 \times \left(\frac{120 + 26 + 10}{84.36 + 10}\right)^{1/1.2} = 4.58\,\text{m}^3$$

Allowing a factor of safety of about 20 per cent raises this value to $5.5\,\text{m}^3$ for the total volume of the vessel.

The design charts enable estimates to be made of the initial air volume and total capacity of the vessel (or vessels). These must be translated into physical shapes. One rule of thumb that has been used for cylindrical vessels installed with their longitudinal axis vertical is that their height is about 2.5 times their diameter.

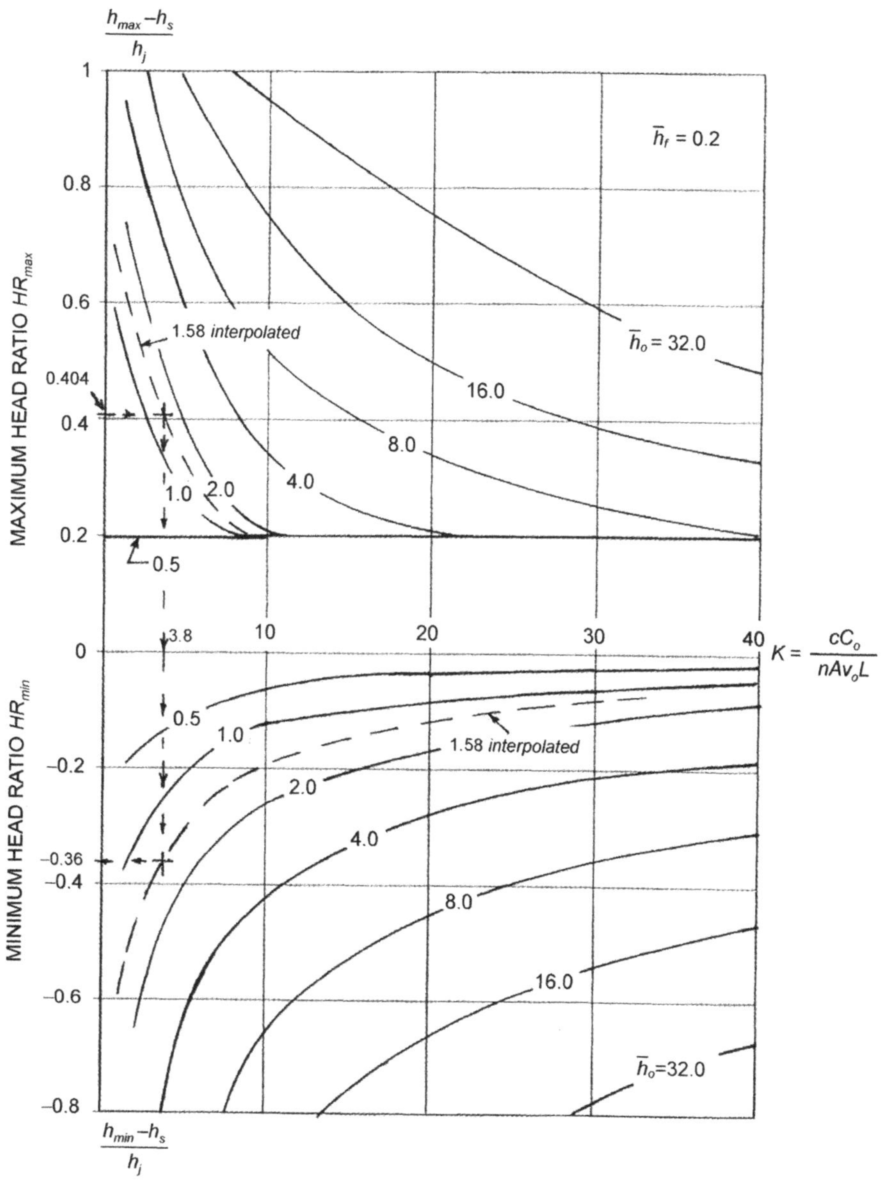

Fig. 2.7 *Air vessel design chart for Phase 2*

Hence, having a provisional value for the total volume, the corresponding dimensions can be estimated, together with the initial depth of water inside the vessel. Nevertheless, there is no firm rule on the proportions, it is a question of what is available at a reasonable cost and complies with pressure vessel standards.

If the diameter for a single vessel proves to be so large that the cost would be excessive, two or more vessels having the equivalent air volume and total capacity can be considered. Various alternatives are presented in Subsection 1.3.3.1.

The next stage is step 9, in which the specification for the computer study is prepared.

The basic data describing the system should be collated. The analyses required should be based on the outcome of the preliminary assessment that has been developed above. Assuming that on economic grounds a decision has been taken to opt for an air vessel, the specification of the problems to be addressed may take the following form:

1. Analyse the system for a full pump trip/power failure from an initial flow of 35 L/s. Check for subatmospheric pressures, especially beyond 8 km from the pump station, and excess pressures near to the pump station. The results should include the predicted maximum and minimum head profiles.
2. Repeat the above analysis, but with an air vessel of 2.5 m^3 capacity, installed close to the pump delivery valves. The air volume under initial steady flow conditions is estimated to be 1.41 m^3. A possible diameter of the air vessel is 1.2 m, with a height of about 2.25 m, and the initial elevation of the water required in the vessel above the pump discharge is estimated to be 1 m. These values should be varied, as found necessary, to ensure that the pressure fluctuations remain within the permitted limits. Recommendations can be made for further analyses if the vessel size appears unduly conservative, with a view to reducing its dimensions and, probably, its cost.

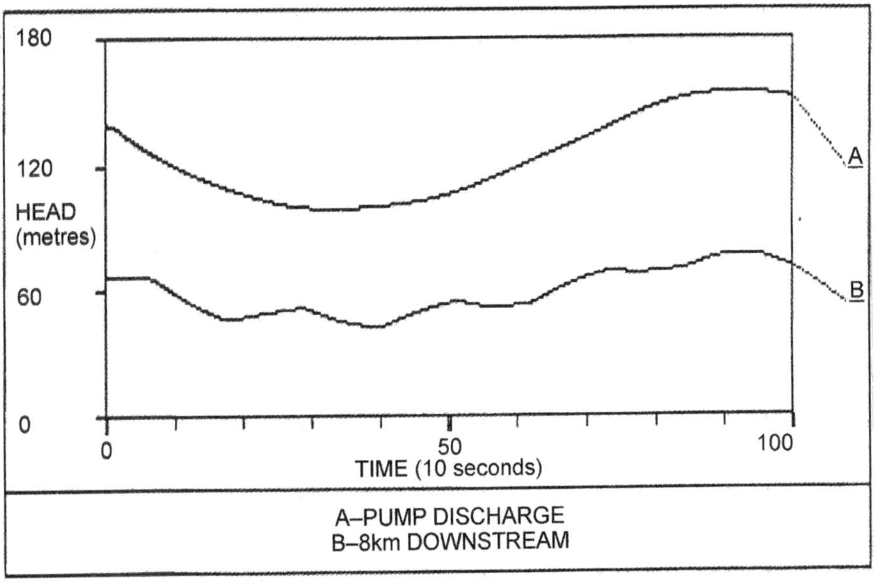

Fig. 2.8 *A pressure–time history at the pump discharge for a full pump trip at a Phase 1 flow of 35 L/s. This indicates that the pressure fluctuations are within acceptable limits with an air vessel of the capacity estimated above. Further analyses could be carried out, if required, to investigate whether a slightly smaller vessel would be adequate*

3. Repeat analysis 2 (with agreed modifications to the air vessel dimensions, if required) to investigate the influence of the closure times of the pump discharge valves. The current proposal is to install gate valves having a linear rate of closure of about 60 s. Is this adequate, can it be reduced, or should it be increased?
4. Analyse the system for a single pump trip from two pumps running in parallel. This should focus on events in the immediate vicinity of the check valve on the pump that has been tripped. Results of interest include the pressures on both sides of the check valve, the maximum reverse velocity of the water and its average deceleration for a range of delayed closures of the check valve after

the flow drops to zero. The suggested range is 1, 0.5, 0.25, 0.1 and 0.05 s after flow reversal occurs.

5. In anticipation of a controlled test during commissioning analyse the system for a pump trip for one pump running on part load (e.g. 25 per cent flow), and provide pressure–time histories for the pump discharge with the air vessel in service.
6. Repeat the second analysis for the anticipated phase 2 flowrate of 55 L/s with the proposed larger air vessel in service. Assume an initial air volume of 3.015 m^3, a diameter of, say, 1.5 m, and an initial water level in the vessel of 1.2 m above the pump discharge.

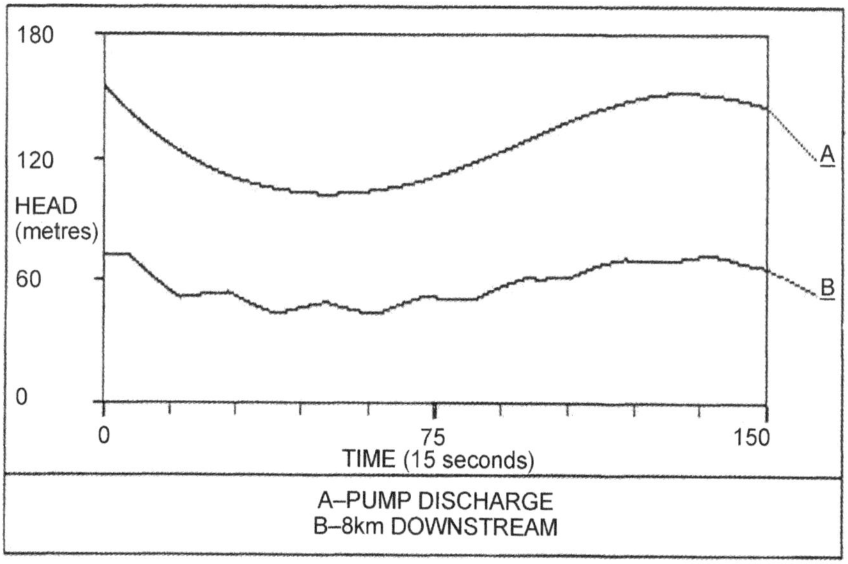

Fig. 2.9 *Similar to Fig. 2.8, but for the phase 2 flow of 55 L/s and the larger air vessel. The pressures are again acceptable. Note the longer period of oscillation of the pressure changes*

When the transient analyses have been satisfactorily completed, the relevant aspects of the system design can be finalized. Any necessary operational constraints, such as the operating instructions for pump starts and controlled shutdowns and pump changeover operations, should also be prepared to ensure compliance with the validated control and suppression strategy. At the same time, a test programme, to be undertaken as part of the commissioning process, should be drawn up and steps taken to have the necessary instrumentation in place.

2.2.2 Rising Main Example 2

This example is the higher route that is proposed as an alternative in the previous case (see Fig. 2.1). The flowrates and major characteristics of the system are exactly the same – only the route changes, and this makes it a different problem with respect to the consequences of transient flows.

The initial steps in assessing whether or not problems may occur are the same as before. A clear longitudinal section of the system must be prepared, the initial hydraulic grade line drawn in and the transient flow parameters of wave propagation speed and Joukowsky pressure change evaluated. These are sketched in as shown in Fig. 2.10.

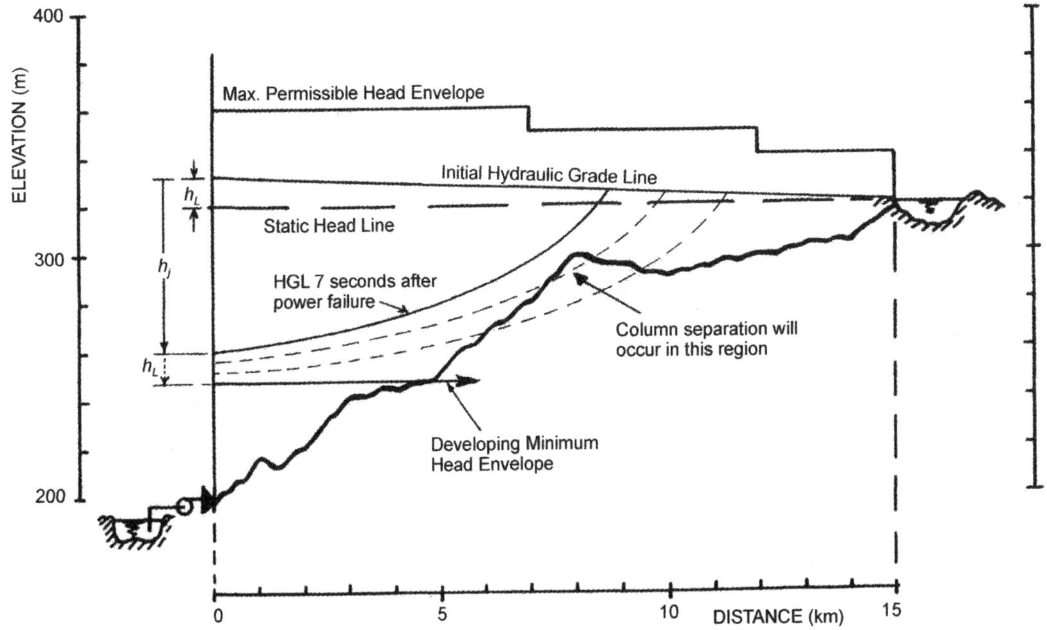

Fig. 2.10 *Longitudinal section of the pipeline following the higher route. A 'snapshot' view of the Hydraulic Grade Line 7 s after the pump trip is shown, together with the developing Minimum Head Envelope*

The operating conditions that are likely to give rise to fluctuating pressures are: pump starts, the stopping of one or both pumps, either through loss of power or a normal shutdown with the discharge valves being closed first, and the tripping of one pump out of two running in parallel.

Of these, the loss of power to the pumps will present the worst case, presuming that a satisfactory time has been adopted for the controlled closure of the pump discharge valves.

To get a feel for the scale of the problem, the changing hydraulic grade line following a full pump trip should be sketched in on the longitudinal section (see Fig. 2.10). The Joukowsky head drop and distance travelled by the leading edge of the wavefront provides two points that can be joined up with the curved line shown for the HGL at 7 s. By 8 s, subatmospheric pressures are predicted at the high point in the line, some 8 km from the pump station, and these pressures continue to fall, leading to separation of the water column. This would be followed by a sudden increase in pressure as vapour cavities collapse. This was illustrated for this system in Fig. 1.8.

Assuming that the operating constraints are that pressures should neither rise above the Maximum Permissible Head Envelope nor fall below atmospheric pressure, some form of suppression and control is again necessary. Since the first effect on the system following the pump trip is the downsurge, i.e. the drop in pressure, the suppression strategy should focus on reducing the rate at which this happens. Among the options that may be considered are:

(a) Use the lower route discussed in the previous example.
(b) Install an air vessel. A larger one will be needed for this route.
(c) Install a feed tank in the vicinity of the peak in the line 8 km downstream of the pump station (in this particular example it will be found that a feed tank will not prevent the pressures exceeding the maximum allowed).

(d) Develop a strategy based on the use of a small air vessel at the pump station, plus a feed tank as in (c).
(e) Alternatively, if acceptable, use air admission at the peak, followed by controlled release, plus a small air vessel at the pump station.

Several of these possibilities should prove capable of controlling the pressure fluctuations within the limits imposed. The final choice will be decided on the basis of cost, the preferences of the system designer and/or operator, perhaps on environmental grounds, and the reliability and maintenance requirements of the remote feed tank or vacuum relief/air release valves.

Meanwhile, some outline designs using logic, aided by the design charts and equations of Part 3, will be necessary precursors to the computer study that will be required if the option of following the lower route is not viable.

If reliance is to be placed on an air vessel near to the pump discharges, the following steps will lead to an outline design requirement. First, sketch in the *practical* curves for the maximum and minimum head profiles, as shown in Fig. 2.11, so that the *actual* maximum and minimum heads at the pump station *as a result of having an air vessel in service* can be estimated. For the present case, these can be read off as 160 and 95 m head respectively.

Since the technical specification of the system is the same as the previous example, the friction and system parameters, $\bar{h}_f$ and $\bar{h}_o$, will be the same, i.e. 0.18 and 2.24 respectively. Hence, use Fig. 3.17a again and interpolate the curve for $\bar{h}_o$ in Fig. 2.12.

The maximum head ratio, HR_{max}, is also the same as before, i.e. 0.634, but the changed profile for the pipeline leads to a new minimum head ratio of

$$HR_{min} = \frac{95 - 120}{63.07} = -0.4$$

This is the critical design case since it points to a higher K value, of 6.75, in

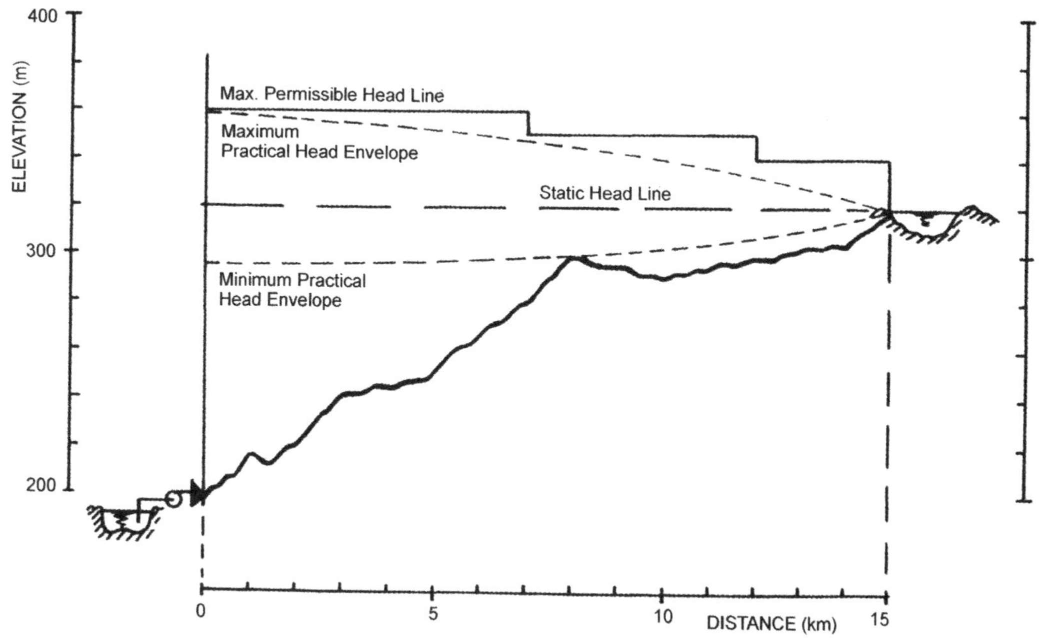

Fig. 2.11 *Longitudinal section with the estimated practical maximum and minimum head envelopes sketched in*

Fig. 2.12 *Air vessel design chart appropriate for the second rising main example*

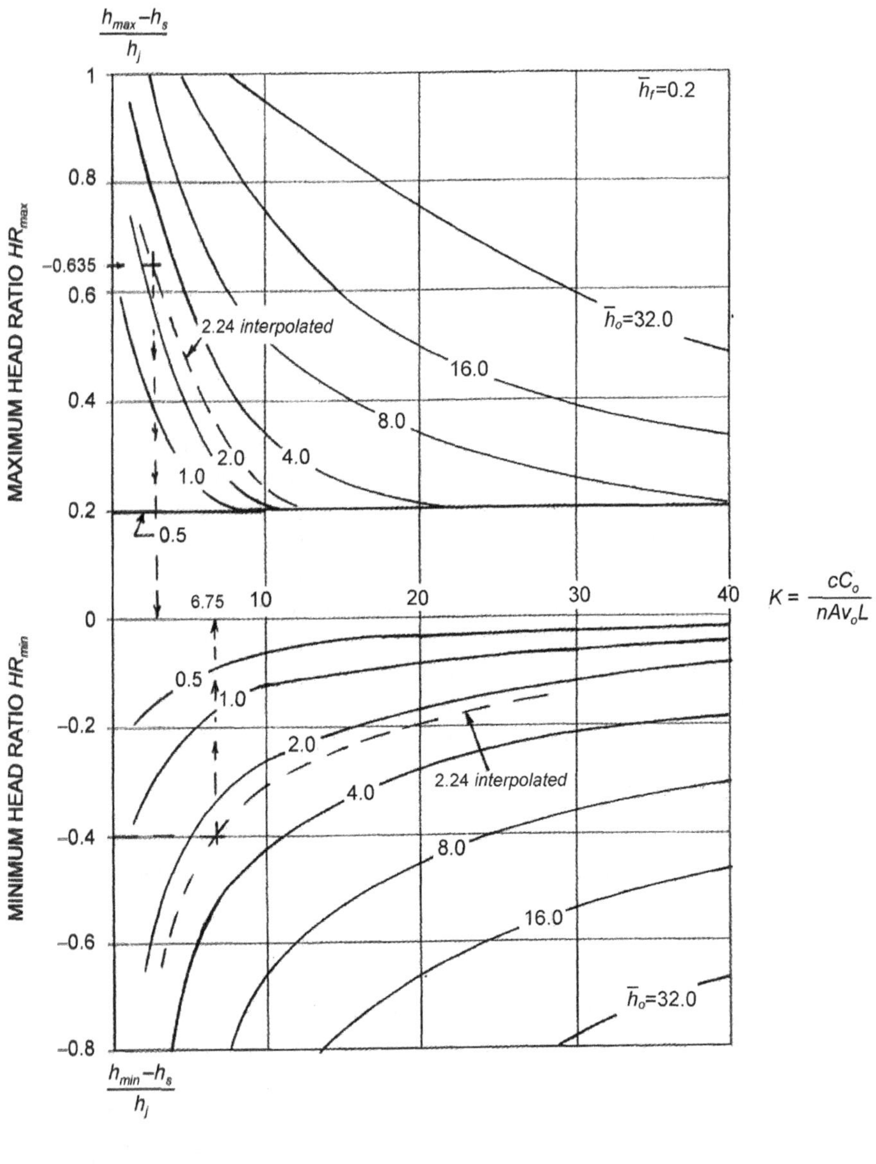

Fig. 2.12. The initial and final air volumes, C_o and C' respectively, are calculated from equations (3.42) and (3.43) as

$$K = 6.75 = \frac{cC_o}{nAv_oL} = \frac{1250}{(\pi/4) \times 0.3^2 \times 0.495 \times 15\,000}\,\frac{C_o}{n} = 2.38 \times \frac{C_o}{n}$$

and letting $n = 1.2$, say, leads to $C_o = 3.40\,\text{m}^3$

$$C' = 3.40 \times \left(\frac{120 + 11.55 + 10}{95 + 10}\right)^{1/1.2} = 4.36\,\text{m}^3$$

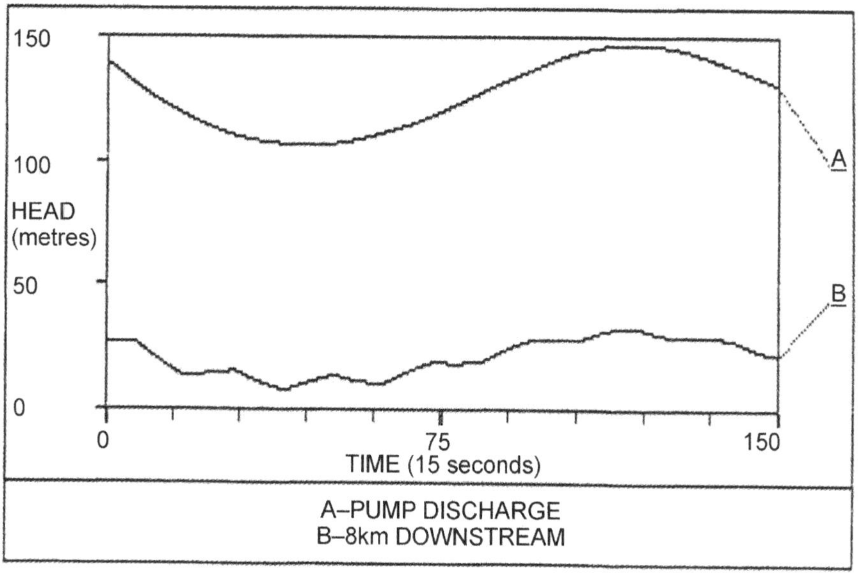

Fig. 2.13 *Pressure head–time histories following a pump trip for Example 2 with a 5.25 m^3 air vessel in service*

Increasing this by a safety margin of 20 per cent gives a final estimate of 5.25 m^3 for the total volume of the air vessel needed for this system.

An air vessel of 5.25 m^3 capacity might have a diameter of 1.4 m and be 3.4 m tall. Such a vessel is not unduly large and is certainly adequate, as indicated by Fig. 2.13 from a computer analysis, though some might prefer to seek a reduction if it can be shown to be safe to do so. Such optimization can only be effected through validated computer studies.

To develop a first estimate for the capacity of a feed tank at the high point 8 km from the pumping station, the equations developed in Subsections 3.2.2 and 3.2.3 can be employed. They will, however, only provide estimates with respect to controlling the pressure reduction. For this example a modest feed tank of under 1 m^3 capacity would suffice to maintain a positive pressure in the system, but a computer analysis shows (see Fig. 2.14), that the subsequent pressure rise exceeds the permitted pressure levels. Recourse to an air vessel is still necessary, therefore, either on its own or in conjunction with a feed tank. A policy decision on the preferred strategy is required!

The specification formulated for the computer study will be broadly similar to that for Example 1 and, briefly, may be as follows:

1. Analyse the system for a full pump trip/power failure from an initial flow of 35 L/s.
2. Repeat this analysis with the surge protection devices in service – either an air vessel on its own or a smaller one in conjunction with a feed tank, as described above. In the latter case it will probably be necessary to perform several analyses to establish the best combination.
3. With the air vessel, or air vessel plus feed tank, whichever is adopted and shown to be satisfactory from Step 2 above, investigate whether the proposed closure time of 60 s for the pump delivery valves is suitable. Advise on an alternative closure time if 60 s is not adequate.

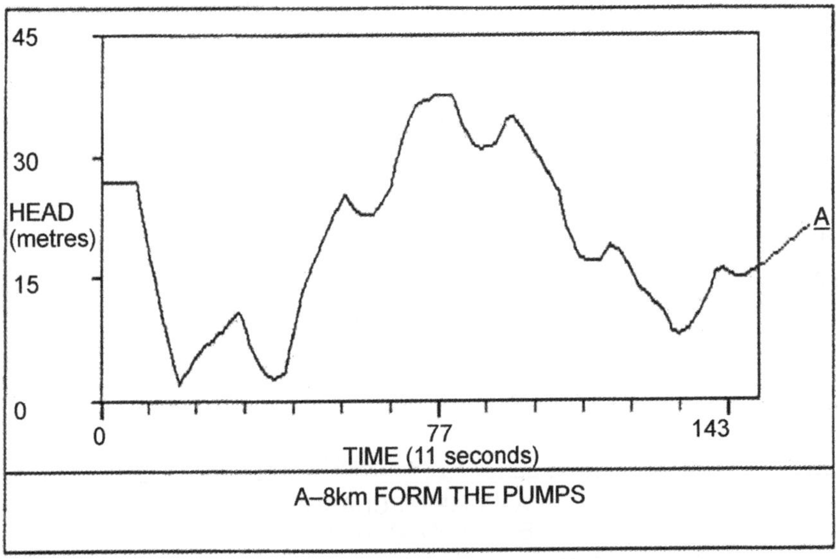

Fig. 2.14 *Pressure head–time history at the site of the feed tank, 8 km from the pump station, when in service on its own. The pressures here exceed the permitted level*

4. The single pump trip and the test programme should also be studied in exactly the same fashion as for the first example, together with the preparation of any necessary operational guidelines or constraints that are deemed prudent with respect to the control of transient pressures. This would include pump start-up and changeover procedures, and controlled shutdowns.

If air admission at peaks in the line is regarded as acceptable, perhaps in lieu of the feed tank, Step 2 should be revised accordingly. Combinations of suppression techniques such as this are quite normal, but it is less easy to develop estimates of capacities and sizes prior to computer studies.

2.2.3 A Pumped Outfall

Figure 2.15 illustrates a proposed sewage outfall. The pumping station is scheduled to discharge 350 L/s of treated sewage 2 km offshore through a 7 km

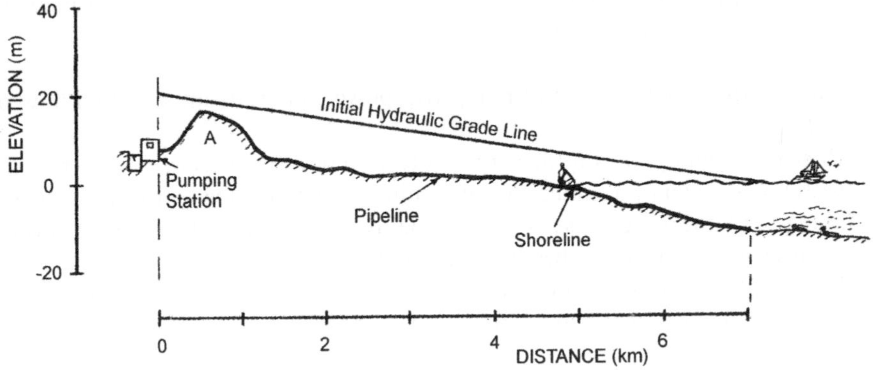

Fig. 2.15 *Longitudinal section for the pumped sewage outfall*

pipeline of Glass Reinforced Plastic (GRP). The downstream end is 10 m below the sea surface at Mean Low Water Spring (MLWS) tides, as shown in the sketch.

The task is to assess the risks associated with transient flows. These will be initiated by pump starts and stops. The former will comprise the start-up, separately, of two 50 per cent duty pumps installed in parallel and the opening of their delivery valves. The latter will comprise controlled pump trips (i.e. one by one, preceded by closure of the delivery valves) and uncontrolled trips caused by loss of electrical power. This summarizes the response to step 4 in Subsection 2.1.2.

This outfall is a low-head system, and consequently the events that will give rise to the most serious risks are the uncontrolled power failures and closure of the delivery valves before tripping the pumps. The valve closure can be regulated, and it will be necessary to specify a suitable valve closure time, but the power failure situation will require closer examination. Not only should the rate of flow reduction be controlled but the rapid collapse of vapour pockets should be avoided.

To assist in resolving Step 5 of the assessment procedure, Fig. 2.15 illustrates the longitudinal section complete with the initial hydraulic grade line. The frictional head drop is 20.4 m, based on a Darcy friction factor of 0.009, a pipeline diameter of 500 mm and the data listed above. The wall thickness is 14 mm, so, from Fig. 3.3, the wave propagation speed, c, is 910 m/s. The flow velocity at 350 L/s is 1.78 m/s, for which the full Joukowsky head change is 165 m.

Since the pipeline profile is largely downhill, there is very little positive pressure in the system. As soon as flow reductions occur, either through valve operations at the pump house or through pump trips, the pressure head at the upstream end will fall. Subatmospheric pressures will occur quite rapidly at the peak, A, approximately 750 m from the pump station. Given the magnitude of the Joukowsky pressure drop for a full flow stoppage, column separation is inevitable together with the risk of pipeline collapse.

To limit the pressure reduction, it is necessary to provide a flow into the system, either of liquid or air. Air vessels are not usually recommended for effluent systems although they are occasionally used. However, the geometry of the system is of significance here. Under shutdown conditions, parts of it would be at pressures that are considerably below atmospheric. The height of the hill just downstream of the pump station is about 10 m, which would render a bypass around the pumps ineffectual. This leaves vacuum relief/air admission as being perhaps the most practical and suitable option for a line with this profile.

By installing a vacuum-breaking valve (or valves) at A, the bulk of the liquid column in the pipe can drain down until it reaches an equilibrium level, by which time all the downward sloping section of pipe over the land will be empty. Assuming adequate air admission, so that the pressure in the cavity is very close to atmospheric, there will be no returning pressure wave to cause a recombination of the liquid column and problems of high pressure – the flow simply dies away.

The overall requirements, i.e. ventilation and surge protection, for air valves at A will be to allow for controlled venting of air on start-up, controlled release of air and gas coming out of solution during normal operation and relief of the subatmospheric pressures in the event of loss of electrical power to the pumps either owing to a fault in the electrical system or owing to the operation of a safety trip (e.g. thermal overload of the motor or high bearing temperatures).

Elsewhere along the line, additional air valves will be required for ventilation purposes, with appropriate ones being selected from the models available at the time and determined by checking with manufacturers' catalogues or through a check on the World Wide Web.

Having allowed the liquid column to split at the high point A, there will be a short, perhaps 10 m, length of pipe immediately adjacent to the pump containing liquid. Backflow through the pumps could be prevented with check valves, although they are not always installed in sewage systems because of problems of poor sealing. Either discharge valves that close automatically can be installed or the liquid can be allowed to return to the sump.

Following the pump trip, the system will ultimately require restarting. This can follow the normal procedure, i.e. against a closed discharge valve which is opened slowly to charge up the system and establish normal flow.

In preparing for the computer analysis, it is necessary to make some estimate of the air valve capacity required. This procedure will be illustrated with data from Subsection 3.6.7, supplied by Glenfield Valves Ltd. Equation (3.80) gives the flow through a single large-orifice valve, for which Table 3.5 provides examples of representative valve coefficients (note that they are not dimensionless). Alternatively, charts such as the lower part of Fig. 3.65a may be used. Other air valve manufacturers provide data in a similar format, though, when using them, always check the unit systems in use.

Assume the design constraint is that the pressure in the pipe should not fall by more than 0.15 bar below atmospheric. Will one large orifice be adequate, or will more be required?

Since the pipeline is 500 mm in diameter, the most appropriate valve is the one designated as suitable for pipes *up to* 600 mm in diameter, i.e. the middle column in Table 3.5. The large orifice in the valve is, of course, much smaller. The valve coefficient is in the range 53–74, so, for the preliminary calculations, assume a value of, say, 60. The inlet pressure, p_1, is 1 bar absolute, as the flow is into the pipe from outside. Solving equation (3.80) gives

$$Q = C_s \times \sqrt{\Delta p \times 1} = 60 \times \sqrt{0.15} = 23.24\,\mathrm{m^3/min}$$

This is equivalent to 387 L/s, well in excess of the initial liquid flow in the line. The actual pressure drop will therefore be less than 0.15 bar and hence acceptable.

However, will the liquid column now accelerate under gravity or will frictional resistance cause it to slow down? The quick answer to this is to estimate the terminal velocity of the liquid column in the pipe, i.e. its maximum velocity when the gravitational force downwards is balanced by the wall friction resisting flow. The terminal velocity, v_t, can be calculated from

$$v_t = \sqrt{\frac{DgS}{2fL}} = \sqrt{\left(\frac{0.5 \times 9.81 \times 17}{2 \times 0.009 \times 6500}\right)} = 0.84\,\mathrm{m/s}$$

Note that S is the elevation difference between the peak where the vacuum-breaking valve is fitted and the downstream 'reservoir' – sea level in this case, but L is the total length of the moving liquid column, i.e. to the end of the pipe.

As the terminal velocity is so much lower than the initial flow velocity of 1.78 m/s, the liquid column will decelerate quite quickly initially, and so a single 600 mm nominal size valve should be adequate. This should really be confirmed through a computer analysis.

In addition to confirming and, perhaps, refining the vacuum-breaking capacity, the computer study should review the effect of varying the closure times of the pump delivery valves on the line pressure and lead to a recommendation for the minimum closure time.

This particular system will always drain down owing to air admission at the peak, but data from a valve closure, ultimately repeated in practice during commissioning, with appropriate instrumentation, will help confirm the strategy for alleviating the effects of the more extreme case of a full pump trip.

2.2.4 A Gravity-Fed Main

The flow in gravity-fed mains is regulated by control valves at the discharge end. In very long lines break tanks may be installed, also with control valves immediately upstream, as a means of limiting the maximum pressures in the pipelines.

Figure 2.16 shows the longitudinal section of a typical system. This pipeline comprises 2.3 km of 450 mm ductile iron pipe, having a wall thickness of 10 mm, and through which water is to flow from an upstream storage reservoir down to a temporary holding reservoir at a treatment works. The static head, S, on the system is 20 m. The intended flowrate is 185 L/s, equivalent to a flow velocity of 1.16 m/s in this pipe, and the control valve at the downstream end is a circular gate valve.

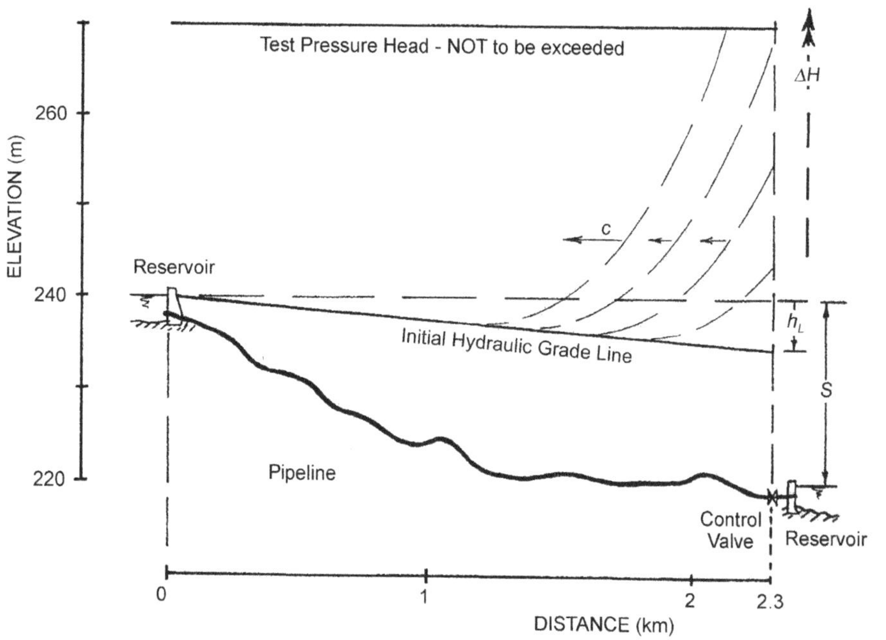

Fig. 2.16 *A longitudinal section of a gravity-fed main. Also shown are the maximum permitted pressure head envelope (i.e. from hydraulic tests), the initial steady flow hydraulic grade line and the changes it will experience as the downstream valve is closed*

Although the pipe is rated at 9 bar, the anchor blocks have been designed for 5 bar. The system will be pressure tested to the level shown in Fig. 2.16, which corresponds to this pressure at the lowest point in the line. Transient pressures should not exceed this. The questions that must be posed are: what risks exist and how should they be contained?

The only source of transient events in this system will be the opening and closing of the control valve. The normal practice with most valves is to crack them open very slowly to initiate flow, and then continue the opening more quickly. Alternatively, where bypass valves are fitted around large and/or high-pressure valves, they would be opened first.

A pressure reduction wave will travel upstream, but since there are no elevated peaks in this particular line, cavitation is unlikely. Ideally, the initial opening phase, which will be some 15 per cent of the valve spindle travel, should be extended over several pipeline periods to ensure smooth development of the flow.

The greater risk is posed by the valve closure. To build up a picture of how transient pressures will develop, the initial hydraulic grade line should be shown on the longitudinal section (see Fig. 2.16) where the frictional head drop, h_L, is 5.8 m.

As the valve closes, it initially has a negligible effect on the flow but then begins to reduce it at an ever-increasing rate. The dashed lines in Fig. 2.16 indicate in general terms how the pressure head builds up as the transient waves are propagated upstream at a speed c. The test pressure head line will soon be exceeded.

From the data given above, the D/e ratio for the pipe is 45, and so, from Fig. 3.2, the wave propagation speed, c, is 1160 m/s. The Joukowsky pressure head, cv_o/g, is 137.5 m which, if imposed on top of the initial pressure head, would certainly lead to a rupture. A transient suppression strategy is therefore required.

Practical options include: a slower rate of valve closure, strengthening of the pipe or the installation of pressure relief valves, an air vessel or a surge shaft.

By far the cheapest option is to use a suitably slow rate of valve closure. This is also the most direct solution, since it goes to the root of the problem and is therefore inherently safer. To explore this further, use can be made of the charts in Section 3.5, and specifically Fig. 3.49 for circular gate valves. It is reproduced here as Fig. 2.17 and used as follows.

First, evaluate the Initial Condition Parameter, α, from equation (3.79), to establish which curve to use

$$\alpha = \frac{H_o}{cv_o/g} = \frac{S - h_L}{cv_o/g} = \frac{20 - 5.8}{137.5} = 0.103 \text{ (say 0.1)}$$

Also, the pipeline period

$$T = \frac{2L}{c} = \frac{2 \times 2300}{1160} = 3.97\text{ s}$$

The next step is to estimate the maximum change, H_{max}, in the pressure head that can be permitted. Although the increases begin from the initial hydraulic grade line, as indicated in Fig. 2.16, some of the friction head will also be added on as the flow decelerates. The amount depends on the ratio of the pipeline period to the

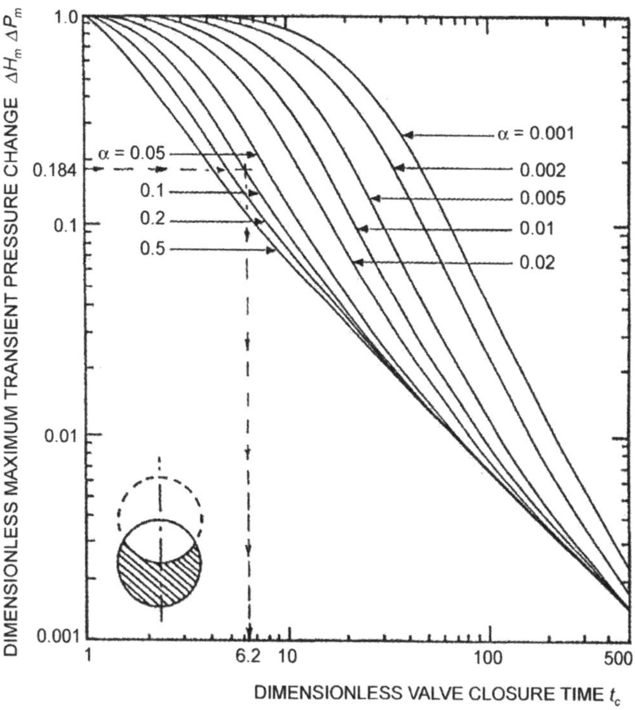

Fig. 2.17 *Valve closure chart for circular gate valves*

anticipated valve closure time. To be conservative, assume in this case that 80 per cent of the friction head is added to the pressure change, i.e.

$$H_{\max} = H_P - (S + 0.8 \times h_L)$$

where H_P is the maximum head permitted (5 bar in this example), i.e.

$$H_{\max} = 50 - (20 + 0.8 \times 5.8) = 25.36\,\text{m}$$

Non-dimensionalize by dividing through by the Joukowsky head of 137.5 m to give $\Delta H_{\text{m}} = 0.184$.

Interpolate this value for the non-dimensional permitted head change on the y axis of Fig. 2.17, read across to the curve for $\alpha = 0.1$, then down to take off a value of 6.2 for the Dimensionless Valve Closure Time, t_{c}, which is defined by equation (3.78). To obtain the *actual* valve closure time, multiply this by the Pipeline Period to yield 24.6 s as the minimum acceptable estimated valve closure time. This then forms part of the initial data for the computer study.

For gravity-fed systems in water resource projects, this timescale would normally be acceptable. If the pipeline were longer, the closure time would also have to be longer, and closure times of 60 min are not unknown.

Once the estimated closure time has been confirmed by a computer analysis, the operational and control procedures should be designed. They should ensure that the closure of the valves cannot occur in a time less than that specified. If the flowrate is subsequently increased beyond the initial design value, the valve closure time must be reviewed.

2.2.5 A Line to an Offshore Oil Terminal

Some of the ports and harbours close to oil fields are too shallow for large modern tankers, and so offshore facilities are used for loading and unloading. These comprise loading platforms connected to the shore by submarine pipelines. Figure 2.18 shows a typical arrangement, in which, over the years, the capacity of the system has been increased in stages.

In the latest phase, it is being proposed that the existing system be uprated to give a flow of 2.25 m^3/s of oil. The system will start at the existing Tank Farm, which serves as a collecting point for the oil to be exported. A battery of pumps, working in parallel, delivers oil to two Main Pumps that pump the oil to the existing facility. New booster pumps at a station on the shoreline are to be added to increase the flowrate to the offshore loading terminal.

The pipe lengths, together with the locations of the pump stations, are shown in the figure. The pipe nominal diameter is 812 mm and the flow velocity is 4.34 m/s. The hydraulic grade lines for this flow and the maximum permissible head lines are shown on the schematic layout (Fig. 2.19). Minimum pressures should not fall below atmospheric.

A particularly vulnerable part of the system is the pipework on the loading platform. This includes flexible hoses and sections of pipe connected by swivel joints for connection to the shipboard facilities. The permissible pressures in these fittings are lower than for the main pipeline.

The valves shown on the diagram represent tanker loading valves, i.e. they are on the tankers themselves. They are an example of a flow control device outside the direct control of the system designer or operator, but which they must nevertheless consider. Flowmeters mounted on the platforms, and through which the oil must flow, are restricted to maximum pressures of 15 bar.

Existing arrangements for the control and suppression of transient pressures consist of three pressure relief valves and a collecting vessel adjacent to the loading platform, and the bypass lines on the pumps.

It is intended that a computer-aided assessment of transient flow conditions, associated with the proposed higher flows, be undertaken. To help focus this study, it is desirable to review what transient problems can occur and the extent to which the existing suppression and control strategies may need to be reinforced.

Following the procedures suggested in Subsection 2.1.2, the next step, step 4, is to consider the likely causes of unsteady flow. These may be summarized as:

Fig. 2.18 *General layout of the supply to an offshore oil loading terminal*

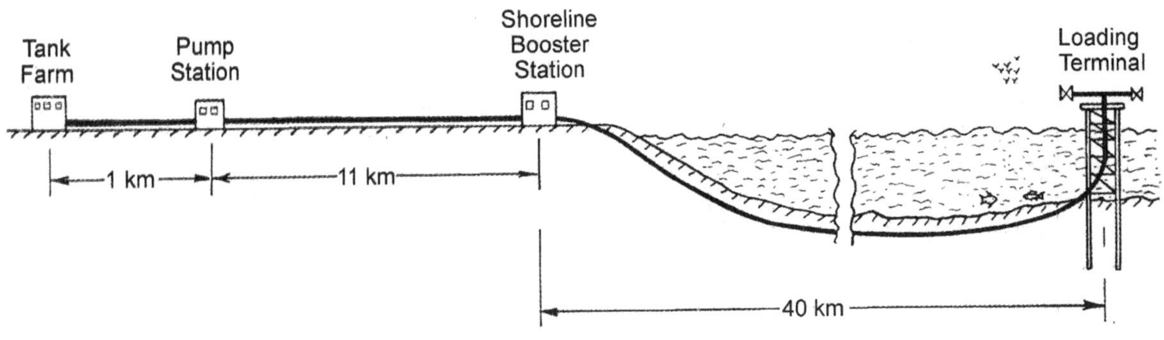

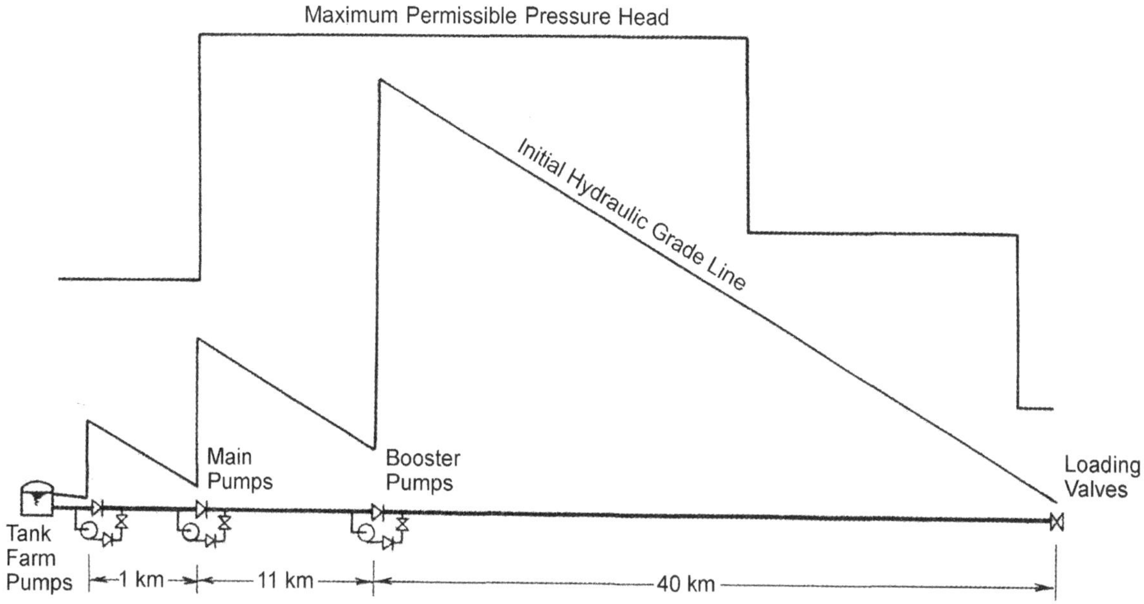

Fig. 2.19 *Schematic of the key pump, pipeline and valve arrangements for the offshore loading terminal, including the hydraulic grade line and the permissible maximum head envelope*

- starting up the system;
- valve closures on the tanker;
- pump trips – controlled and uncontrolled.

By considering these events in turn, it is possible to build up a picture of how the pressures and flows will develop – the point being to try to get a 'feel' for how the system will behave so that, in step 5, the more serious, i.e. critical design, cases can be identified.

The normal starting routine for the pumps is to begin with the Tank Farm Pumps, and then the Main Pumps, to establish a flow in the system. The arrangement of pumps and their bypass lines ensures that the Main and Booster pumps do not inhibit flows from the pumps upstream. Once flow is established the Booster Pumps can be brought on line. The starting up of the pumps, especially the Booster Pump, should not occur without the tanker loading valves being open, otherwise the permissible pressure at the downstream end of the line will be exceeded – this can be seen by examination of Fig. 2.19.

Of the flow reductions, consider first the closure of the tanker loading valves and refer to Fig. 2.20. The valves on modern tankers can close in as little as 15 s, though 30 s is more usual. Rapid closure is required to avoid oil spills when the tanks on the ships are full. Closure of these valves is under the control of the tanker crew. It is quite divorced from the operating requirements of the pipeline and the staff who operate it. Control strategies for the safety of the line with respect to valve closure must therefore be based on an assumed worst case.

Whether closure is in 15 or 30 s is rather academic on a long line such as this – it will be a 'Rapid' event since the Pipeline Period, T $(= 2L/c)$, will be much longer than the valve closure time and the full Joukowsky pressure rise will occur upstream of the valve. On this line, the speed of propagation of pressure waves in the oil will be of the order of 1150 m/s [see Table 3.1 and equation (3.6)], and, with

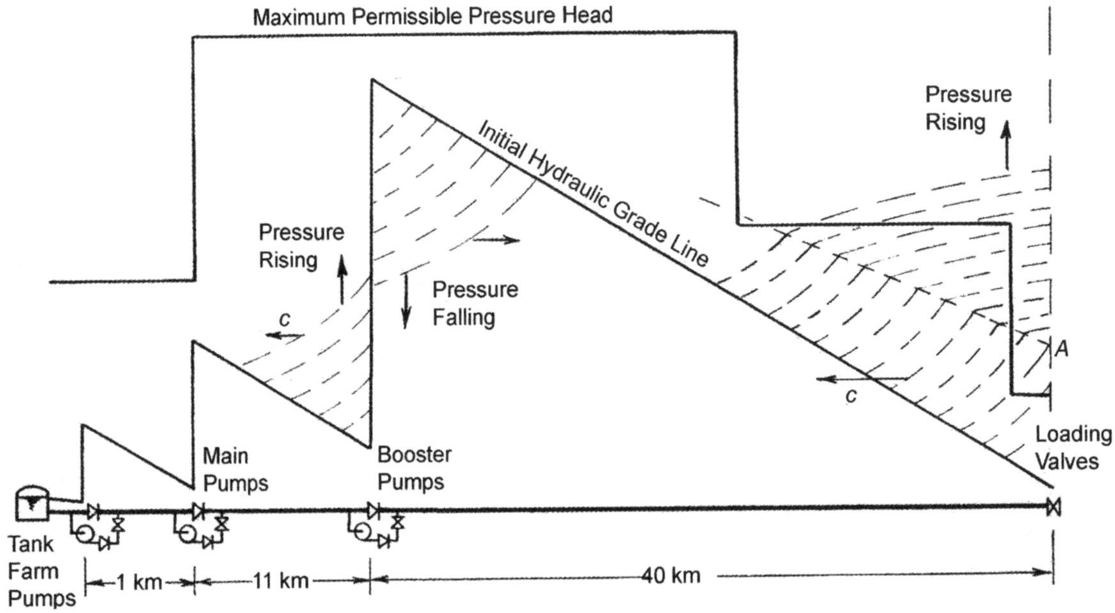

Fig. 2.20 *As Fig. 2.19, but with indications of how the pressure heads will tend to change after a valve closure and loss of power to the booster pump*

the flow velocity of 4.34 m/s given above, the Joukowsky head rise will be in the region of 509 m head.

The instant the valve has closed fully, and if no suppression devices are in place, the pressure on the upstream side of the valve will rise by the Joukowsky head to point A in Fig. 2.20, as the leading edge of the pressure wave moves upstream at 1150 m/s.

On this type of system, especially with a very low static head, the bulk of the energy input from the pumps is to overcome friction and the hydraulic grade line is quite steep. The pressure wave will tend to advance up this hydraulic grade line, bringing the flow to rest. However, if the full Joukowsky head simply moves up the hydraulic grade line there will still be a pressure gradient – which is physically impossible if the flow is also at rest – and it will tend to even out. It can only do this by moving forward a little and further compressing the liquid in front of it – but, by moving forward its velocity has therefore not been reduced to zero, and hence it cannot experience the full Joukowsky pressure head.

Consequently, as the main pressure wave travels upstream, its magnitude decreases from the full Joukowsky value, in the manner indicated in Fig. 2.20, although the pressure behind it continues to rise significantly. The pressure head will build up as the system moves towards a no-flow shut-off condition. At the valve itself, the pressure could become many times greater than the Joukowsky head, tending ultimately to a maximum value in excess of the pump shut-off head.

This process, of more fluid being accumulated in the system and greatly increasing the pressure rise, is known as 'Line Packing', and can pose serious threats to the safety of a system such as the one shown here. Some form of transient suppression strategy is clearly required. To be most effective with respect to this particular factor, it should be based on relieving the pressure at the downstream end of the line where the transient event is initiated by closing the valves.

Turning to the pump trips, they can either be controlled events, as for normal shutdowns, or uncontrolled events due to fault conditions or power failures. In the former case the usual procedure would be to shut down the pumps in the reverse order to that in which they were started, so the first to be shut down would be the booster pumps. As the discharge valves are closed, the pressures either side of the pump will gradually come into balance and the bypass line will open. Following this, the Main Pumps may be progressively shut down, followed by the tank farm pumps. Since there can be two Main Pumps running in parallel, an acceptable strategy would be to shut down one of these and one or more of the Tank Farm pumps. The remaining Main Pump would then be shut down, followed by the last of the Tank Farm pumps still in service.

Insofar as guidance on the controlled shutdown will be required with respect to ensuring pressure head changes are within limits, the minimum closure time for the discharge valves should be determined to avoid creating subatmospheric conditions downstream of the Booster Pump.

If, on the other hand, the Booster Pump was simply tripped out, just as with a power failure, the pressures upstream and downstream would tend to change more rapidly, as indicated in Fig. 2.20. When the pressure upstream exceeds that downstream, the bypass line will open and ultimately the flow will settle down to whatever the Main Pumps can provide. As described above, the Joukowsky pressure change will be modified by line packing, and the pumps will ride up their characteristics to settle eventually at a new steady state value.

When the Main Pumps are tripped, intentionally or otherwise, and experience loss of power, negative pressure waves will propagate downstream as they do when the Booster Pump is tripped. Although not shown in Fig. 2.20, subatmospheric pressures will soon occur on the suction side of the Booster Pump. This is unacceptable and must be prevented. Similarly, if the Tank Farm Pumps are tripped and the others continue to run, both the Main and Booster Pumps will soon cavitate.

To summarize progress thus far, the main problems that have been identified are associated with reductions to the flow, either by closure of the tanker valves or by the pumps being tripped out. More specifically:

1. Valve closures lead to excess pressures – these must be relieved in a controlled manner, so the existing pressure relief system should be investigated and uprated as necessary, including the temporary storage facility (see Section 3.7). Tripping out the pumps by remote control could help reduce the storage capacity required at the loading terminal.
2. Pump trips – the principal hazards seem to be cavitation conditions at the Main and Booster Pumps when other pumps upstream are tripped. This suggests that, when a particular set of pumps trip, other pumps downstream should also be tripped.

These interim deductions can now be used to formulate the specification for a computer analysis of the system. The following list is a suggestion of the form this might take:

(a) Assume complete closure from full flow in 15 s of the tanker loading valves and assess the extent to which the existing relief system limits the pressure rise upstream. If the permitted pressure is exceeded, how much additional relief capacity will be required? The relief valves can be increased in size and/or number. Increasing the storage capacity of the relief system will be expensive and must be kept to a minimum. It may be assumed that the Booster Pumps can be tripped within 5 s of the closure of the tanker valves being initiated. What additional capacity would be required if the delay were 10 s?

(b) Model a power failure at the Booster Pump Station. Determine the pressure head profiles upstream and downstream, assuming the other pumps continue to run. Do subatmospheric pressures occur downstream and, if so, for how long? Is the maximum permissible pressure exceeded upstream?

(c) Investigate the risk of cavitation at the Booster Pump suction following power failure to the Main Pump Station. Can cavitation be avoided by intertripping the Booster Pumps within, say, 5 s of power loss to the Main Pumps? If not, for how long would it last, and how much of the line would experience low pressures?

(d) Investigate the effects of power loss to the tank farm pumps. What is the risk of cavitation at the pumps downstream, and can it be avoided by rapid intertripping of the Main and Booster Pumps?

(e) For a controlled shutdown of the system from full flow, determine the closure time required for the discharge valves on the Booster Pumps to avoid cavitation in the pipeline downstream.

These suggestions have been developed for the tanker loading system used as the basis for this example. They are strongly influenced by the physical details of the particular system, such as the pipe lengths, the relative positions and power of the three pumping stations, etc. In other systems the significant problems may be associated with different components, but the basic approach will be the same – look at what could give rise to unsteady flows and try to decide which will be the more serious. Alleviating the undesirable effects of these will usually protect the system from similar, but less drastic, consequences of other events.

This example also illustrates the importance of understanding the physical processes taking place and the benefits that can then be derived from looking at the whole system. For instance, by representing events and their consequences on a longitudinal section, e.g. how the use of bypass lines provides a simple yet effective way of limiting the effects of booster pump tripping, the design engineer does not need to be particularly skilled in computer modelling but is nevertheless able to pose the important questions and appreciate the significance of the results.

2.2.6 A Process System Supplied by a Ram Pump

Positive displacement pumps, such as ram pumps, are often used in chemical and process plants. Figure 2.21 is an example of a typical arrangement and, in this case, comprises a stainless steel pipeline 800 m long and 150 mm in diameter.

A mean flowrate of 16 L/s will be provided by a three-cylinder ram pump. It is proposed that the crankshaft of the ram pump will run at 22.5 r/min, being driven through a gearbox by an electric motor.

At the downstream end, a control valve will regulate the flow into the reaction vessel and, in the event of a fault developing downstream, will close automatically in 0.5 s. At the same time, the power to the ram pump will also be cut off. It has

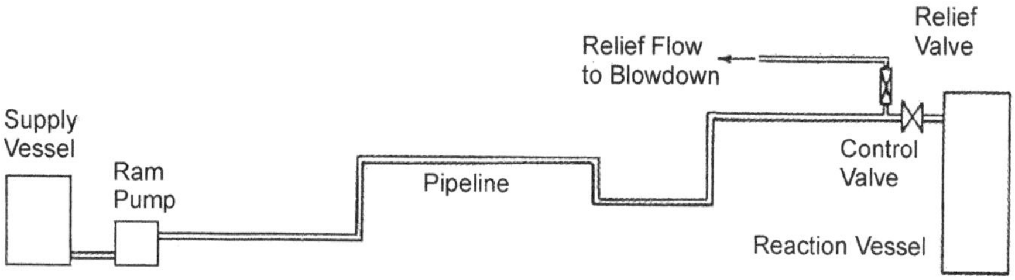

Fig. 2.21 *A typical arrangement of a positive displacement pump supplying liquid to a reaction vessel in a process plant*

been suggested that a pressure relief valve be installed upstream of the control valve to relieve excess pressures following the rapid closure of this valve. The plant Safety Officer has asked what transient events might occur and whether the system is adequately protected.

The causes of unsteady flow will include the start-up and shutdown of the system, and the emergency closure of the control valve at the downstream end. In addition, the pulsatile nature of the flow from the ram pump should not be overlooked.

The regular start-up and shutdown operations are controlled events and, provided standard procedures for systems fed by positive displacement pumps are followed, should not present significant problems. However, the emergency shutdown and the pulsatile flow do merit closer examination.

Consider first the emergency closure of the control valve. Using data from Subsection 3.1.3.9, the transient propagation speed is 1200 m/s for the liquid in this system, and hence the pipeline period, T, is 1.33 s. At 0.5 s the valve closure is therefore a 'rapid' event.

If the single pressure relief valve fails to operate, the full Joukowsky pressure head will occur. This amounts to

$$\Delta H = \frac{cv_o}{g} = \frac{1200 \times 0.91}{9.81} = 111.3\,\text{m}$$

where 0.91 m/s is the liquid velocity for a flowrate of 16 L/s in a 150 mm diameter pipe. Most of the line will be exposed to this increase in pressure.

A pressure rise of this magnitude within 0.5 s has to be viewed in the context of the consequences of line failure, bearing in mind the risk of fire, explosion, environmental contamination, etc. from the fluid in the system.

Even if, in principle, a straight length of uniform pipe may be able to sustain this increase in pressure, shock loads on the many bends and junctions that occur in process plant piping systems can enhance the risk of pipeline failure. It is therefore highly desirable that the pressure relief arrangements should have some built-in redundancy. They might consist, for example, of three 50 per cent duty valves set at slightly different pressures. Normally, two would operate, the third one being on standby and to provide cover during maintenance. This arrangement will help protect the line if the system to trip the pump automatically should also fail.

Alternatively, to ensure a fast response consideration should also be given to the use of a bursting disc as this would rupture faster than a relief valve would open. The downside is that, before the system could be restarted, it would be necessary to replace the disc. Such a delay, and possibly the Health and Safety implications for staff installing a new disc, may not be acceptable.

Turning to the pulsatile nature of the pump discharge, the mean flow is 16 L/s but data supplied by the pump manufacturer indicate that the actual flow will oscillate either side of the mean by ± 2 per cent with a frequency governed by the speed of the driveshaft and the number of cylinders in the pump – 22.5 r/min and 3 respectively in the present case. Is there a potential resonance problem?

The Period, T_1, of the first harmonic of the liquid column in the pipe is

$$T_1 = \frac{4L}{c} = \frac{4 \times 800}{1200} = 2.6667\,\text{s}$$

The Periods for the third, fifth, seventh and ninth harmonics will therefore be 0.8889, 0.5333, 0.381 and 0.296 s respectively. To be sure of avoiding a resonance situation developing, the frequency of the pulsations from the pump should not be close to any of these or the higher harmonics of the system.

For a three-cylinder pump running at 22.5 r/min, the pulsation frequency, f, will be

$$f = \frac{3 \times 22.5}{60} = 1.125\,\text{Hz}$$

which is equivalent to a Period of 0.8889 s. This corresponds to the third harmonic of the liquid column in the pipeline – and implies that, although the amplitude of the flow oscillations may only be 2 per cent, significant pressure fluctuations could occur. A more thorough assessment is therefore required.

From the discussion thus far it follows that the computer assessment of the system should be formulated to include the two major issues of the pressure relief capability and the risk of resonance. For example, it should provide assistance in:

(a) establishing the size, number and set pressures of the relief valves to suitably control the pressure following an emergency shutdown and
(b) checking the implications of delays in the opening of these valves *or*
(c) if acceptable, establishing the relief requirements in terms of bursting (rupture) discs to be fitted in lieu of the relief valves shown in Fig. 2.21;
(d) testing the response of the system to a slightly pulsatile supply from a three-cylinder pump running at 22.5 r/min and
(e) if the pulsations are unacceptable, determining a safer speed, or the characteristics of a suitable alternative pump.

In this latter respect, Fig. 2.22 is a plot from a computer assessment of the pressure fluctuations in this system for the condition described above, with the pump running at 22.5 r/min. Fig. 1.10 in Subsection 1.2.12 is a similar plot for a pump having a shaft speed of 53.3 r/min, which is close to the seventh harmonic of the system.

To reduce these fluctuations the pump speed should be mismatched with these harmonics. Figure 2.23 is one such case, where the pump shaft has been set to run at a speed of 26.7 r/min, equivalent to a cyclic period of 0.75 s. There will be other acceptable speeds, and the one selected should be chosen so that the required flows are provided without generating a dynamic excitation of the system.

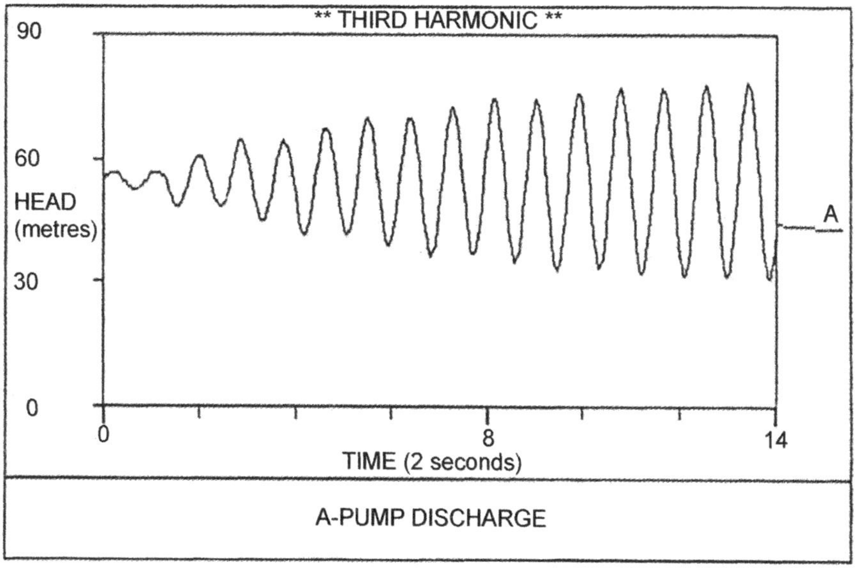

Fig. 2.22 *Pressure oscillations building up after starting up the ram pump at a speed of 22.5 r/min*

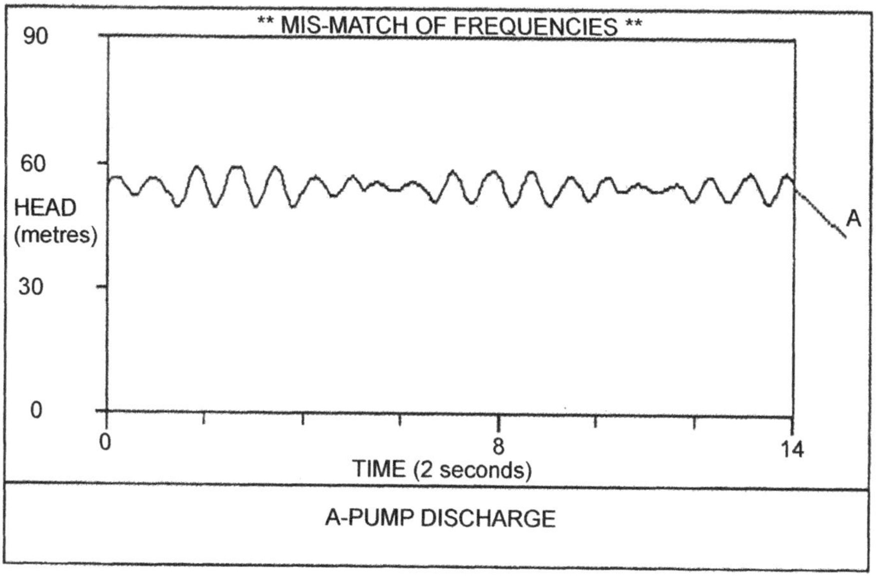

Fig. 2.23 *As Fig. 2.22, but with the pump speed changed to 26.7 r/min to mismatch the frequency of the flow pulsations and the natural frequencies of the system*

If there are still slight ripples in the pressure and flow, these can usually be damped out by installing a pulsation damper on the pump discharge. Pulsation dampers may also be required in the suction line of positive displacement pumps if they are not mounted close to the source of the liquid to avoid damaging cavitation occurring in the suction line and the pump itself during the suction stroke. Even if a relatively small degree of cavitation and pulsatile flow occur, the repetitive shock loads can, over a period of time, give rise to cavitation and fatigue damage to the pumps and their components [Collier (1983), Taylor and Harrison (1990) and Vetter and Schweinfurter (1990)].

In some process systems there may be a requirement for a check valve to be fitted to prevent a reverse flow from the reaction vessel and as a second line of defence should the control valve actuator fail to operate. If a check valve or, alternatively, an Emergency Shut Down Valve (ESDV) is to be used, it will be necessary for it to respond adequately, and an assessment of this should also form part of the computer study.

2.2.7 A High-Pressure Feed System

When pipelines are simple systems, consisting of a single source and a single discharge point connected by one pipeline, it is usually not too difficult to predict the response of a system to the early stages of transient propagation. For instance, the previous examples have demonstrated how it is possible to get a 'feel' for how a system will respond to various fluid transient events and to form estimates of the magnitudes of peak pressures, to identify the principal hazards and to formulate strategies for the control of unacceptable conditions.

As soon as other pipes are introduced, to create branches, loops and networks, it rapidly becomes much more difficult to predict, with any degree of certainty, how transient pressures will develop following a disturbance to the flow. Nevertheless, it will still be necessary to interpret predictions made by computer analyses, and so some attempt must be made to understand how a system will respond to a transient event.

Consider the high-pressure boiler feed system shown schematically in Fig. 2.24. This is not the complete system but, in the context of transient flows, can be bounded by the constant pressure source on the pump suctions at one end and by the boilers, another constant pressure point, at the other. This is one practical and realistic example of system idealization.

Two pumps deliver feed water through check valves, pump isolating valves and a heat exchanger to a common main. Four pipelines lead off from this common main to the boilers, each line containing three heat exchangers, an Emergency

Fig. 2.24 *Schematic layout of part of a high-pressure boiler feed water system. When considering the transient response of complex networks, sections may be assessed independently, provided they can be isolated from adjacent regions by reservoirs or similar constant pressure points*

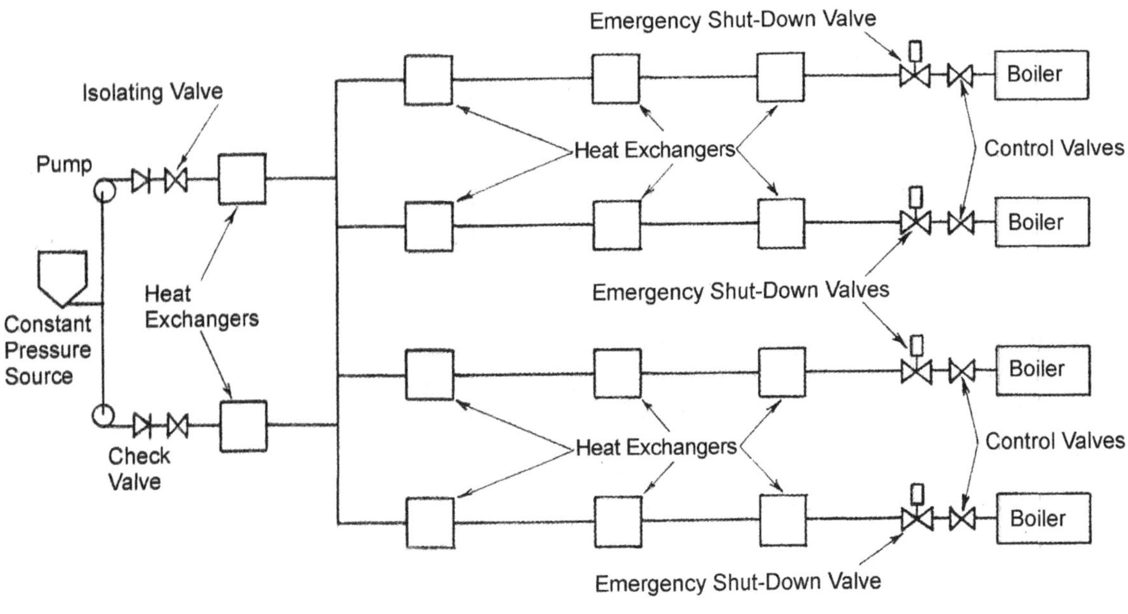

Shut Down Valve (ESDV) and a control valve. The control valves regulate the flow into the boilers according to steam demand.

At the pump discharge, the normal operating pressure is 81 bar and at the ESDVs it is 75 bar, the difference being due to changes in elevation, plus head losses in the pipes and fittings and across the heat exchangers. The flowrate corresponding to these pressure heads is 3.5 m^3/s, which gives a flow velocity of 7 m/s in the main pipes – typical for these high-pressure systems.

The pipelines on the pump discharges are 575 mm internal diameter and have an average length of 35 m. In the four separate legs to the boilers they reduce to 400 mm diameter and are all in the region of 215 m long.

The principal transient events in this system will include the tripping of both pumps, the tripping of one pump when two are initially in service, the rapid closure of all the ESDVs and of one valve on its own.

The two most severe transient situations will be associated with the tripping of both pumps and the rapid closure of all four ESDVs. For these two cases, some degree of idealization of the system is reasonable since it is fairly symmetrical, e.g. the four lines to the boilers are of similar lengths and properties, as are the two pump discharge lines. Being a high-pressure system the pipes will be thick walled and so the wave propagation speeds will be quite high. For water with an average temperature of, say, 80 °C in thick-walled pipes, it would be in the region of 1300 m/s.

Take the case of the four ESDVs all being closed in a time of 1.5 s, while the two feed pumps continue to run. The pressure in the system will, over a period of time, change from its initial steady state to a final steady state and, in the process the system will experience a series of rapid transients. This concept is illustrated in Fig. 2.25, which indicates *in outline* the events as they will occur on the upstream side of one of the ESDVs.

The final steady state pressure will be higher than the initial one because the pump will ride up its operating characteristic (see Fig. 2.26) to the shut-off head, or close to it if some form of leak-off arrangement is in service. Since check valves are fitted, there could be a risk of even higher pressures being locked into the

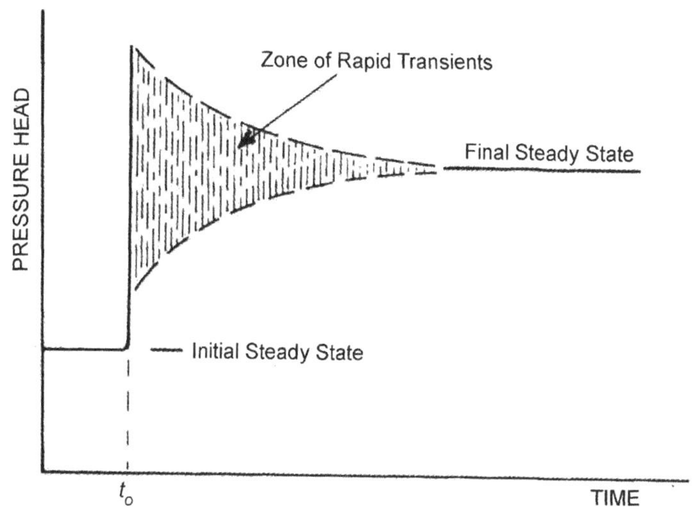

Fig. 2.25 *Following closure of the ESDVs, the pressure will change from an initial to a final steady state condition, passing through a period of rapid transients*

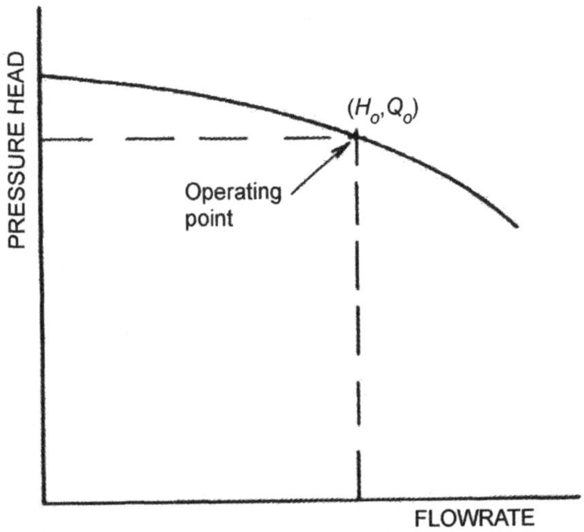

Fig. 2.26 *Typical pump performance characteristic. The shut-off head may be 15–20 per cent higher than that at the normal operating point*

system between the check valves and the closed ESDVs, as sometimes happens in fully charged fire-mains.

It is, however, the 'zone of rapid transients' that is of major concern. Initially, the fluctuations in pressure will be quite large, but over a period of time they will be damped out in an exponential fashion owing to viscous effects. The details of the pressure changes must be predicted through a computer-based study, though it should be possible to get a general idea of how the initial pressure rise develops near the ESDVs.

As a result of the ESDVs slamming shut, the pressure will rise on the upstream side and a pressure wave will be propagated towards the pumps. Reflections will occur at the interfaces with the heat exchangers and at the various bends and junctions, some decreasing and some increasing the general trend in which the pressure will change. Figure 2.27 is an example of how the pressure–time history at the ESDV may develop immediately after closure.

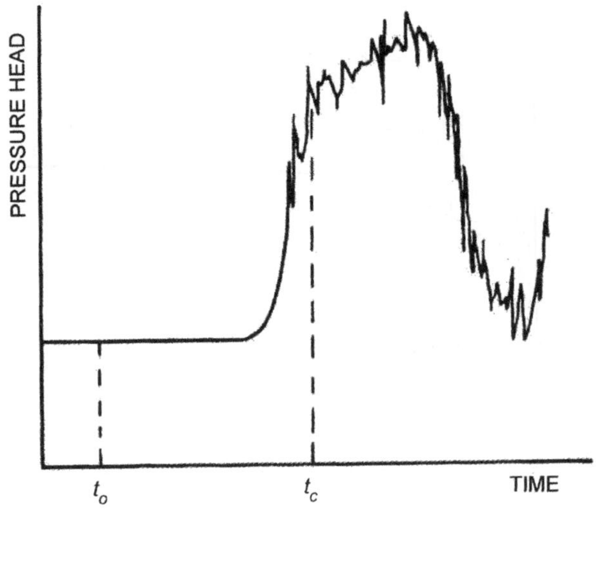

Fig. 2.27 *A predicted pressure–time history, with a much expanded timescale compared with Fig. 2.25, on the upstream side of an emergency shutdown valve, at the time it closes*

Closure commences at time t_0, and is completed at time t_c, 1.5 s later. The valve is of the circular gate variety and so does not really start to reduce the flow until it is about 80–85 per cent closed. Hence, the pressure does not start to rise until closure is well advanced. The Pipeline Period, T $(= 2L/c)$, can be based on the distance from the ESDVs to the pumps, owing to the symmetrical nature of the system, and has a magnitude of 0.385 s. Although the valve closure is almost 4 times the pipeline period, the initial pressure rise can be expected to be close to the full Joukowsky head because of the highly non-linear reduction in flow.

This can be explained by reference to Fig. 3.49 in Section 3.5, which shows the closure characteristics of circular gate valves. The ESDVs are normally fully open and will have a fairly small frictional head loss across them, so the value of the initial condition parameter, α, will also be small. Reading up from a dimensionless closure time of 3.9 (i.e. 1.5/0.385) to the curves for low α values indicates that virtually the whole of the Joukowsky head rise will occur at the valve.

The magnitude of the Joukowsky head in this system is $1300 \times 7/9.81 = 928$ m, i.e. approximately 93 bar. On top of this should be added a large part of the frictional head drop, perhaps 4 bar, before major reflections arrive back from the upstream boundary of the system. This suggests that the first peak pressure near the ESDVs will be in the region of 97 bar on top of the initial pressure head of 75 bar, i.e. an actual pressure of about 172 bar.

This can only be viewed as a rough estimate, an outline of the general trend of events in the immediate aftermath of an emergency shutdown. No allowance at all has been made for the effect of Fluid–Structure Interaction (FSI) which can also introduce additional pressure spikes similar to those generated by pressure waves reflected at bends, junctions and heat exchangers. However, by establishing a plausible explanation of the results from a computer analysis for one point in a system, some credence can be extended to others.

If only one ESDV closes, rather than all four, the initial response in the line in question will be exactly the same as already described. However, instead of four pressure waves arriving more or less simultaneously at the junctions in the pump discharge main there will be only one. The discharge main now becomes the location of the principal reflection point at the upstream end, although the pressure will change slightly as the pumps adjust to a combined discharge of around 75 per cent of the previous flow.

The effective Pipeline Period, T, based on the line to one boiler, is 0.33 s, instead of 0.385 s. Hence, although the initial response at the valve will be the same as before, it will rapidly depart from it owing to the earlier return of reflected pressure waves from the upstream end.

Irrespective of whether one or several of the ESDVs close, the initial peak pressures upstream will rise to the same magnitude, and the strategy to withstand or control them can be the same. The consequences of high-pressure hot water being released if the system were to rupture would be quite disastrous. One control strategy is to ensure that the pipeline and all the components are strong enough to withstand the high pressures. Another suitable option is to install pressure relief valves or bursting discs, if acceptable, connected to a blowdown vessel to receive the water discharged from the system.

The initial consequence of the pumps being tripped out will be a drop in pressure. If both pumps are tripped together, an impression of how the system responds may be deduced by considering one of the symmetrical halves, i.e. a pump feeding two boilers. A further idealization would be to combine the lines to two boilers into one having an equivalent cross-sectional area. The result is a single-pipe system, like the rising mains already discussed. However, this model can only indicate the general trend of the initial pressure fluctuations. It will not include any partial wave reflections from the various components in the real system.

The loss of all the boiler feed pumps will, in fact, create a much wider problem, and the rapidly changing pressures in the feed lines and heat exchangers may be of less concern than boilers starved of water. The strategy for dealing with the loss of the pumps will depend on the nature of the plant, the speed with which the pumps can be restarted and whether other sources of feed water may be available. In some circumstances it may be necessary to shut down the boilers, which starts to become a major incident.

The transient analysis could therefore become quite involved and form only part of a larger risk management problem, and one certainly requiring computer-aided investigations. That said, the foregoing discussion does indicate how one can develop an insight into how events could unfold and help raise confidence in the predictions that a computer model of the system can yield.

Finally, the loss of one pump from a running pair must be considered, but in a well-designed system this need not present any major problems. The initial drop in pressure downstream of the check valve on the tripped pump will be similar to when both pumps trip, but only briefly so. The principal reflection point is again the pump discharge main where the pressure will remain high owing to the other pump continuing to run. The water column in the line between the pump and the common main will decelerate rapidly, come to rest and then reverse.

The extent to which a reverse flow develops depends upon the speed with which the check valve responds to the changing flow. If the valve responds quickly, a negligible reverse velocity will occur prior to closure (see curve A in Fig. 2.28). If, on the other hand, the valve is rather sluggish a much greater reverse velocity will develop (see curve B), leading to a high-pressure rise when the valve does finally close – the phenomenon of check valve slam.

This preliminary assessment of the four potential fault conditions confirms that computer studies will be necessary and provides some indication of the specific issues that need to be addressed. Furthermore, it gives indications of the kind of results that should be anticipated and of measures that can be taken to render the plant safe from untoward events.

The analysis of the valve closures should concentrate on the initial high and low pressures in the pipelines and especially in the heat exchangers as they may well be the weak link in the chain. If the use of pressure relief valves or bursting discs is being considered, the computer model will have to be modified to include the valves and their discharge lines.

If low pressures in the heat exchangers, which might give rise to vaporous cavitation, are to be avoided, the use of pressure relief valves to alleviate the initial rise could help reduce the subsequent drop in pressure. A detailed consideration of the closure of just one ESDV could be optional, since the significant pressure changes will be similar to those when several close.

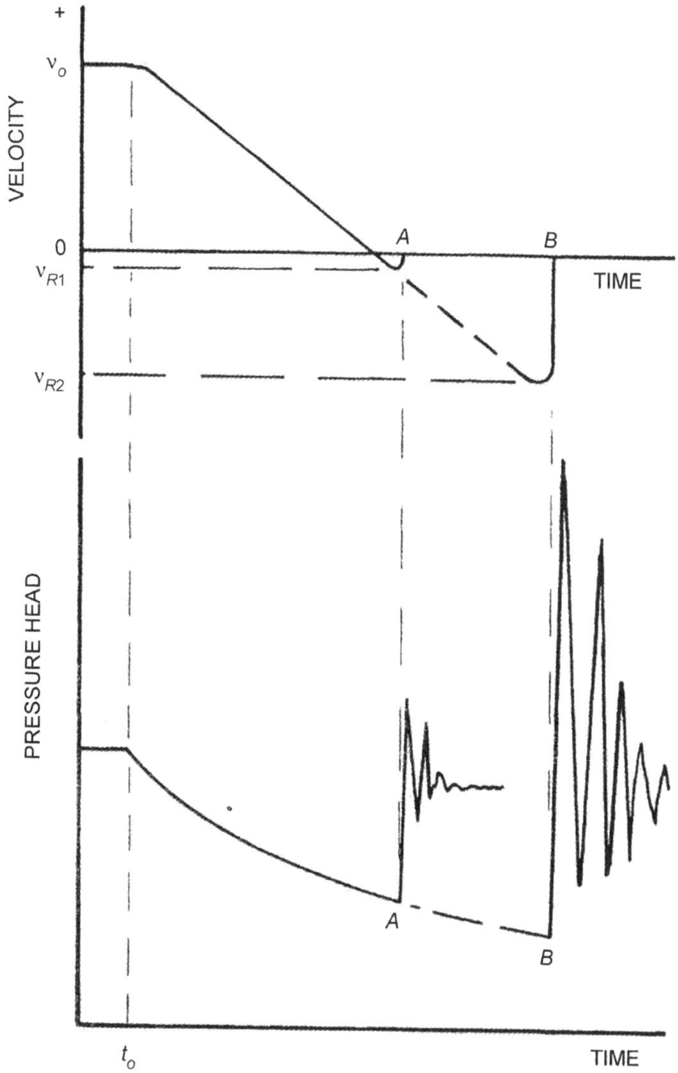

Fig. 2.28 *Plots of flow velocity and pressure head against time immediately downstream of the check valve next to the pump that has been tripped. If the check valve is able to respond quickly to changing flow conditions, the maximum reverse velocity, v_{R1}, when it closes is quite small. If the valve response is slow, a much higher reverse velocity, v_{R2}, will develop and the associated pressure rise will also be very large*

The extent to which a single pump trip is a serious risk will depend on several factors. Among these are the inertia of the pumps, their driving motors and the shafts and couplings between them, the distance between the check valve and the junction with the discharge main, and the pressure in the system.

The dominant factor will, however, be the type of check valve, and especially whether it is capable of closing with a minimum reverse velocity of water through it. If the results of the computer analyses of the system indicate that the deceleration of the flow adjacent to the tripped pump is very high it will be essential to select check valves that have a fast dynamic response. For more details on this, see Subsection 3.8.2.

2.2.8 Looped Networks

As the complexity of piping systems grows it becomes increasingly difficult to follow the detailed interactions of pressure waves propagating to and fro within the system. Networks, and looped networks in particular, consisting of relatively short lengths of pipe (i.e. less than, say, 500 m) are usually inherently safer than the longer, single-line systems discussed in the previous examples. This is because wave reflections from junctions and reservoirs, etc. usually tend to counteract the initial change in pressure, and limit further changes in pressure at the point where a transient is initiated.

This generalization breaks down as lines become longer. The following example helps illustrate the need for computer studies and how difficult it can be without them to do other than predict general trends. For these more complex systems it is therefore essential that computer-based analyses be undertaken to determine how they will respond to transient events.

Figure 2.29 shows the layout of the basic supply grid of a water distribution system for a small community. The various pipes in the two loops are between 200 and 300 mm in diameter and their lengths vary from 300 to 700 m. The principal supply is from the pumping station, with the reservoir and elevated storage tank providing additional flow, especially at periods of peak demand. The end-users are supplied from a network of small-diameter pipes in the two distribution networks fed from the nodal points indicated in Fig. 2.29.

There is usually no particular merit in incorporating every little pipe in a system – it does not really improve the model, and computer run times become much longer than is really required for an acceptable result, especially for a demand-led system such as in the water supply industry.

In systems such as these the usual cause of significant transient events tends to be pump trips, with the complete loss of power to the pump station being the most severe. This scenario is reviewed below for the network shown, for two situations. In one case, case A, the line from the pump station to the first junction in the network is 100 m long; in the second case, case B, it is 4 km in length. This is to illustrate a point made above, namely that systems with shorter pipes tend to have smaller pressure head changes than longer-piped systems.

The net result of all the pumps tripping out is that the supply to the two distribution networks will be taken over, to some extent, by the reservoir and the

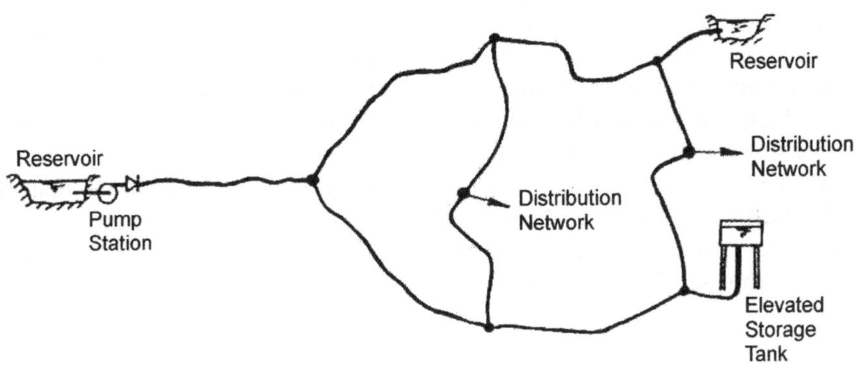

Fig. 2.29 *Layout of the principal pipes in a water supply grid for a small community*

elevated storage tank. The system will switch eventually from one steady state situation to another after passing through a period of rapid transients. The pressures and flowrates will adjust to whatever the remaining sources can provide.

Figures 2.30 and 2.31 show the variation in pressure head at two locations for Case A. The first figure refers to a point near to the pump discharge, while the second (Fig. 2.31) is for Node 6, which is at the connection to the distribution network located between the elevated storage tank and the reservoir, shown on the right-hand side of Fig. 2.29.

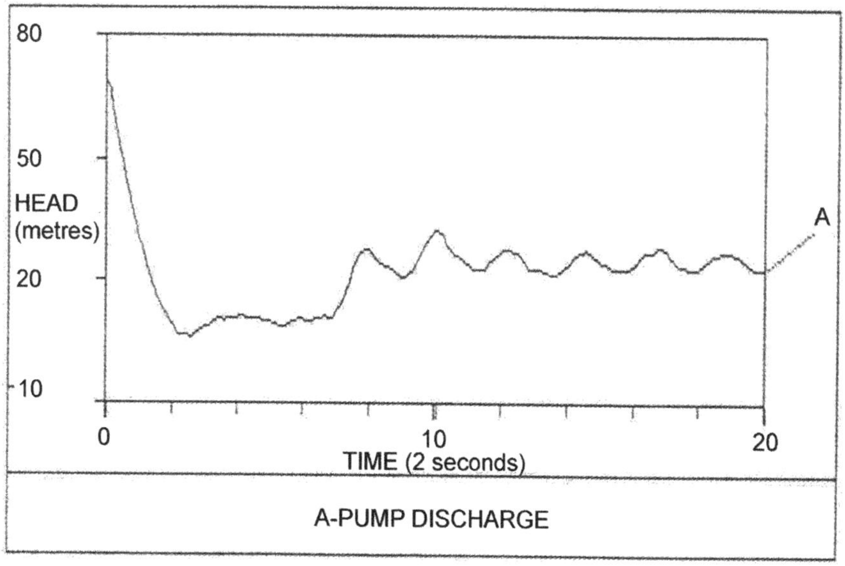

Fig. 2.30 *Changes in pressure head at the pump discharge, for Case A, following loss of power*

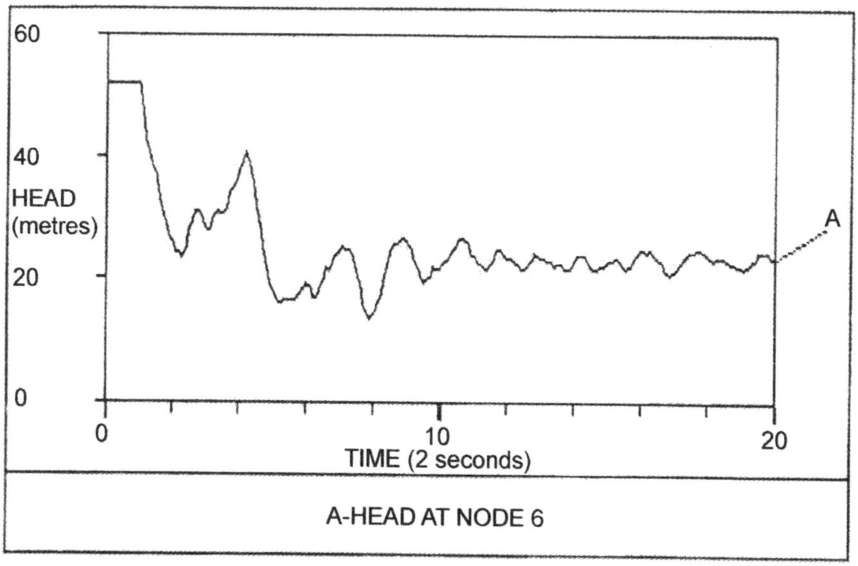

Fig. 2.31 *Changes in pressure head at the connection to one of the Distribution Networks for Case A, following loss of power to the pump station*

The influence of numerous interacting pressure waves is evident in these two pressure–time histories. The initial transient pressure wave is rapidly attenuated on account of the many reflections, and the system is seen to be settling down to a reasonably steady state some 10 s after the pump trip.

Compare these results with Figs 2.32 and 2.33 for similar locations in Case B, in which the pumping station is 4 km from the network. The fluctuations in pressure are rather more violent and take much longer to die down since the Periodic Time for the pump discharge line is now in the region of 8 s, compared with 0.2 s for Case A. This is the only difference between these two systems, yet it makes a considerable difference to the way they respond.

Fig. 2.32 *Pressure–time history part-way along the long pump discharge main of Case B following a pump trip. Compare this result with Fig. 2.30*

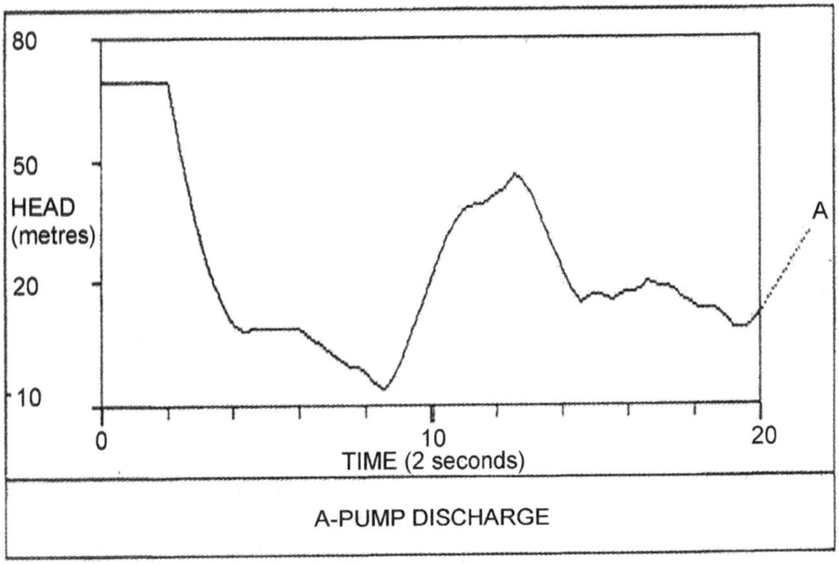

Fig. 2.33 *Pressure–time history at Node 6, the connection to a Distribution Network, following a pump trip, when the pumping station is 4 km from the main grid*

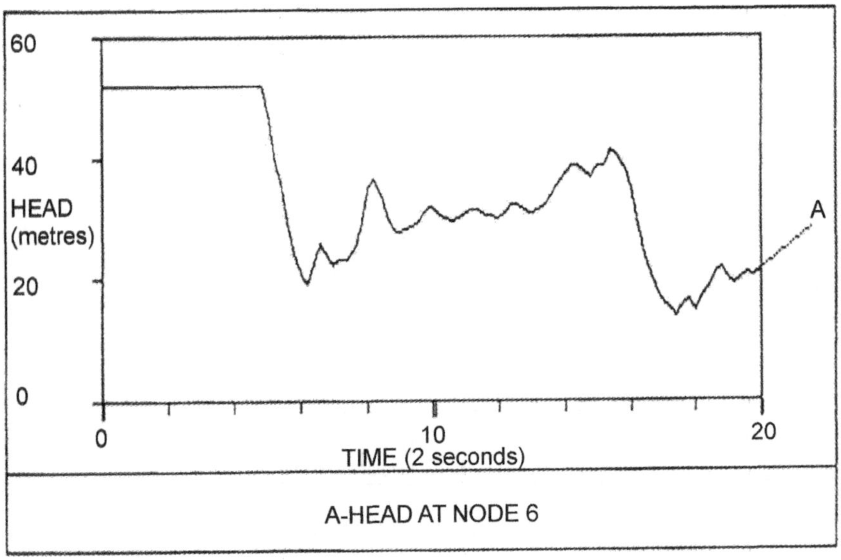

It is the networks containing longer lines that are more vulnerable to problems associated with fluid transients. Note how, for example, in Fig. 2.32 the pressure becomes subatmospheric. In the shorter discharge line of Case A, 'beneficial' wave reflections have returned from the other end of the pipeline to counteract the initial transient, and the consequences are felt throughout the system. Compare Figs 2.31 and 2.33 for Node 6. Even though all the other pipes between this point and the end of the pump discharge main are quite short, the pressure changes at node 6 have increased.

In developing the computer model to produce the results used in the discussion above, a number of idealizations were made. These include combining two pumps running in parallel into one pump, representing the associated valves as a single (check) valve, and collapsing the small-scale distribution systems into two demands leaving the main trunk network. Are such simplifications valid?

The answer is – it is a matter of engineering judgement as to how faithfully the detail of a physical system should be modelled. The computer model should be good enough, fit for the accurate prediction of transient events within a reasonable amount of computing time and commensurate with the reliability of the basic data. In the present case the demand on the system is driven by the end-users and, at the very local level, will be sporadic and random. However, once getting back towards the large-bore pipes, there will be a fairly steady flow and the fluctuations occurring in the very small bore pipes will become of minor consequence. For a longer discussion of this issue, turn to the discussion on idealizations in Subsection 2.3.4.

2.2.9 An Ash Slurry Line

Coal-fired thermal power stations produce vast quantities of ash that has to be removed from site prior to subsequent reuse in, for example, the manufacture of building blocks or in road construction. As part of this process, the ash is transported as a slurry to settling lagoons prior to being reclaimed in a more concentrated form. Usually, there are several lagoons in reasonably close proximity. Slurry is pumped into one until it is full, when it is then isolated so that the water can be allowed to drain away prior to the ash being reclaimed. Meanwhile, slurry is pumped into other lagoons and the cycle is repeated.

Figure 2.34 shows the longitudinal section for a representative system. It comprises a 400 mm ductile iron pipeline 15 km long from an ash slurry sump to a settling lagoon. The static head is 6 m. At the upstream end the slurry is pumped

Fig. 2.34 *An ash disposal system at a coal-fired power station*

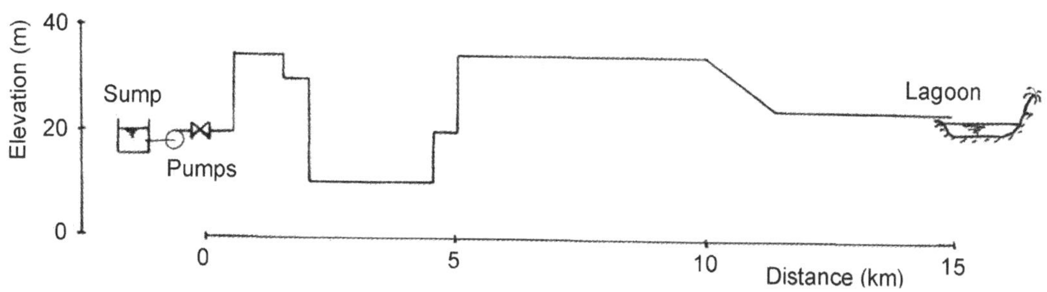

into the system by three parallel banks of pumps, each bank comprising four pumps in series. Normally, only two banks of four pumps operate, the others are on standby to ensure continuous availability of the ash disposal system.

Other salient data are: the maximum permissible pressure in the pipeline and fittings is 35 bar (equivalent to 330 m head of slurry), the frictional head is estimated to be 154 m at the required flow of 2.4 m/s and the plant operator does not wish to have check valves fitted owing to previous bad experiences of blockage and poor sealing. The density of the slurry is 1060 kg/m^3, and the required flow velocity is 2.4 m/s.

The time-varying flow scenarios are the usual ones for a pumped system, namely: controlled starts, stops and changeover of the pump combinations while continuing to maintain flow of the ash slurry to the settling lagoon. In addition, there are the uncontrolled events associated with power failure and automatic trips due to sensors responding to faults.

In terms of sequencing an assessment of the hazards due to fluid transients, the worst-case scenarios will be the power failure and other fault conditions provided, as always, that valve operations, etc. to achieve the flows associated with the planned and controlled operation of the system are so arranged that they cause less serious problems and can be carried out safely.

The additional derived data required to consider events further are: the propagation speed of pressure waves may be estimated as about 1160 m/s (the wall thickness for the pipes is 9.52 mm and the bulk modulus for this slurry can be assumed to be very similar to that for water). This enables the Joukowsky pressure change to be determined as 284 m head and the pipeline period $T\ (= 2L/c) = 26$ s.

The delivery pressure at the pump discharge valves will be the static head plus the friction head, amounting to 160 m, so, following a loss of power to the pumps, the fall in pressure will lead to subatmospheric conditions and cavitation as the Joukowsky change far exceeds this. Given the very low static head on the system, this is one of those cases where, if it can be arranged, the best strategy is to allow the liquid column gradually to come to rest without a serious rebound. How can one achieve this? Sketch the initial hydraulic grade line on the longitudinal section and proceed as for the early examples in this section to examine the sequence of events.

If no action is taken, a full vacuum would tend to occur initially either at the pumps or, more probably, at the first high point about half a kilometre downstream from them. This must be relieved to help ensure forward flow continues and the rate of flow reduction is reduced. Two options are available: install either a special air vessel or vacuum-relief valves at the high points.

The 'special' air vessel is of a construction used very successfully in many low-head slurry systems. It comprises a vessel with a full-bore pipe connecting it to the main pipeline and having a vacuum relief valve mounted on its top. As the pumps are started and flow commences, air is driven into it and, as the slurry starts to fill the pipe and enter the vessel, the air is compressed to form an energy source downstream of the pumps. Following a pump trip, the air expands and slurry flows out of the vessel, reducing the deceleration of the slurry in the main line. In that way it reduces the surge pressures and the flow gradually dies away to rest.

However, for this approach to be used, a non-return valve must be fitted to the pump discharges. The operator has specified that there should be no non-return valves in the system. This precludes the use of air vessels since they would induce a significant reverse flow through the pumps and so vacuum-breaking valves at key locations become the preferred option.

The suggested key locations are at the elevated sections 0.5–1.5 km downstream of the pump station, and at intervals along the section from 5 to 10 km. Bearing in mind that vacuum-breaking valves are normally closed when the pressure is balanced across them, air release valves, suitable for the *controlled* release of air from sewage and similar effluent lines, will also be required at similar locations for start-up. This operation must be undertaken slowly to ensure shock loads are not imposed on the pipes as slugs of liquid encounter sudden changes in direction, although the pipe is strong enough to contain the Joukowsky pressure change.

The pump start-up sequence should be that, after opening the suction valve, the first pump, i.e. the one closest to the sump, is started against a closed discharge. This is then opened to supply the second pump. The third pump is then started similarly and the discharge valve opened. As these pumps are very close together, there is no particular hazard associated with the opening of the discharge valves. However, after starting up the fourth pump, its discharge valve *must* be opened very slowly otherwise the flow would be initiated too quickly.

It will be recalled that, when a pump initiates flow against zero, or very low, static head, it will accelerate very rapidly to the extreme right of its H–Q curve, unless throttled by the discharge valve. The potential consequences of this are twofold: firstly, the air pocket that will have formed over the first elevated section could be compressed too quickly, leading to an excessive pressure within the cavity as the high kinetic energy from the liquid column is converted into a pressure head, and secondly, the column of slurry that will have been left in the low section 2–4.5 km from the pumps will be accelerated through the several bends with the risk of local shock loads on their supports (NB not shown on the sketch are various expansion loops).

Over what period, therefore, should the final discharge valve from the pumps be opened to charge up the system and initiate flow? Although we know the Pipeline Period is 26 s and could, in principle, use the design charts in Section 3.5, these do not directly apply to a line containing air pockets. A feasible and careful approach could be to adopt the following strategy, to be confirmed by a computer analysis and during commissioning tests:

1. Knowing the line length is 15 km, assume that flow is initiated at, say, 0.5 m/s and that the air release valves specified can cope with the release of the entrapped air. Then it could take up to 8.33 h on *initial* start-up for the flow to reach the far end and the line to be fully primed. Following this, the flow can be increased gradually to full flow over a much shorter period. However, the critical issue is to get a full liquid flow up to the high point some 5 km from the pump station, which would take only 2.7 h at the same flowrate.
2. For subsequent start-ups, the crucial point is to endeavour to shut the system down with minimum air ingress. There will be a tendency for the long high elevated section, beyond the 5 km point, to drain down, so, provided there is very little air in the first elevated section, the start-up can be faster, extending over, say, $15 \times 26/60 = 6.5$ min, but with the initial cracking open of the final

discharge valve being done very slowly – bearing in mind the highly non-linear characteristics of valves.

3. This suggests one aspect of an analysis that could be specified for a computer assessment of the system and/or an experiment that could be conducted during commissioning to investigate how long it will take *in this system* for the air to be purged. In the commissioning test, observers should be posted at the air release valves to time when air release was completed on start-up. In this way the start-up process can be optimized.
4. A shutdown process can be proposed that is the reverse of the start-up process. The pump delivery valve into the system could be closed by about 85 per cent quite quickly, but it must be followed by a very much slower closure of the last 15 per cent, i.e. a two-stage closure must be adopted, with the final phase taking several pipeline periods, e.g. perhaps $6 \times 26/60 = 2.5\,\text{min}$.

For a pump changeover sequence the aim will be to effect this as smoothly as possible. The approach should be to start up the standby pumps to the point where it is time to open the discharge valve. The discharge valve on the running pumps can be shut in by the 85 per cent or so referred to in proposal 4 above first, and then opening of the delivery valve on the new pumps can be initiated. As they start to pick up flow, further closure of the discharge valve on the pumps coming out of service can take place. Finally, when it is closed, the pumps can be tripped out in sequence.

2.2.10 A Sub-Sea Recharge System

A technique that is sometimes used to maintain production from gas and oil wells in order to extend their useful economic life is to pump water down into them. As the water is denser than both the gas and the oil, it gets beneath them and maintains the pressure in the well, thereby continuing to force them towards the surface. This case is included as, although it is not a frequent example of a transient flow analysis, it demonstrates that the assessment principles are exactly the same and some of the issues it raises are applicable to many other systems.

Clearly, there is much skill, and perhaps some art, in establishing the most useful places into which to inject water, but it has been found to be a useful practice in a number of installations. Several holes are drilled into the well field, some of which can be quite deep. As these are all high-pressure pumped systems, they have the potential to experience fluid transient problems and appropriate risk assessments must be undertaken. One of the unusual features is that the water is being pumped downhill, often against a significant back pressure.

Figure 2.35 is a schematic layout of a sea water injection system for a sub-sea gas field. Pumps mounted on the platform deliver sea water from buffer tanks (also on the platform) into a steel 'riser' which drops 245 m to the seabed where, after a short and almost horizontal run, it splits at junction J to feed two flexible hoses running along the seabed to remotely activated control valves at A and B. These valves control the flow of sea water into steel-lined boreholes leading from A and B to C and D respectively approximately 3 km below the seabed.

The boreholes themselves are considerably longer and terminate within the

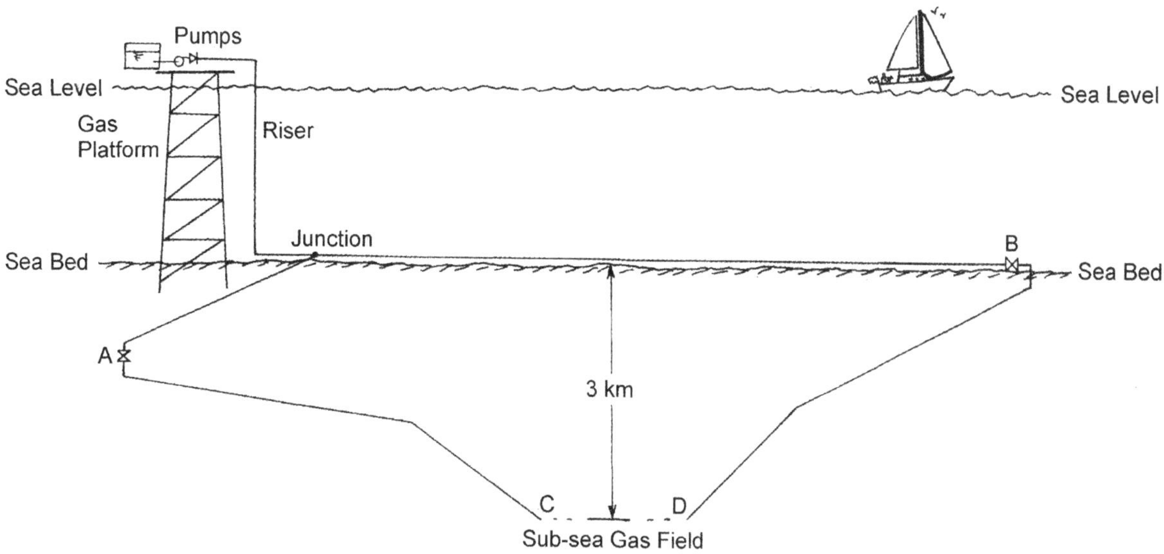

Fig. 2.35 *Schematic layout of a sea water recharge system for a representative gas field*

fractured material of the gas field. Also, for this discussion example, several nominally parallel boreholes from manifolds at A and B have been idealized into two of equivalent flow cross-sectional area.

A feature of this type of example is the uncertainty with respect to some of the data that will be required for a transient assessment. It is a dynamic situation in that conditions over the lifetime of the application will change as the gas field becomes further depleted and so the 'steady state' operating conditions change. However, an assessment of the risks must still be undertaken in the interests of the safety of personnel and the integrity of the system. It also illustrates the advantages of involving all concerned, analyst, client, owner/operator of the system, etc., in the assessment so that everyone understands where the uncertainties in the data lie, and the assumptions that have to be made in order to perform a risk assessment.

In designing the recharge system, use will have been made of data gathered over the early stages of operating the well field. For example, the back pressures from various parts of the well field will enable some estimates of the pressures at C and D to be made.

The sizes of the tubes and pipes, and often their lengths, are governed by what is feasible, in terms of the equipment available, and the ground through which the drilling occurs. As the figures in Table 2.2 imply, the economics of drilling are such that relatively high flow velocities may be involved, leading to high friction losses in the pipes and injection tubes. For this example, using data from the table below, the friction losses in the lines from junction J near the base of the platform are approximately 365 m head from J to C, and 781 m head from J to D.

To add further typical data that would have been recorded from site information, the pressure heads at C and D are assumed here to be of the order of 640 and 680 bar at C and D respectively, i.e. 6155 and 6540 m head of sea water. The corresponding pressures at A and B will be just over 3000 m head.

By allowing for this, the friction losses and the drop from the platform to the

Table 2.2 Data for a Sea Water Recharge System for a Typical Gas Well Field

Pipe	Material	Length (m)	Diameter (mm)	Flowrate (L/s)	Velocity (m/s)
Pumps – J	Steel	285	250	185	3.77
J – A	Flexible hose	2500	200	100	3.18
A – C	3 × Steel-lined tubes	5500	125	100	2.72
J – B	Flexible hose	12 000	200	85	2.71
B – D	2 × Steel-lined tubes	6000	125	85	3.46

seabed, the normal delivery pressure from the recharge pumps on the platform will be in the range 250–280 bar when all the recharge lines are in use.

The wave propagation speeds will be of the order of, say, 1200–1300 m/s in the steel pipes and steel-lined boreholes, and 900–1000 m/s in the flexible hoses. The closure time for the valves at C and D is specified as being 10 s.

This scenario of flows and pressure distributions gives a background against which to consider the impact on the system of fluid transients, bearing in mind that, with these operating conditions, the pipelines specified must be capable of withstanding very significant pressures.

The likely causes of flow changes will be:

1. Normal start-up.
2. Normal shutdown of the whole system.
3. Normal shutdown of the flow to C.
4. Normal shutdown of the flow to D.
5. Closure of the valves at both A and B simultaneously.
6. Tripping out of the pump under fault conditions.

A key feature of this system is the length of many of the pipelines. This can mean that even the normal controlled events may be ‘rapid’ in fluid transient terms. Although the Pipeline Period for the line from the pump to the junction on the seabed will be 0.44 s, from J to A it will be 4 s and from J to B it will be 26.7 s – this latter one is of particular significance.

Using the various data items for the system and the flow velocities given in Table 2.2, the Joukowsky pressure head changes can be estimated (see Table 2.3).

The significance of the various change events can now be explored. The normal start-up and shutdown of the whole system are both controlled events. Pumps

Table 2.3 Joukowsky Head Changes in the Recharge System

Pipe	Velocity (m/s)	Wave speed (m/s)	Pipeline period (s)	Joukowsky head (m)
Pumps – J	3.77	1300	0.44	500
J – A	3.18	900	5.6	292
A – C	2.72	1250	8.8	345
J – B	2.71	900	26.7	250
B – D	3.46	1250	9.6	441

discharging at the pressures involved here normally have a leak-off arrangement and, provided the system is designed to withstand the shut-off head of the pump, these operations should remain within the normal envelope of permitted pressures. A pressure pulse injected into the system on start-up and opening of the delivery valve will be reflected after about 0.44 s from the junction, with reduced pulses being propagated towards the valves at A and B.

The more critical cases are likely to be the closure of the valves at A and B. Closure of the valve at A in 10 s will not be a 'rapid' event in the strict sense of the definition. However, with respect to the boreholes downstream the pressure drop is likely to be virtually the full Joukowsky head owing to the non-linear nature of effective flow changes. Does this matter?

Bearing in mind that the maximum head change is a drop in pressure of 345 m on an initial operating pressure of the order of 3000 m head, column separation will not occur, especially as the pressure wave is moving into a region of even higher pressure – so 'no', it is unlikely to be a problem.

Meanwhile, on the upstream side of the valves at A, an increase in pressure travels 2500 m to junction J, where some will travel up the riser to the pump and some towards the valves at B. At the pump, there will be a slight increase in pressure as it rides up its characteristic to settle eventually at the new flowrate. The reflection of the pulse going to B will depend on whether the valves there are open or closed. If open, it will continue through the valves and eventually come back as a reduction wave; if closed, the pulse will be doubled in magnitude as it is reflected back towards J. Some of the reflection will be transmitted up the riser and some towards the closed valve at A. Some line packing will also occur at A as the frictional head builds up against the closed valve.

For a valve closure at B the sequence of events will be quite similar, although time-scales will differ owing to the length of the line from J to B now being 12 km rather than 2.5 km. However, closure of the valves at B is a 'rapid' event as the closure will be completed within the Periodic Time of this pipe and the full Joukowsky pressure head change will occur. Added to this, line packing will increase the pressure further as the friction head builds up.

If the valves at A and B are closed simultaneously, the effects described above will occur together. The more severe effect is likely to be felt at B where, in addition to the pressure changes described in the previous paragraph, an increase in pressure will be transmitted into the line J to B from the closure of the valve at A.

Finally, tripping out of the pumps under fault conditions will send a pressure drop of up to about 500 m into the riser if the flow stops within 0.44 s. However, as the pressure is much in excess of this column, separation will not occur and the issues to address will be to ensure an appropriate check valve is fitted to prevent shock loads on the structure.

This preceding discussion now sets the scene for defining the specification for a computer analysis of the system. After confirming that the computer model reproduces initial steady state conditions, the key scenario to investigate will be the closure of all valves at A and B together as this is the likely worst case in this system owing to the long lines. This may lead to a recommendation for intertripping of the pumps to send a negative wave into the system to help mitigate the pressure increases, depending on how close the ensuing changes in pressure come to the permitted maximum values.

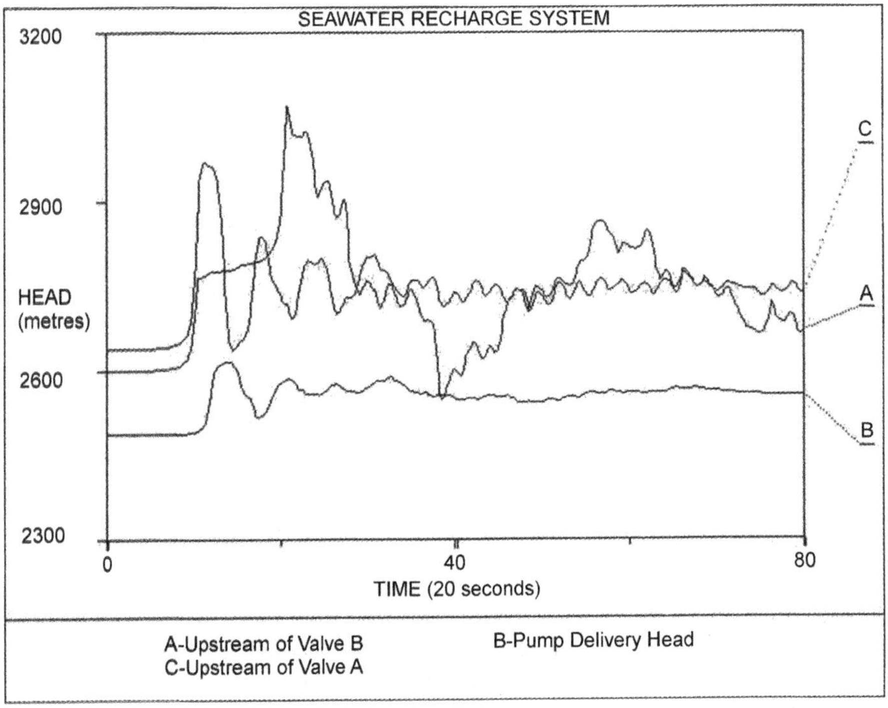

Fig. 2.36 *Pressure–time histories in the sea water recharge system. Note how the initial pressure rise at Valve B is increased further as the effect of the closure at A is added, together with some line packing. Meanwhile, the pump settles at a new operating point*

Following this, any or all of the other 'controlled' events can be analysed, together with the pump trip with delivery valve open with a small selection of delayed closures of the check valves.

2.2.11 Cooling Water Systems

Cooling water systems are a feature of thermal power stations, petrochemical plants and a number of other industrial applications. They usually involve significant flows through large-diameter pipes or tunnels, e.g. 0.5–2 m in diameter or more. The water source is either the sea or a large river or, in some cases, it is a recirculation system using cooling towers with make-up water being drawn from the sea or a river. When water is drawn from the sea or a river, it is returned some distance downstream to avoid warm water being drawn back into the intake.

A distinguishing characteristic of these systems is that the static head is normally zero, or very close to it, and the function of the pumps is primarily to overcome frictional resistance. Figure 2.37 illustrates the general layout of two common types of system that are encountered in practice.

The upper example of the two is typical of those found in thermal power stations. The condensers are mounted in the basement area of the power station beneath the low-pressure turbines. Their role is to condense the exhaust steam, condensate then being drawn out of the base of the condenser and returned to the boilers. The cooling water is drawn in from a source, which may be as much as 16 km away, and passes through screens to remove debris before entering the pumps. The relatively cold water enters the water box at the base of the condenser. It will have two or three passes through the several hundred tubes

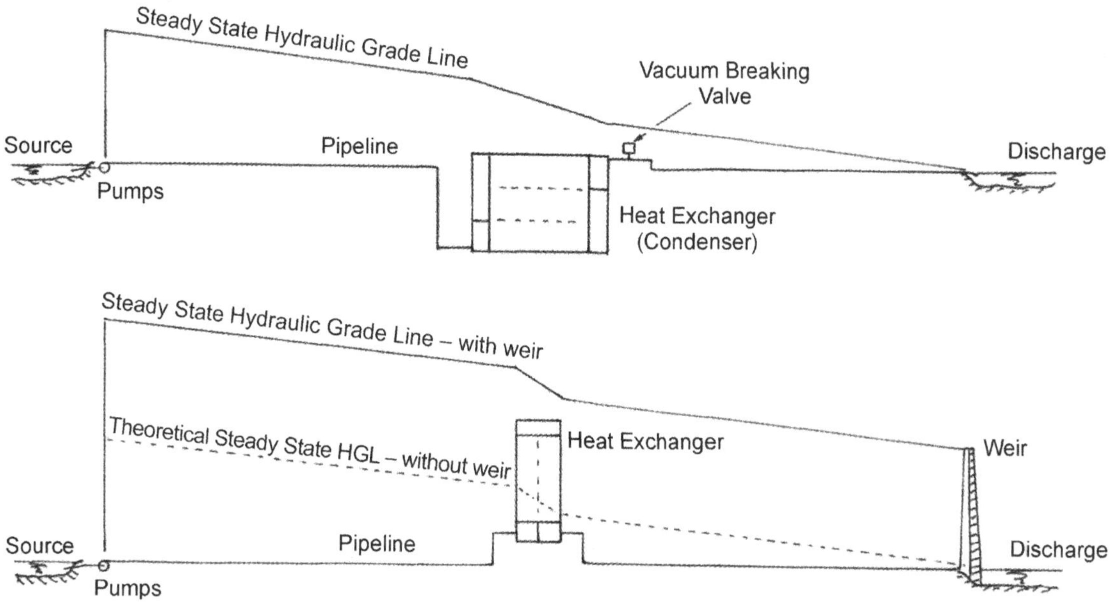

Fig. 2.37 *Longitudinal section of two types of cooling water system that draw water from a river or the sea. The upper diagram is a typical arrangement for the condensers in thermal power stations; the lower example is found in the oil and petrochemical industries*

within it before leaving the top section of the discharge water box to enter the outfall pipeline and flow back out to sea or the river.

In some installations there are separate pumps, supply and delivery lines for each of the generating units, albeit with cross-connections at the pump station that are normally closed except during maintenance periods or emergency use in the event of breakdowns. In other installations the supply and discharge lines will be common, with shorter lines within the confines of the power station itself to the separate condenser units. For the purpose of this general discussion, leading up to issues relevant to surge suppression in such systems, it is assumed that the installation is of the former type. The steady state hydraulic grade line is also shown on the longitudinal section.

The lower example is taken from the oil and gas industry. Since the nature of the physical plant is much lighter than that found in generating stations, it is usually not necessary to excavate deep foundations, and so the various heat exchangers are mounted at, or above, ground level. In practice, the cooling water flow may pass through more than one heat exchanger, and some will be mounted horizontally and some vertically. The orientation may be influenced by their function, but if the choice is optional it is better, with respect to alleviating fluid transient effects, for the heat exchangers to be as low as possible, which will generally mean being installed horizontally.

The case shown above is a simplified, i.e. idealized, version of what can be encountered in practice and is shown this way here to highlight the point that it is useful to think ahead to the potential risks associated with fluid transients before firm decisions are taken with respect to plant layout. It repeats a point made earlier in Part 1 – that the choice of pipeline route and, by the same token, the location and orientation of components can impact on the risks of unacceptable conditions developing.

The main fluid engineering difference between these two examples is the location and orientation of the heat exchangers. First, compare the steady state hydraulic grade lines and then contemplate the consequences of changes to the initial steady flows.

Under initial shutdown conditions in the power station example, the energy and hydraulic grade lines would be horizontal, and most of the pipeline would be at, or close to, atmospheric pressure. Once commissioned, the condenser water boxes and tubes would be full of water, possibly with small air cavities in the top of the water boxes. On start-up, any small pockets of air could be vented easily and the hydraulic grade line would assume the position shown.

Similarly, with a controlled shutdown, involving a timed closure of the pump discharge valves pressure reductions in the system can be limited. A 'critical point' common to these systems is the exit point from the top water box of the condenser where, as shown in this example, a vacuum-breaking valve is often fitted to protect the condenser and the line downstream against subatmospheric pressures. Complete draining of the outfall line can be avoided by installing a small weir at the end over which the water has to flow to reach the river or sea.

Loss of power to the pump motors or tripping out under fault conditions is the worst-case scenario here, followed by a controlled closure of the pump delivery valves, for which an acceptable closure time must be specified. Controlled vacuum breaking followed by delayed restart with controlled release of air are often the preferred surge protection methods and should be included in the computer analysis to be undertaken to confirm the number and location of appropriate valves. Occasionally, air vessels will be preferred owing to features of the system ruling against air valves, but they would tend to be quite large as there will be very little back pressure to decelerate the forward flow.

Some quite sophisticated computer models have been developed for this class of example, particularly with respect to the condenser. The more detailed ones enable predictions to be made of cavity growth within the water box, the pipeline downstream and even the number of tubes within the condenser itself that drain down, although full validation of these is an open question. When specifying a computer analysis, this would be a point to explore.

Turning now to the second example, there is an immediate issue with respect to the design conditions: the pumping capacity required merely to overcome the friction head is unlikely to have sufficient power to prime the heat exchanger. Therefore, unless it can be mounted horizontally or at a lower level, it will be necessary to generate a back pressure for two reasons: firstly to avoid subatmospheric pressures in the top of the heat exchanger to ensure it will function efficiently, and secondly to provide scope to withstand the pressure reductions that will follow a pump trip.

Back pressure can be created by either a choke or a partially closed valve at the end or, as shown here, by a weir. This will require the use of a more powerful pump, although this will increase both the capital and running costs. There may, however, be mitigating features owing to the operating requirements of the system, as well as some benefit in relation to surge control.

It is often the case that cooling water systems in the oil, gas and petrochemical industries are rather more complex than those for power stations or the discussion example being used here. They will be supplying several heat exchangers and consequently they will have flow and return lines running in parallel, with heat exchangers being connected between them, rather like a building heating system.

The fluid transient risk assessment will still need to review exactly the same conditions as for the power station – controlled pump starts and shutdowns and trips due to loss of power or under fault conditions. For the simple, one heat exchanger scenario shown here it may be possible to use a vacuum-breaking valve provided the weir generates a reasonable back pressure. However, for the more complex systems, the use of vacuum-breaking valves will introduce a number of uncertainties owing to the potential number of air cavities created in the system, and so air vessels become more attractive, especially if the integrity of the system is important to the availability of the plant as a whole.

2.2.12 A Phosphate Ester Pipeline

Although the vast majority of published work on fluid transients refers to systems conveying water, many other liquids, e.g. oil, fuel petrochemicals, food products, etc., are also transported by pipeline – and they can all experience problems due to changes to their steady flow. The general approach to dealing with them is, in principle, the same, although there may often be limitations on strategies that can be used for surge control owing to the nature of the fluids involved. For example, contact with air may be prohibited owing to the volatility or toxicity of the fluid. It may be corrosive when in contact with certain materials, so care must be taken to ensure materials compatibility in additional components such as pressure relief systems, etc.

The following example is a system for unloading a phosphate ester liquid from a tanker berthed at a jetty and conveying it some 2.7 km to a set of storage tanks at a chemical works. A longitudinal section of the system, with the pumps and storage tanks idealized into an equivalent pump and storage vessel, is shown in Fig. 2.38. The discharge into the tank is above the liquid surface, which is to avoid liquid siphoning back in the event of a pipeline rupture.

The design flow is 700 metric t/h which, given a density of 1295 kg/m^3, is 150 L/s. The pipes are 280 mm in diameter, so the flow velocity will be 2.35 m/s. The bulk modulus is 0.9 GPa, the wave speed is 775 m/s and the Joukowsky head will be 186 m.

As with all pipelines, it will be necessary to commission it and, ultimately, to drain it, especially if it is used for conveying different materials to different storage

Fig. 2.38 *Pipeline conveying a volatile phosphate ester from a tanker at the jetty to storage tanks over a distance of 2.7 km*

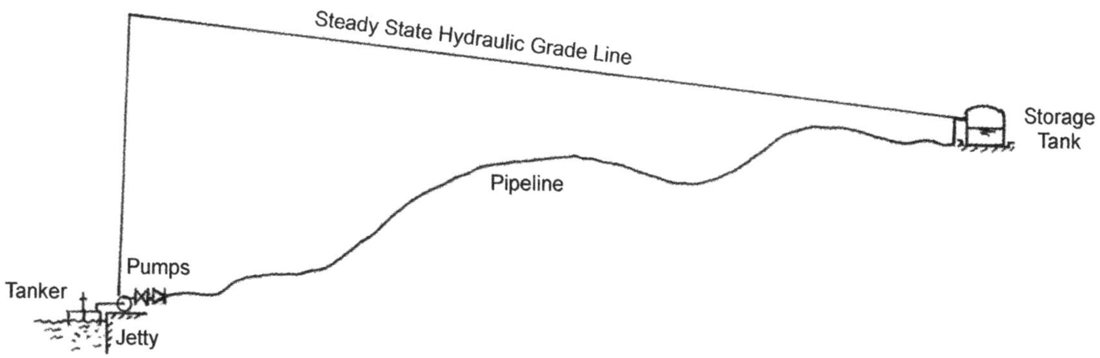

facilities related to the manufacture of various products, and so venting of the line will be required. Special venting valves are available for use in fuel and chemical lines and these should be the type recommended in systems such as this. It may be necessary to purge first with, say, nitrogen before admitting either air or the fluid to be transferred.

Having primed the line it will experience the usual routines of controlled starts and stops, so questions will arise with respect to safe valve operating times. This applies to the valves on the pump discharge as well as to the isolating valves at the storage tank end of the line. Knowing the permitted maximum pressures for the pipes and various components, estimates can be made with the aid of the charts in Section 3.5, as already demonstrated.

As with most systems, it is the unplanned and uncontrolled events that cause the most problems and generate the critical design case. The electricity supply can fail, and a slip in maintenance procedures, perhaps, can cause a bearing or motor to overheat and the pump to be tripped out on fault. With the predicted Joukowsky pressure head change, and assuming a pump rundown of maybe 1 s, cavitation conditions will definitely occur in this line. Note that volatile fluids may vaporize at pressures well above atmospheric, the actual pressure being dependent upon the prevailing temperature of the liquid which may be preheated to reduce its viscosity during transfer. Protection strategies will need to account for these effects.

Clearly, the fluid must be contained within the system to prevent a major environmental disaster. Can vapour cavities be permitted and, if so, what may be the consequences for the fluid and for the system. If permitted, can restart procedures and the system accommodate shock loads as the cavities move through the system and collapse? It will be recalled that the shock loads will give rise to pressure spikes as well as the risk of pipe movement. Is the pipe strong enough? Are the pipe supports adequate?

Although conventional air vessels could not be used, devices known as 'bladder tanks' that fulfil the same function are available (see Fig. 2.39). A bladder in the

Fig. 2.39 *Bladder-type vessel for use with volatile liquids. The bladder is filled with an inert gas, usually nitrogen, and the vessel behaves in a similar fashion to a conventional air vessel. Picture courtesy of Charlatte, France*

upper part of the vessel can be precharged with nitrogen or air. Such vessels are used in some sewage systems and, provided the bladder material is compatible with the fluid in the system, they can be used here. The key points to clarify from a computer-aided surge analysis are the capacity of the vessel and the size of the precharge required.

2.2.13 Key Points Developed in Sections 2.1 and 2.2

The preceding sections have:

- Introduced and demonstrated the application of a procedure for identifying and managing risks that arise in pipeline systems from fluid transients.
- Suggested that, even if the causes of unacceptable fluid transients are outside the direct control of the designer, operator and owner, they still have a moral and legal responsibility to take steps to ensure the safety of their personnel, the plant and the environment.
- Examined twelve typical situations to illustrate methods and procedures from Part 3, e.g.:
  - to estimate transient pressures following pump trips and valve closures, etc.;
  - to size air vessels;
  - to determine safe valve closure times;
  - to establish where vacuum relief and air release may be used with relative confidence;
  - to illustrate that, usually, systems with long pipe lengths are generally those most at risk;
  - to show how to avoid check valve slam;
  - to illustrate the cause of resonance and how it can be minimized.
- Reinforced the message that every system must be considered unique, at least with regard to fluid transients, and treated on its merits, keeping an open mind with respect to strategies that may be used for the alleviation of unacceptable conditions.
- Emphasized the importance of identifying:
  - the critical design cases, which may not necessarily be the peak flow case;
  - the sequence of events as this is the key to reducing fluctuations in pressure.
- Described how one can approach defining the specification for a computer-aided analysis of a system to confirm initial predictions of the changes in pressures in the system and the risks it faces, then proceeding to confirm and optimize the strategies proposed to reduce unacceptable conditions to within safe limits.
- Suggested how one can specify tests during commissioning to help validate the computer model of the system and its response to transient events without exposing the system to the full magnitude of a power failure or pump trip under fault conditions.

A basic theme that runs through these two sections is the importance of understanding the physical processes taking place when a transient event is initiated. The discussion of the various examples has been undertaken with the minimum of complex equations and relied on the basic ideas developed in Part 1.

SECTION 2.3

Computer modelling of transient flows

2.3.1 Introduction

Several computer codes for the analysis of transient one-dimensional 'liquid only' flows are available for purchase at a modest cost, or for use through specialist consulting companies. This is generally far cheaper than developing a new program from scratch, although it may be necessary for systems in which the following are important factors:

- complex boundary conditions (e.g. large heat exchangers, hydroelectric plant);
- when multiphase and multicomponent flows occur;
- when two- and three-dimensional flows occur;
- when heat and mass transfer effects are significant;
- when fluid–structure interaction (FSI) must be modelled.

The basis of the computer models for the analysis of transients in systems conveying liquids is the numerical solution of the equations describing the conservation of mass and of linear momentum. In their most general form, these are a pair of non-linear hyperbolic partial differential equations in which pressure and flow are variables that are dependent upon position in the system and time after some event has occurred [see, for example, Chaudhry (1987), Fox (1989), Swaffield and Boldy (1993) and Wylie and Streeter (1993)].

The development and manipulation of these equations are too well documented in these references to justify a full rehearsal here, and readers interested in program development are urged to consult them. However, a brief explanation of the relevant equations, their numerical solution and the assumptions and approximations involved may help the reader appreciate the issues to be considered when preparing for a computer-based analysis of specific systems.

2.3.2 Brief Outline of Solution by the Method of Characteristics

The most popular technique for solving the transient flow equations is the method of characteristics. Gray (1953) was among the first to apply this method to fluid transient problems, but the method was really brought to prominence through the work of Streeter and Lai (1962) and Streeter and Wylie (1967) at the University of Michigan. This method has also been exploited by other major writers in the field, such as Chaudhry (1987), Fox (1989), etc.

The two conservation laws referred to above may be expressed in the following form:

Linear momentum

$$\frac{\partial Q}{\partial t} + gA\frac{\partial H}{\partial x} + RQ|Q| = 0 \tag{2.1}$$

Mass (or continuity)

$$\frac{\partial H}{\partial t} + \frac{c^2}{gA}\frac{\partial Q}{\partial x} = 0 \tag{2.2}$$

where A is the duct cross-sectional area, g is acceleration due to gravity, c is the wave propagation speed [given by equation (3.5)] and R represents frictional resistance [$= f/(2DA)$ in terms of the Darcy–Weisbach friction factor for circular pipes]. The parameters of interest are the piezometric head (i.e. the elevation of the hydraulic grade line above an arbitrary datum), H, and the flow, Q (pressure p and flow velocity v could be used instead, but usually H and Q are preferred). They are the dependent variables, expressed with respect to position x in the system and time t.

As closed-form solutions to these partial differential equations are not possible because of the non-linear friction term, a numerical solution is required. One of these is the Method Of Characteristics (MOC) in which these two equations are first converted into ordinary differential equations and then solved by a finite difference method. Full details are given in the texts to which reference has already been made. Suffice it to say for present purposes that, in their simplest possible form, 'compatibility equations' can be obtained from equations (2.1) and (2.2) as

$$H_{Pi} = C_P - B \times Q_{Pi} \qquad C^+ \tag{2.3}$$

valid along the characteristic line defined by $\Delta x = c \times \Delta t$, and

$$H_{Pi} = C_M + B \times Q_{Pi} \qquad C^- \tag{2.4}$$

valid along the characteristic line defined by $\Delta x = -c \times \Delta t$.

Equations (2.3) and (2.4) are generally referred to as the C^+ and C^- (see-plus and see-minus) equations. They are a pair of simultaneous equations which, in principle, can be solved as easily as, for example, $y = 1 - 2x$ and $y = 3 + 4x$, where 1, 2, 3 and 4 are known quantities. They can therefore be solved to give $y = 5/3$ and $x = -1/3$.

Here, H_{Pi} and Q_{Pi} correspond to x and y and represent the head and flow at a point i in the system and are to be *predicted* after some short time interval Δt into the future, and C_P, C_M and B are known quantities corresponding to the 1, 2, 3 and 4 from the trivial example above. True, C_P, C_M and B are more involved (see below), but the principle is the same.

Note that a number of simplifying assumptions have been made and yet more will follow. These will be recalled in detail later as they influence the validity of computer codes based on them. Meanwhile, to take the discussion further, the 'known quantities' are

$$C_P = H_{i-1} + B \times Q_{i-1} - R \times Q_{i-1} \times |Q_{i-1}| \tag{2.5}$$

$$C_M = H_{i+1} - B \times Q_{i+1} + R \times Q_{i+1} \times |Q_{i+1}| \tag{2.6}$$

in which

$$B = \frac{c}{gA} \quad \text{and} \quad R = \frac{f \times \Delta x}{2gDA^2} \tag{2.7}$$

Now consider what these various equations mean and how they are applied.

As stated above, H_{Pi} and Q_{Pi} are the (piezometric) pressure and flow to be *predicted* at a point i in the system at a new time $t + \Delta t$ in the future. Insofar as C_P, C_M and B are known, they are known *now* at time t for positions in the system either side of point i, namely at $i - 1$ and $i + 1$.

Equation (2.5) takes information on changing pressures and flows from the upstream point $(i - 1)$, projects it into the future through a time interval Δt, and over a distance Δx (from i to $i - 1$) at the wave propagation speed c, to help enable future conditions at i to be predicted.

Similarly, equation (2.6) takes information from a position downstream, namely $i + 1$, and projects it through time and distance towards i so that, with C_P, C_M and B now being known quantities, equations (2.3) and (2.4) can be solved simultaneously to give both H_{Pi} and Q_{Pi} at i.

This process is often illustrated in a diagram referred to as the x–t plane, in which the horizontal axis represents the pipeline and time is on the vertical axis (see Fig. 2.40).

The distance from point 1 on the left to $N + 1$ on the right is the full length L of the pipeline which is divided into N equal lengths of distance Δx $(= L/N)$. The point about this diagram is that the two characteristic lines denoted by C^+ and C^- represent lines in space and time along which information travels from the two points either side of i. Note that this information is travelling at the propagation

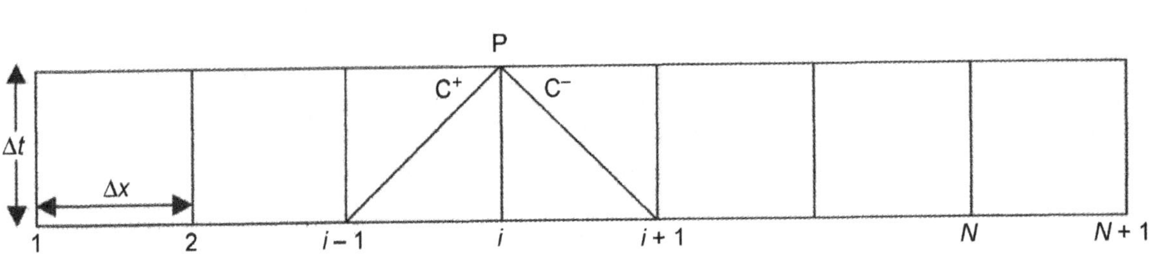

Fig. 2.40 *An x–t plane showing characteristics from i − 1 and i + 1 converging at point P (in time) which is at the intermediate point i along the pipe itself*

speed c and, in this specific case, the link with Δx is that $\Delta x = c \times \Delta t$. For this reason this particular procedure is known as 'the method of specified time intervals' – more later!

Now, *provided* it can be assumed that the wave propagation speed is constant for the liquid along the whole of this pipe and that the initial conditions along the pipe are known, the two compatibility equations (2.3) and (2.4) can be used to predict new pressure heads and flows at all the points from 2 to N at the end of the time interval Δt (see Fig. 2.41). Note that the C^+ equation (2.3) conveys information in the $+x$ direction, the C^- equation (2.4) conveys it in the $-x$ direction. Between them the two equations can be solved for the two unknowns, H_{Pi} and Q_{Pi}, for all these 'intermediate' or 'internal points' as they are sometimes called.

At each of the two ends, or 'boundaries', of the line, only one compatibility equation containing two unknowns is valid, C^- at $x = 0$ and C^+ at $x = L$. Some other link between head and flow must therefore be identified to combine with the relevant compatibility equation to enable conditions at the ends to be evaluated. These are usually referred to as the 'Boundary Conditions'.

A few simple examples, namely a reservoir, closed end and a junction between two pipes with different properties (e.g. material, wall thickness), will serve to illustrate the process. In principle, it is exactly the same for each case, it simply is necessary to identify as many equations as there are unknown quantities and solve them simultaneously.

Suppose the left-hand end of the pipe, where $x = 0$, is connected to a reservoir. By definition, this means that the pressure head there will not change with time and will remain at, say, $H_{p,1} = H_{res}$. The relevant compatibility equation is the C^- equation (2.4), i.e.

$$H_{Pi} = C_M + B \times Q_{Pi} \qquad C^-$$

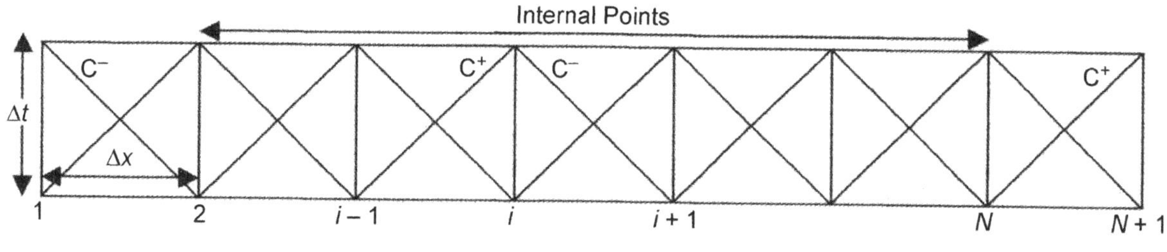

Fig. 2.41 *Similar to Fig. 2.40 but with more characteristic lines included*

Fig. 2.42 *The two ends of the line, extracted from Fig. 2.41*

Combining this with $H_{p,1} = H_{res}$ enables the new flow $Q_{P,1}$ to be determined, i.e.

$$Q_{P,1} = \frac{H_{res} - C_{M,2}}{B} \tag{2.8}$$

Note that, over a period of time, $C_{M,2}$ may not be constant since it is derived from the flows and pressures at grid point 2, at a distance Δx from the reservoir. It must be evaluated after each time interval Δt and, if conditions are changing elsewhere in the system, they will eventually be felt here and at the reservoir by way of a change in the flow to or from it.

If, on the other hand, the reservoir is at the other end of the pipe, where by definition $x = L$ and $H_{p,N+1} = H_{res}$, it is necessary to use the C^+ equation (2.3), i.e.

$$H_{Pi} = C_P - B \times Q_{Pi} \qquad C^+$$

Combining this with $H_{p,N+1} = H_{res}$ leads in the same way to

$$Q_{P,N+1} = \frac{C_{P,N} - H_{res}}{B} \tag{2.9}$$

in which $C_{P,N}$ incorporates flows and pressure heads from the position N adjacent to $x = L$, i.e. a distance Δx to the left in Figs 2.41 and 2.42. These various locations, as shown on the x–t plane, are referred to as 'grid points'.

Consider another simple example, a closed end at $x = L$. The flow, assuming that a vapour cavity does not form (an example of a more complex Boundary Condition), will remain constant at zero, i.e. $Q_{P,N+1} = 0$. Combining this with the C^+ equation (2.3), gives simply

$$H_{P,N+1} = C_{P,N} - B \times 0 = C_{P,N} \tag{2.10}$$

Finally, consider the junction of two pipes, denoted as pipe 1 and pipe 2, connected in series. From the Laws of Conservation of Mass and of Energy:

Conservation of Mass

$$Q_{P2,1} = Q_{P1,N1+1} \tag{2.11}$$

Assuming no losses due to friction at the junction, the law of Conservation of Energy gives

$$H_{P2,1} = H_{P1,N1+1} \tag{2.12}$$

Equations (2.11) and (2.12) contain the four unknowns $Q_{P2,1}$, $Q_{P1,N1+1}$, $H_{P2,1}$ and $H_{P1,N1+1}$. The two compatibility equations also contain the same four unknown quantities, so the four equations together enable a solution for all four unknowns to be found. Firstly

$$H_{P1,N1+1} = \frac{C_{P1}/B_1 + C_{M2}/B_2}{1/B_1 + 1/B_2} \tag{2.13}$$

This gives the new head in the junction and it can be substituted into either of the compatibility equations to give the new flows.

There are now, of course, two pipes. The additional subscripts denote pipe 1 on the upstream side and pipe 2 on the downstream side of the junction. There is also an implied assumption in the development above, namely that the second pipe can be divided up into an integer number of incremental lengths of Δx. These need not be the same as for the first pipe, *but* the time interval must be the same and be related to the incremental distance through $\Delta x = c \times \Delta t$, in which c is the wave propagation speed for pipe 2. If this is not the case, the discrepancy must be accommodated within the modelling process to help maintain the accuracy of the predictions.

These simple examples have been included to illustrate one of the main approaches to solving the partial differential equations that describe unsteady flow in closed conduits. The Method of Characteristics is a very powerful tool and many programs have been written around it. However, a number of assumptions are made in their development with regard to how multiple pipe systems and the various Boundary Conditions are modelled. It is important that these are recognized when using the programs as they can undermine the validity of the results.

To summarize:

- The Method of Characteristics provides a means for solving the compatibility equations for all the internal points in a system conveying liquids and, together with auxiliary equations, also enables conditions at the boundary conditions to be evaluated.
- Networks involving numerous pipes and complex boundary conditions can also be analysed provided the codes adequately model the network and the various components that are installed.

Several other numerical strategies have also been explored, especially for two-phase and two-component flows. Some have been found to be equally reliable and convenient, although they sometimes use MOC techniques to deal with boundary conditions. One strategy that has survived is the wave-plan method [Boulos *et al.* (1990) and Wood *et al.* (1966)] which forms the basis of the SURGE series of programs from the University of Kentucky.

An excellent review has been compiled by Niessner (1980), but see also Ames (1979), who writes for engineers.

Irrespective of whichever numerical basis is used, the end-user is only really interested in whether or not the programs are reliable and reasonably efficient, and whether they can adequately model events in the system being designed.

2.3.3 Idealizations and Assumptions

The starting point in setting up a computer-based solution for most time-varying flows is a set of partial differential equations derived from the application of the Conservation Laws of Mass, Linear Momentum and, sometimes, Energy. Except for a few cases, the approach used assumes the flow is 'one-dimensional', meaning that any changes in direction perpendicular to the axis of flow are negligible, i.e. the flow velocity and pressure are uniform over the flow cross-section, although they can vary with both time and axial position. Further development to the form given in equations (2.1) to (2.4) above involves additional assumptions, including:

- The flows are of low Mach number, i.e. $v \lll c$ so that $\mathrm{d}x/\mathrm{d}t = v \pm c \cong \pm c$.

- Although density changes occur, they are small, i.e. the flow is only slightly compressible.
- Inertial effects in the pipe and duct walls are negligible.
- The wave propagation speeds are constant within given sections of pipe or duct at least for each individual time-step if not for the duration of the computer run.

The implication of these assumptions and approximations means that the equations and their solution apply to Newtonian liquids and some homogeneous slurry flows, and that the effects of Fluid–Structure Interaction (FSI) are small. In developing the numerical solutions to the equations, including assigning numerical values to various parameters, yet more assumptions and approximations are normally necessary, for example:

- Steady state values for friction factors, Young's modulus for the pipe wall material, bulk modulus and viscosity of the fluid will normally be the only ones available.
- Loss coefficients and various parameters used in equations to define the behaviour of boundary conditions such as pumps will be steady state values and may involve extrapolations into areas not covered by available data.
- To accommodate the various pipes and ducts in a network into a solution by the Method of Specified Time intervals, the computer program will have to be structured so that the analyst can:
  - decide a strategy for defining an acceptable time interval Δt for the whole system; it must be considerably shorter than the time for a 'rapid' event to occur but without incurring excessive computational time to achieve accurate and complete results;
  - make use of a suitable interpolation procedure to ensure that the data used in the compatibility equations originate from the correct starting points at time t and arrive at the various points in the grid at the correct time, $t + \Delta t$; and/or
  - set the distance increments $\Delta x = c \times \Delta t$ for each pipe, then include any sections left over as a lumped inertia if the measured or intended length is to be retained; or
  - note that, at best, the wave propagation speed c will be known to only about $\pm$8–10 per cent and use engineering judgement to divide the various pipe sections up into a reasonable number of incremental lengths so that the characteristic lines pass through the various grid points at wave propagation speeds that are plausible.
- If the software is to make provision for the occurrence of vapour cavities and column separation, and for these cavities to move within the system, suitable boundary conditions will have to be devised. In some codes this may mean that, when a cavity forms, it occupies an incremental length, Δx, or lengths, which can introduce significant errors, or provision will need to be made for the interface between liquid and gas or vapour to move. This is similar to the moving boundary problem when modelling the movement of trains in tunnels.

In deciding how to resolve these issues, the analyst has to balance the need for a reliable output from a computer code that is 'user friendly' and making use of data that are sometimes unreliable for a variety of understandable reasons. Nevertheless, this is a common facet of engineering design and, if it is recognized that approximations, assumptions and idealizations are made, it is often possible to use parametric studies to assess whether or not these may be significant or may even to some extent balance out.

Likewise, the end-user employing a consultant or computer bureau to undertake an analysis of one-dimensional liquid flows will want to be satisfied that:

- A full range of boundary conditions (pumps, valves, junctions, transient control devices, etc.) is available.
- The behaviour of components that are causing transient flows, such as pumps running down, valves closing, etc., is adequately modelled.
- A suitable strategy for incorporating line friction is included. This applies particularly to long oil pipelines.
- The technique used to yield numerical solutions does not itself distort the fluctuations in the pressure and flow being predicted in the system.
- When pressures fall to levels conducive to separation of the liquid column, or the formation of vapour cavities, either these are satisfactorily modelled or a message to the effect that further results may no longer be valid is output. In systems where such low pressures are to be avoided, by suppression and control strategies if necessary, it is sufficient to be aware of if, when and where they occur.
- The program has been validated as far as possible against experimental data or other recognized criteria.
- Adequate results are produced and they are in a suitable format.

If the end-user intends purchasing, or authorizing the purchase of, software for these analyses, he will also wish to know:

- What equipment is required in order to run the software?
- How easy is the software to use, especially with respect to data input and editing, and for the output of results in a convenient graphical and tabular format?
- Can the software be interfaced with other programs (e.g. word processors, spreadsheets, etc.) for data entry, post-processing and reporting?
- Is the documentation comprehensive and intelligible?
- Is tuition available if needed? What level of technical support and back-up is provided?

The responsibility for providing all the necessary data for the system to be modelled rests with the design engineer requiring the transient flow analyses to be undertaken. A comprehensive list that embraces the majority of pipeline systems will be found in Subsection 2.3.4.

2.3.4 Preparation for Computer-Aided Analyses

A lot of time can be saved if all the relevant data can be prepared before initiating an analysis or, perhaps, presenting a fluid transient problem to a specialist. The following list should be completed as far as is appropriate for each system to be analysed.

The list includes data for transient control devices, since it is assumed that, for most cases, some form of preliminary assessment would have been undertaken. For a few special cases, additional data items may also be required.

2.3.4.1 System Data

General description of the system, giving an overview of its intended function, normal operating conditions, demands on the system, commissioning dates, planned extensions, etc., and a summary of the likely hazard scenarios identified in the preliminary risk assessment.

An overall plan showing the pipeline routes and the locations of all junctions, pumps, valves and other components.

A longitudinal section showing pipe lengths (chainage) and the elevations of the pipeline and all components, junctions, changes in pipe slope, hydraulic grade lines, maximum permissible head profiles, etc.

A summary sheet listing the type and number of the pumps, valves and other components.

2.3.4.2 Fluid Data

The key data needed are the physical properties at all the relevant operating conditions of pressure and temperature for the intended system, and especially:

- Density.
- Viscosity.
- Working temperature (and likely range).
- Wave propagation speed, bulk modulus of compressibility or the acoustic velocity.
- Vapour pressure.
- Dissolved air, gas and/or solids content, by volume, if appropriate.
- Density and elastic properties of the solids, if appropriate.

2.3.4.3 Pipes and Tunnels

- Lengths.
- Diameters.
- Wall thicknesses.
- Roughness or friction coefficients.
- Material of manufacture.
- Elastic (Young's) modulus.
- Poisson's ratio.
- Pressure rating.
- Method of jointing, support and anchoring.
- If buried, depth of burial, nature and quality of backfill material.
- Maximum permissible pressure (pipes, components, joints, supports).
- Minimum acceptable pressure (pipes, components, joints, supports).

2.3.4.4 Junctions

- Locations.
- Elevations.
- Loss coefficients.
- Any special features, such as change in pipe materials or dimensions, flow into or out of the system, etc.

2.3.4.5 Pumps

- Type, number and location.
- Performance characteristics, with operating curves and the following Rated Conditions:
  - head,
  - flow,
  - speed,
  - power (or torque),
  - efficiency.
- Inertia of rotating elements (impeller, motor and coupling).

2.3.4.6 Valves

- Type, number and location.
- Head loss characteristics.
- Pressure rating and maximum permissible pressure.

Additional data required for special-purpose valves include:

- Air relief/vacuum-breaking valves:
  - orifice sizes and head loss characteristics.
- Check valves:
  - dynamic performance characteristics,
  - anticipated delay in closing after flow reversal,
  - characteristics of damper, if fitted.
- Control valves:
  - area–, head– and flow–time characteristics, as appropriate.
- Pressure relief valves:
  - set pressures for opening and closing,
  - time taken to open and close,
  - losses in the discharge pipework.
- Bursting discs:
  - make, type (metallic, graphite, other), set pressure, losses in discharge pipeline, etc.

2.3.4.7 Reservoirs, Sumps and Tanks

- Maximum, minimum and normal levels of the liquid surface.
- Elevations of pipe (or tunnel) connections.
- For sumps and tanks:
  - cross-sectional areas,
  - pressure above the liquid surface.

2.3.4.8 Air Vessels, Accumulators and Surge Shafts

- General layout, showing type, dimensions and orientation.
- Point of connection to the pipeline.
- Length and diameter of connecting pipes and associated head loss coefficients.

- Elevations of the vessel and contained liquid in relation to the pipe centre-line.
- For air vessels and accumulators:
  - initial volume of the air (or gas) trapped above the liquid under design flow conditions,
  - ratio of specific heats for the air (or gas),
  - initial pressure at the design flowrate,
  - vessel diameter or cross-sectional area,
  - head loss characteristics for inflow and outflow.

2.3.4.9 Feed Tanks

- Location.
- Contained water volume.
- Cross-sectional area.
- Elevation relative to the main pipeline.
- Length and diameter of the connecting pipe.
- Type, head loss, time delay in closing and dynamic characteristic of the check valve in the connecting pipe.
- Time taken for the tank to refill after use.

2.3.4.10 Bypass Lines

- Location.
- Length.
- Diameter.
- Type, head loss and dynamic characteristic of the check valve.

2.3.4.11 Transient Event Data

The precise nature of the data required will depend upon the cause of the transient events to be studied. These will be from the list of the various 'impending' events discussed earlier and may, for example, include an unscheduled pump trip or start-up, or a valve closure or opening.

For a pump operation, in addition to the data outlined above, the timescale over which the events are intended, or expected, to occur should be given. If the operating range under transient flow conditions extends outside that of the data usually available, some assumptions will be required about how this will be handled. Recourse to the standard 'Suter' characteristics [Marchel *et al.* (1965) and Suter (1966)] may be appropriate.

Similarly, for valve movements, the timescales involved should be added to the data referred to above, together with the rate of valve spindle movement, the head loss (or loss coefficient) versus valve position, and the initial and final positions of the valve.

If oscillatory motion is involved, e.g. from a valve or a positive displacement pump, etc., the amplitude, frequency, pump speed and number of chambers will be required.

2.3.4.12 Aims and Objectives

A clear statement should be prepared and agreed with the analyst, setting out the purpose and aims of the analyses to be undertaken. It should define the transient flow conditions that it is expected may arise, and indicate the order in which some, or all, of them are to be investigated.

A schedule of progress meetings should be agreed to provide an opportunity to revise the scope of the investigation as it progresses and especially if preliminary results indicate that this might be desirable. This is, perhaps, stating the obvious, but it is surprising how often transient analyses are ill-defined.

2.3.4.13 Expectations on Completion

It is important to have mutual agreement, between the analyst and the system designer, on the outcome expected, before the analysis commences. Details will vary from case to case. On the one hand, all that may be required is confirmation that, say, the overall size of an air vessel, based perhaps on a preliminary assessment, will provide adequate protection. On the other hand, the analyst may be required to look at a number of possible fault conditions, identify the critical cases and propose, specify and perhaps even design the necessary control and suppression strategies.

The essential point is that, to avoid subsequent disappointment, a clear understanding is reached at the outset on how far the analyst is expected to go down the path of investigation and towards the actual detail design of control devices and components.

Ideally, the whole risk assessment should be an exercise in which the analyst, designer, operator and owner are all involved, sharing awareness of the approximations and assumptions being made to progress the analysis, and sharing responsibility for the decisions made to ensure the safety of personnel, the system and the environment.

On completion, a full report should be prepared, including a concise summary and cover sheet, signed and dated by all the participants. A specimen is given at the end of this section.

2.3.4.14 Idealizations and Assumptions

Although the system should be described in full to the analyst, some idealizations and approximations will be both necessary and desirable. There are two reasons for this. Firstly, some of the basic data required for a transient flow analysis will not be available and may not be reliable even if they are. Secondly, some simplifications can avoid unnecessarily long computer run times. This can lead to a more manageable problem without compromising the validity of the analysis.

While it is quite possible to build a computer model that faithfully represents the physical layout of a pipeline system, the improvement in the accuracy of its predicted behaviour can be far outweighed by the considerable increase in computer run time over that of a slightly simpler version. There is normally a satisfactory compromise that is computationally efficient and yet will provide answers well within the accuracy of the basic data used to describe fluid properties, component operating characteristics, etc.

Some examples of valid system idealizations referred to above are:

1. Equivalent pumps. If two or more similar pumps are running in parallel or series, they may be modelled as a single equivalent pump – unless, of course, it is the effect of just one of them being started up or tripped that is being looked at. This avoids the necessity of coping with a variety of valves and very short connecting pipes.
2. Variation in pipe lengths and wave propagation speeds. The shortest individual pipe in a model is one of the dominant influences on the computer run time. Consideration may be given to ignoring extremely short pipes if the analyst deems their effect on transient behaviour will be negligible, e.g. a short suction line from a sump to pumps discharging to a pipeline several kilometres in length. In other situations it could be more acceptable to combine short pipes to make a longer one. Associated with this process may be the adjustment of wave propagation speeds from their theoretical values. Provided these are kept to within about ±8–10 per cent they will be within the usual error band associated with the accuracy of the data used for their estimation.
3. Transient events. Several events may occur in sequence, but perhaps only one of them is significant in the context of initiating serious transient flow problems. For example, a system start-up may comprise one or more pumps in series being run up to speed and then the final discharge valve being opened. Since it is usually only this final valve opening to the system that is likely to cause problems, the analysis could start with the pumps already running.

Deficiencies in at least some of the basic data available for the fluid and material properties, pump and valve performance characteristics, etc., at the time they are required for a transient analysis are quite common.

This is not necessarily as big a problem as it might seem. The computer model, and its associated database, can usually be constructed with representative values culled from sources such as the present text or from the analyst's own experience. Parametric studies will indicate whether or not the accuracy of that data to within, say, 5, 10 or even 15 per cent is important. If necessary, the model can usually be rerun when more reliable data become available.

Some judgement will clearly be required, and it will depend upon the circumstances. Suppose, for example, that a pump trip is being considered in a rising main. Many modern pumps will run down and the flow will reverse within 1–3 s of being tripped, depending on the discharge head and the inertia of the rotating parts, of which the latter is usually unknown. If the discharge pipeline is more than about 4 km in length, the exact details of the rundown time are not really that important, since the full Joukowsky pressure changes will be experienced by most of the pipeline. Hence, a notional inertia will be quite adequate. On the other hand, if the pump discharges into a nearby network, the converse would be true.

2.3.4.15 Confirmation and Testing

It is surprising how seldom the results of fluid transient studies are validated on prototype systems. Even when the specification for transient analyses is being prepared, some thought should be given to a likely test programme, and in particular what measurements could be taken. By devising a suitable test procedure at an early stage, the output from the computer model can include predictions for the places in the system where measurements can actually be made.

It may be deemed prudent not to test the most extreme fault conditions, such as loss of power at full load, though even a part-load shutdown of a section of a complex system can be extremely serious. However, there is every reason to seek the confidence that can be gained from undertaking less demanding tests as these can still provide reassurance that the system has been adequately protected against the more serious risks. This should be an essential part of the normal commissioning procedure and included in the computer-based fault assessment.

Fluid Transient Risk Assessment – Summary Sheet

NB The attached report and tables summarize the results of a risk assessment for the effect of fluid transients on the system specified. The recommendations that follow from this are only valid for the system as defined and the conditions under which it was assessed. If any changes are made to the physical layout of the system or its components, or to the operating conditions or procedures, the risk assessment *must* be reviewed

System	A name that identifies the system
Function	Statement of its function, e.g. fuel transfer, rising main to treatment works, including, say, the design flowrate
System and product data	List of data sheets attached: System data and layout Pump and valve data Fluid data Safety Information, COSHH data sheets, etc.
Design criteria	Statement of the operating requirements that must be met
Design constraints • **Maximum permitted pressure** • **Minimum permitted pressure** • **Maximum permitted flowrate** • **Minimum permitted flowrate** • **Shock loads to be avoided**	Statement of relevant numerical data e.g.: • 25 m head • minimum acceptable pressure is 2 m gauge • not applicable in many cases • this may apply in slurry pipelines
Fluid transient criteria	Statement of safety and other criteria to be met following transient 'events', e.g.: • Maximum head must not be exceeded anywhere • No air admission except when draining – pressure must always be above the minimum permitted value • Check valve slam must be avoided • Pump delivery valves *must* close within 30 s, but two-stage closures are allowed • Flows diverted out of the system via relief valves limited to 20 t owing to capacity of blowdown vessel • Air vessels, if recommended, must not exceed a height above ground level of 5 m
Key transient events	List of the events (to be) investigated e.g.: • Loss of power to all pumps – no protection • Loss of power to all pumps – with protection • Normal shutdown procedure – with protection • Normal pump changeover sequence • One pump trip (on fault) from three running • Normal closure of emergency shutdown valve

Assumptions/idealizations List of key assumptions and idealizations of the system; could also refer to data sheets attached for longitudinal section and if significant assumptions concerning fluid data are required

Recommendations for surge control
The following are suggested headings that should be modified as appropriate

Equipment requirements
A statement of changes to the system layout and/or additional equipment required, e.g.:
- air vessels having an initial air (or inert gas) volume of $x\,m^3$, a total capacity of $y\,m^3$ and a liquid level when the system is in operation of z m above the pipe axis;
- check valves fitted to pump discharges must have a dynamic response better than x;
- pump delivery valves must close in not less than (say) t seconds;
- if a positive displacement pump is used, its speed of operation should *not* match the following frequencies, etc.

Operational requirements
A statement of operational constraints that should be included in the Operations Manual for the plant. It may include any or all of the following plus additional items related to the specific plant and fluid being conveyed:
- Normal start-up procedure/sequence
- Normal shutdown procedure/sequence
- Pump changeover:
  (a) one pump running,
  (b) two (or more) pumps running
- Start-up procedure after emergency shutdown
- Priming/draining the system

Maintenance, service and inspection requirements
A statement of the requirements for the surge control devices fitted. Some, such as pressurized air vessels, will be subject to statutory inspections, e.g. annually, but good practice to help demonstrate compliance with Health and Safety legislation could include:
- Pressure relief valves should be floated once per month
- Liquid levels in surge vessels must be inspected visually daily/weekly
- Automatic flow control valves should be inspected for wear every 1, 2, 6 (?) months
- All vacuum relief and air valves must be inspected visually every 1, 2, 6 (?) months
- Bursting discs must be inspected visually for evidence of corrosion at 3, 6 (?) months

A similar list for service and replacement can be made as part of the annual (?) overhaul.

Following an emergency shutdown, the following should be considered:
- Delay restarting until the system has stabilized, feed tanks have been refilled, bursting discs replaced, blowdown vessels emptied and product recovered
- Pipe and equipment supports should be visually inspected for evidence of movement
- Emergency shutdown valves, etc. should be inspected for evidence of damage

This Risk Assessment was:	Name	Signature	Date
Specified by			
Undertaken by			
In consultation with			
Approved by			
Read and agreed by			
• **Safety Officer**			
• **Operations Manager**			
• **Plant Operator(s)**			
• **Etc.**			

Fluid Transient Risk Assessment – Summary Sheet

NB The attached report and tables summarize the results of a risk assessment for the effect of fluid transients on the system specified. The recommendations that follow from this are only valid for the system as defined and the conditions under which it was assessed. If any changes are made to the physical layout of the system or its components, or to the operating conditions or procedures, the risk assessment *must* be reviewed.

System:

Function:

System and product data:

Design criteria:

Design constraints:

- **Maximum permitted pressure**
- **Minimum permitted pressure**
- **Maximum permitted flowrate**
- **Minimum permitted flowrate**
- **Shock loads to be avoided**

Fluid transient criteria:

Key transient events:

Assumptions/idealizations

Recommendations for surge control

Equipment requirements

Operational requirements

Maintenance, service and inspection requirements

This Risk Assessment was:	Name	Signature	Date
Specified by			
Undertaken by			
In consultation with			
Approved by			
Read and agreed by			
• Safety Officer			
• Operations Manager			
• Plant Operator(s)			
• Etc.			

SECTION 2.4

Accidents and incidents

Most of this book has been dedicated to identifying and preventing unacceptable conditions arising from unsteady flows in piping systems. This particular section could also be subtitled 'Tales of the Unexpected', since it contains descriptions of a number of incidents that have suddenly confronted rather surprised system operators.

Even with the most carefully prepared design and supervised operation, there is always the risk that, just occasionally, something unforeseen can go wrong. This is as true in the field of fluid transient control and suppression as anywhere else. Perhaps an oversight on someone's part, an incomplete set of design data provided to the analyst, a maintenance inspection deferred or simply a change in operating conditions can lead to a 'transient event' that was not anticipated.

The following examples give a rather broader perspective to the more obvious Potential Causes of Fluid Transients in a Pipe System listed in Table 2.1.

2.4.1 The Case of the Lightweight Anchor Blocks

A sewage scheme was being uprated by laying a 900 mm ductile iron pipeline in parallel with two existing mains of smaller diameter. The total length of the system was some 1080 m and the new pipeline was to be supplied by two pumps running in parallel. Being sewage, the design engineer preferred not to use check valves on the pump discharges, choosing instead automatic valves that would close over a predetermined period.

A fluid transient study was undertaken on the following lines:

- a site investigation, including an instrumented pump trip, of the existing system to obtain physical data such as actual wave speeds, etc.;
- a computer study of the existing system, including calibration of the program with the aid of the test data;
- a computer analysis of the proposed new system, with the twin objectives of specifying the optimum closure time for the automatic valves on the pump

discharges, and assessing the safety of the overall system with respect to transient flow events.

The analyses yielded an optimum valve closure time of around 17 s, to be fine-tuned during commissioning in view of assumptions concerning wave propagation speeds in sewage. It was also predicted that the peak transient pressures in the system following a pump trip would be about 5.5 bar, well within the permitted level of 10 bar quoted in the specification.

The pipeline was duly constructed and commissioning tests undertaken. During the first test the system ruptured, with a reported maximum pressure of approximately 4 bar.

The subsequent investigation revealed two things of note. Firstly, while it was confirmed that 10 bar was indeed the permitted pressure level in the pipe itself, the anchor blocks had been designed to withstand a mere 2 bar, but this had not been communicated to the analysts!

Secondly, being naturally cautious, the commissioning engineer had decided to trip the pumps from low flowrates during the initial tests, working up to higher flowrates for the later tests. The significance of this is that, with low initial steady flows, flow reversal at the pumps occurs sooner than for higher flows and hence the optimum closure time for the automatic valves under these conditions should be shorter. Consequently, much higher reverse velocities than expected were allowed to develop, leading to higher transient pressures as the valves finally closed.

The pipe itself could withstand the higher pressures. However, the principal weakness was the anchor blocks, and hence the joints failed.

Lessons from this incident are to ensure that:

- a complete set of data is provided for the transient analyst;
- the actual site tests have been checked beforehand as part of the detailed risk analysis.

2.4.2 The Dancing Feed Range

Five boilers in a coal-fired power station draw their feed water from a 200 mm bore ring main. The ring main in turn is supplied by two constant-speed pumps and one variable-speed pump, all of which pump water through high-pressure heat exchangers before joining the feed range. The three connections to the feed range are all in close proximity to one another, so the water will tend to split fairly evenly to flow in opposite directions at that point.

Normal practice at the time was that, when steaming under a fairly steady load, all of the boilers except one were set to run at a constant output, and the remaining one would be set to an automatic mode for feed water supply to compensate for minor changes in steam demand from the turbines. In this mode the flow of feed water into the boiler was controlled by two automatic valves, one on each side of the boiler at entry to a heat exchanger known as an Economizer.

One evening, quite suddenly, while operating in this fashion, the whole feed range began shaking violently, being restrained only by its supports and the connections to the boilers. The amplitude of the pipe vibrations was in the region of 10–15 cm. The noise and vibration were transmitted to adjacent parts of the building and they also shook.

Fortunately, the engineer in charge recognized the symptoms of hydraulic resonance, though the exact cause was not immediately apparent. Since it seemed to be associated with the feed range, and something was happening at a frequency that appeared as though it could match a natural harmonic of the system, his first thought was that it could be connected with the variable-speed pump. The speed of this pump was therefore increased by about 5 per cent, and the vibrations died away as rapidly as they had begun.

On reflection, it was realized that the pump had run at the previous speed many times without any problems of resonance or vibration, so, while changing its speed provided a rapid cure, it seemed unlikely that the pump itself could have been the cause of the problem. It then transpired that a different boiler than usual had been put onto automatic feed mode.

This plant was quite old and maintenance was not always as thorough as perhaps it should have been. The moving parts on the automatic valves on this particular boiler had become quite badly worn in the region of their normal position when operating at a constant load, although there was much less wear elsewhere in the mechanism.

The cause of the problem was that the looseness in the mechanism had allowed the control element to hunt and a self-excited motion had been set up, corresponding to one of the natural frequencies of the feed range.

Increasing the speed of the pump had increased the pressure, and hence flow, in the feed range. As the water level in the boiler drum increased, on account of the greater flowrate, the valve opening was reduced by the controller moving to another, less worn, part of the mechanism. At this new position it was not so free to oscillate and the vibrations stopped.

Lessons from this incident are to ensure that:

- inspection, maintenance and test procedures are properly carried out;
- plant operators are familiar with the symptoms of fluid transients.

2.4.3 Where has all the Water Gone ?

This incident concerns an elderly hydroelectric power plant located, like the two previous examples, in the United Kingdom. A 3 m diameter riveted mild steel pipeline, having a wall thickness of 9.5 mm, extends about 600 m from the upstream reservoir, followed by the high-pressure penstocks and the turbine main inlet valves.

During an otherwise uneventful shift one day, the pressure head on the turbines fell away to zero over a period of some 6–8 min. No valves had been operated and no other changes had been made, but still the water pressure fell away. No record is available concerning any noises that may have occurred, but it was realized that the water supply had ceased and it became necessary to shut down the plant as the turbine was then in motoring mode.

In the investigation that followed it was discovered that, near to the upstream end of the riveted steel pipeline, a runaway valve had inadvertently closed. Downstream of this valve the mild steel pipeline had been unable to support what must have been almost a full vacuum and had collapsed inwards over a distance of some 500 m. The cross-sectional shape was closer to a banana than a circle.

The runaway valve had been installed as a protective device. In the event of the pipeline rupturing, this valve was intended to close automatically to prevent the reservoir from emptying and causing damage to land and property in the path of where the water would otherwise have run.

Automatic closure was supposed to be initiated when the flow velocity exceeded about 2.6 m/s, which was higher than the normal velocity when the turbines were on full load.

The closure of this valve at a time when clearly it should not have occurred was attributed, as in the previous example, to a lapse in inspection, maintenance and service procedures.

2.4.4 A Midnight Feast

Holes dug in the ground sometimes have a tendency to fill up with water. Coal mines in South Africa, as in other countries, are examples of deep holes that, if people are to work in them, need to be drained on a fairly continuous basis. This is achieved with pumping plants located near the base of the mines pumping water that has seeped into the mine up to the surface and then to a convenient runaway or reservoir.

The plant that is the subject of this incident was quite typical. Multistage pumps at the base of the mine pumped the water up to the surface through an almost vertical pipe system, which then ran in a fairly horizontal fashion to a storage reservoir. The pipes were steel, coated with bitumen on the inside, and connected by flanged joints.

To cope with anticipated transient pressures, the high-head pumps operating in parallel had been fitted with fast-acting check valves, and the pipes stressed to withstand both high pressures and a full vacuum. The normal operating procedure was for the system to operate intermittently, removing water from a sump, then shutting down for several hours until the sump was again full. It had, in fact, been designed to handle a much greater degree of seepage than it regularly encountered.

The system had operated without any problems for several years. Nevertheless, one day a long section of the almost horizontal section at the surface leading to the storage reservoir burst following a pump trip. The split was a classical hoop stress failure, running axially along the top of the pipe. A curious feature was that examination of the inside of the pipe showed that, along the upper surface, the protective bituminous coating was absent. How had this happened?

Inspection records proved that the pipe had been properly coated when new. Chemical analysis confirmed that there appeared nothing unusual about the quality of the water. The operating strategy of the system had not been changed, so no unexpected transient pressures should have occurred. It was only when the microbiologists began studying some of the little organisms in the mine water that the truth eventually began to dawn.

They discovered that the favourite meal of the particular variety of microorganism in the water was bitumen. During the shutdown periods they migrated to the top part of the pipeline and ate up the bitumen coating. The steel pipe was then left unprotected and suffered corrosion.

Over a period of time the wall thickness was reduced to a point where it could no longer withstand the variations in pressure associated even with normal running conditions. The intended protection against transient pressures, i.e. the strong pipes, was compromised.

The direct cause of this failure is so unusual that it is difficult to argue that it should have been foreseen, unless regular internal inspections of the pipes were undertaken or very detailed assessments of water quality had been obtained. However, it does serve to emphasize the care that is required when reviewing potential hazards that can strike, some of them long after the system has been commissioned.

2.4.5 Green for Danger

Part of a chemical process involved pumping a heavy combustible liquid between tanks. The salient features of the piping system in which this incident occurred are as follows. A 50 mm bore suction pipe, approximately 8.5 m long, took liquid from a stock tank to a centrifugal pump with a duty point of around 7.5 L/s at 30 m head. The delivery line, also 50 mm bore, consisted of a 5.2 m horizontal run followed by a 7.6 m vertical leg to a plug valve. Beyond the valve was a further short vertical section, then a horizontal run, although these are not relevant to this incident.

On the fateful day, an operator started the pump to initiate a transfer of liquid but, after a short while, the pump was tripped out by its thermal overload protection. The operator pressed the green start button again and there was an almost instantaneous explosion and fire at the valve. The studs retaining the valve cover fractured owing to the high pressure, and the valve plug shot out at high speed and disintegrated. The pipeline was also damaged.

This particular system had been in operation for some 50 years without any previous incident of this nature. The inquiry into the accident found that, as often happens, a chain of events was involved [Thornton (1983)]. A principal link in this chain was that the plug valve, which should have been open when the pump was started, had been shut.

However, for explosive combustion to have occurred, air must have been present, and the mixture of air and vapour from the liquid must have been at an elevated temperature for spontaneous combustion to have been initiated. How had air been able to collect upstream of the closed valve and, even more curiously, why did the system not explode when the pump was first started?

As the valve was destroyed in the incident, it can only be presumed that air entered the system through a leaking gland or a weak seal, possibly being sucked in as the liquid in the pipe tended to drain back to its source. The amount of air is also, of course, unknown. The next question is, what pressures and temperatures could have developed in the vapour cavity?

Thornton himself investigated this question for this particular incident [Thornton (1983)], while this and similar incidents prompted other more recent studies [Thorley and Main (1986) and Thorley and Spurrett (1990)].

These all demonstrated that, as a vapour cavity is compressed by a moving liquid column, rather like a piston in a cylinder, both the pressure and the temperature will rise dramatically. In a diesel engine the piston travel is limited by the design of rotating crankshaft, but when the piston is a liquid column in a pipe it is the build-up of pressure in the cavity that stops it. If the liquid column develops a high kinetic energy, a correspondingly high pressure in the cavity develops. The researches quoted have shown that an initial cavity pressure of a

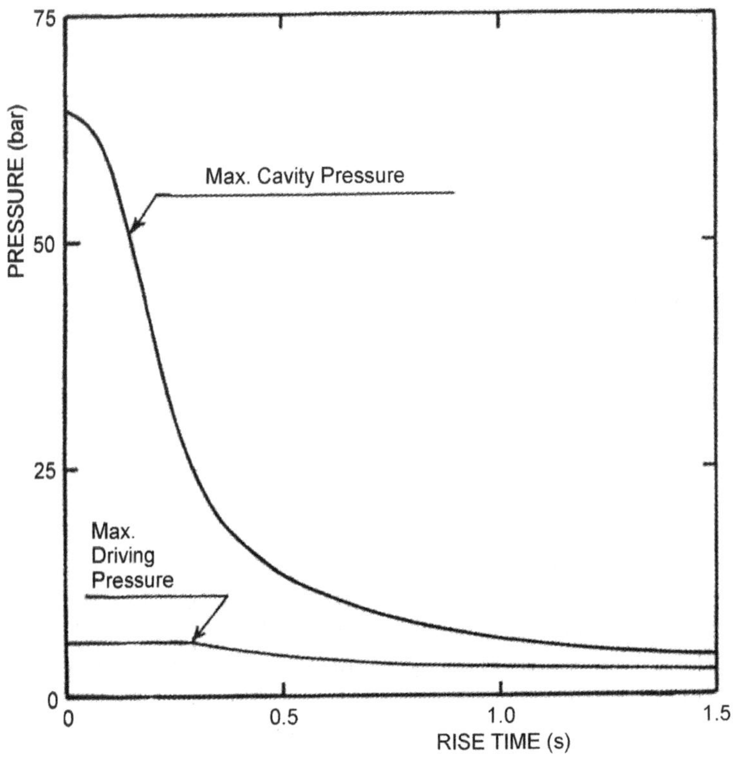

Fig. 2.43 *Even with a pressure as low as a few bar, the resulting pressure in a shrinking vapour cavity can reach extremely high values*

mere 1 bar can rapidly be magnified many times, with temperatures of several hundred degrees being generated. Figure 2.43 is taken from Thorley and Main (1986) and shows the maximum pressures that can occur as a vapour cavity is compressed. The corresponding temperatures will also be very high.

For an air/vapour mixture to ignite spontaneously it must be above its auto-ignition point. Unfortunately, this is not constant, but drops as the ambient pressure increases. A typical mineral oil will have an Auto-Ignition Temperature (AIT) of around 380 °C at atmospheric pressure, but at 100 bar this can drop to 200 °C. Even fire-resistant hydraulic fluids experience a reduction in AIT, which can fall from, say, 580 °C at atmospheric pressure to as low as 320 °C at their working pressures. These temperatures are well below those that can be generated in an air cavity being compressed by a moving liquid column, implying that, with the correct air/fuel, mix, combustion is virtually inevitable.

Figure 2.44 shows two phenomena. The two parallel curves illustrate how the AIT drops quite significantly as the pressure in the vapour cavity increases. The danger zone, i.e. the area on the graph where combustion can occur, is above the relevant AIT curve.

The curve that rises up from near the origin characterizes the pressure and temperature changes within a cavity containing an air/fuel mix such as the one in the incident being described in this example. During a fluid transient event, conditions within the cavity are represented by a point moving up and down this line. Since this line clearly crosses into this danger zone, it indicates that, in principle, spontaneous combustion can occur.

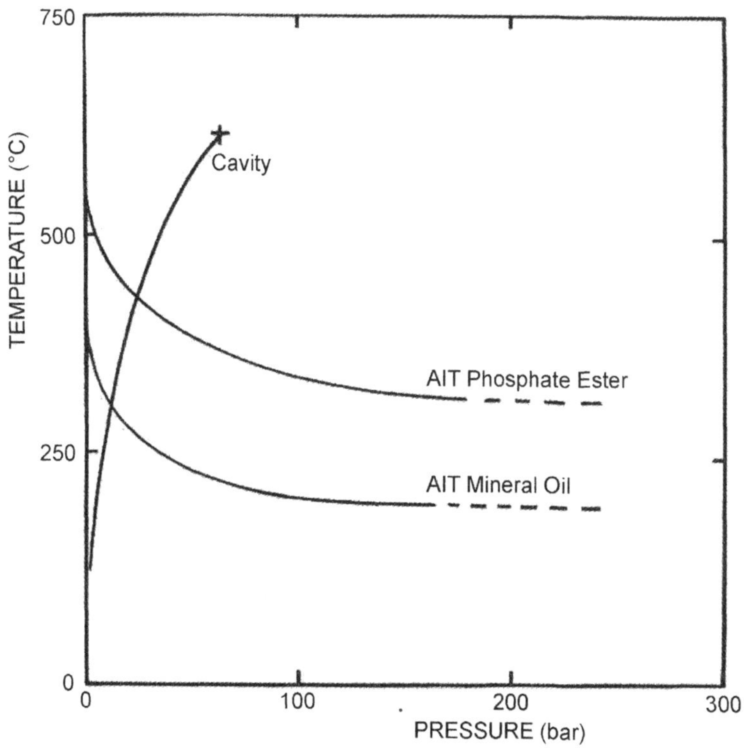

Fig. 2.44 *Curves for the auto-ignition temperature for two types of combustible liquid together with the pressure–temperature curve for a cavity in a duct illustrating how it can extend into the danger zone above the AIT curves*

The remaining question is, if pressures and temperatures conducive to combustion could be created in the air/vapour cavity adjacent to the valve, why did nothing untoward happen when the pump was first started?

The answer to this almost certainly lies in the results of some patient detective work by White (1960) and Main (1985). White investigated the causes of two explosions on US aircraft carriers that led to the deaths of nearly 100 people. Main studied explosions in a number of situations, including offshore oil and gas platforms and land-based oil-hydraulic systems, as well as researching into methods for characterizing the AIT properties of fire-resistant hydraulic fluids.

Basically, these showed that, as well as having an air/fuel mixture at a temperature and pressure suitable for combustion, other factors are also significant in determining whether combustion actually occurs or not. These include the distribution of the air/fuel mixture, the degree of surface wetting of the pipe walls, the surface roughness and contamination of the surface by rust or other corrosion products.

The event that initiated fluid transients here was a pump start, for which the pipeline and its components appear to have been strong enough to withstand the ensuing fluctuations in pressure. However, what may not have been fully appreciated when the line was built was the significance of the auto-ignition characteristics of the liquid being conveyed and the consequence of an air cavity being allowed to form. The situation was compounded by weak inspection and maintenance procedures that did not ensure that the valve seal was properly maintained or pick up the fact that air was leaking into the system.

Lessons that can be drawn from this case are to ensure that:

- inspection and maintenance procedures are followed, even on older plant;
- full details of all relevant fluid properties are known and their significance appreciated.

2.4.6 Minor Change – Major Problem

The Manzanillo II Thermal Power Plant was designed with two 350 MW generating units which were scheduled to begin operation in 1988. The condenser cooling water system comprises an inlet channel through which sea water is fed from Cuyutlan Lake to an intake structure. Located in the intake structure are auxiliary cooling water pumps, travelling screens and the two main cooling water pumps. Each of these pumps supplies water at a rate of 6.42 m^3/s against a head of 8.7 m to similar cooling water circuits. The conduits from the pumps to the condensers are square in cross-section and made of reinforced concrete. Downstream from the condensers, similar conduits lead to a seal well, from which the discharge is a free surface flow.

A computer analysis had been undertaken, which predicted that under all operating conditions transient pressures would be satisfactorily controlled. During commissioning tests the system was started up and ran satisfactorily, with several starts and stops. However, at 1500 h on Friday, 27th November 1987, following a violent thump, the top surface of a section of the concrete duct fractured and jets of water sprayed out of the many holes [Paz Soldan (1989)].

Immediately prior to this, the circulating water pumps had been operating in their backwashing mode to clean the screens. The valve that controlled the flow to the washers had closed accidentally, initiating a transient, and then the pump had tripped on overload, causing another transient.

Reviewing the design for the cooling water system, and the specification for transient flow analyses, it was realized that the system for backwashing the screens had entailed a small change from the original design for the cooling water system. This had been introduced after the transient analyses had been undertaken, and no checks were made subsequently to see how the modification might affect the transient response of the pipelines.

The problem was overcome by providing alternative pumps for the backwashing system, so that the cooling water circuit could be returned to a layout consistent with the original design. Meanwhile, the damaged concrete pipe was dismantled and relaid to a greater thickness.

Moral – ensure that all changes to a design are incorporated in the transient flow risk assessments, especially if they occur late in the design stage.

2.4.7 A Positive Reflection

A fire protection system installed at a certain international airport consists basically of two parallel pipes, with several cross-connections between them, forming a water distribution network. Various external hydrants and internal sprinkler systems are connected to the basic system. In common with other systems of this type, there are some lines that terminate in a dead-end.

The system covers an area of some 520 000 m^2, and the total length of the main piping is 9 km. The pump station is located centrally and is capable of providing a flow of up to 1400 L/s for about 45 min [Herforth and Heuser (1986)]. The pipes are of asbestos cement, having diameters ranging from 600 mm at the pump station down to 300 mm at the extremities.

Following completion of the system, a series of seemingly inexplicable line breaks occurred. In addition, operators noticed that some of the sprinkler systems became pressurized at twice the normal level. These events occurred when the system was idle, i.e. on standby mode, and concern was expressed about how it might function in a real emergency. If failures could occur when the flowrates were minimal, or even apparently zero, the potential for a catastrophic failure under the intended firefighting flows seemed enormous.

As the system was, in effect, a fairly complex network, several possibilities existed for the likely causes of the excess pressures. A programme of carefully controlled tests, combined with computer studies, was devised that enabled the problems to be identified. The essential outcome was as follows.

The high pressure in the sprinkler systems was the first real clue. It was then demonstrated, by closing one of the fire hydrants, not especially quickly, from a modest flow, that pressure waves travelling back towards the pumps met a closed check valve. Being shut, it represented, in fluid transient terms, a dead-end. These give positive reflections, i.e. if the pressure wave meeting it represents an increase in pressure, then the pressure is doubled. Conversely, if the pressure drops, the drop is doubled or a cavity formed if vapour pressure is reached. Basically, pressure waves were being locked into the system in a cumulative fashion.

Several remedies were explored, but the one finally adopted was to install five pressure relief valves at critical points in the system where it had been shown that the higher pressures occurred.

This phenomenon, of high pressures being locked into a system, is a hazard associated with certain types or combinations of automatic valve. Among the systems frequently fitted with such valves are fire protection systems designed to be maintained under pressure, and various service water systems in high-rise buildings. These include the supplies to washrooms and toilets. Vertical pipes run down the service shafts, with a take-off on each floor. Pressure regulating valves at each level are intended to ensure that the supply pressure to taps and faucets is not too high.

Problems can arise when the taps and the flushing devices on toilets close automatically. High pressures can be generated and become locked within sections of the system when the pressure regulating valves behave, in effect, like check valves. Different sections may also interact with one another, and significant pipe movement, and noise, can be a consequence.

Moral – if it is necessary to incorporate several check or pressure regulating valves in a system, it is prudent to ensure that transient pressure waves cannot become locked in. This can be achieved by ensuring that rates of change to the flow are kept as low as reasonable, commensurate with the main function of the system. If this still results in high pressures, a strategy for relieving the excess pressure must be devised. Typically, this may be to install pressure relief valves, a small air vessel or an accumulator.

2.4.8 Hanging Free

To change over from one pump to another supplying water to a system ought to be a trouble-free operation, but this is not what it proved to be in the following incident.

Feed water systems on a nuclear power plant were supplied by one of two pumps fitted in parallel. As is typical of such plants, the pipes followed a contorted route, being supported at various locations by hangers. Some movement was permitted to allow for thermal expansion and contraction. The pumps drew their supply from a common source, and on each pump discharge a counterweighted swing check valve and an isolating valve were fitted. The discharge pipeline continued for several metres before combining into a single line.

During commissioning tests, a pump changeover was scheduled. One pump was running at its design flowrate. The second pump was started and brought into service. Although some pressure fluctuations occurred, they were not of great significance.

Once the pressure and flow had stabilized, the original pump was tripped out – with rather frightening results. A very violent and noisy movement of the discharge piping occurred, causing fractures and other serious damage to the pipe supports. Fortunately, the pipe did not rupture, but it was evident from data recordings that large variations in pressure had occurred.

The incident was judged to have been so severe that the system was carefully instrumented and a series of tests undertaken to gain a better understanding of what had happened. Data recorded included pressures and flows at various locations, pump data and the position of the doors on the swing check valves. Several tests were completed, including the pump changeover that had given rise to the incident described above, though at lower flowrates.

The detailed results confirmed that violent fluctuations in pressure occurred. The basic problem was thought to be the inability of the check valve to respond quickly to the changing flows. After the pump that was running initially was tripped, the flow reversed rapidly in that section of the pipe close to the check valve. This was due to a high pressure being maintained in the discharge manifold by the pump that was still running. When the swing check valve eventually closed, a large step change in pressure occurred on the downstream side. Meanwhile, a vapour cavity formed on the upstream face. The slamming of the valve door onto its seat caused a shock load on the piping system, aggravated by the step change in pressure. When the vapour cavity on the upstream side collapsed shortly afterwards, this also gave another violent shock to the system. This cavity was shown to open and collapse several times.

The hypothesis that the late closure of the check valve was the fundamental problem was reinforced. To verify this, the swing check valve on one pump discharge was replaced with a nozzle type of check valve and the tests were repeated. The start-up of the second pump went more smoothly, but the most dramatic improvement came when the pump running initially was tripped out. The changes in pressure were significantly reduced and no shock loads occurred.

The net result was that 12 swing check valves were replaced with nozzle valves and no further problems occurred. At the same time, a temporary restriction on the plant to 30 per cent of its design output, which had been imposed when the symptoms first manifested themselves, was lifted.

This particular incident was one of several that prompted the studies of check valve slam [Provoost (1980) and Thorley (1989)], leading to recommendations for how such valves should be selected – see Subsections 1.3.2.4 and 3.8.2 for a further discussion of this topic.

2.4.9 The Devil is in the Detail

This is another example where problems that had not been anticipated emerged during start-up and commissioning. It concerns a cooling water system for a process plant. The supply system comprises several main branch lines and a number of smaller loops. Take-offs from the supply lines feed a number of heat exchangers and other components in the plant, the discharges from these then entering the return lines back to the cooling towers [Locher and Wang (1996)].

As is typical of such plants, the various pipes follow fairly tortuous routes and are mounted on racks and supports. These are intended to allow for thermal expansion and contraction and to support the pipes against axial and lateral movement.

The main lines are up to 980 mm in diameter, with branch lines typically being 200–500 mm in diameter. The distances involved between individual bends and junctions are measured in terms of tens of metres, with the overall distances for major sections and loops being in the 100–200 m range. Changes in elevation from the pumps to the top of the racks for the system of interest here are about 13.5 m. When the system is in full operation, with two pumps running, the flow is expected to be of the order of 1290 L/s. For part-load operation, only one pump would be used, with surplus water being returned to the cooling towers via a bypass line. The flow in the bypass line is controlled by a valve that also maintains a set pressure in the line to the rest of the plant. This is to help ensure that the various heat exchangers and other components receive their intended supply.

At the time of the incident, only one pump was in operation. Water was being supplied to two small sections of the plant and the bypass line was open. During a start-up test the operating pump tripped out. After a short period, defined as 2–5 min later [Locher and Wang (1996)], the standby pump was started. This initiated a significant vibration of the pipe, and a section of 450 mm pipe on one of the elevated racks was recorded as having moved 500 mm laterally and 300–500 mm axially. Unfortunately, a workman who had been standing on the pipes was seriously injured.

The subsequent investigation established that, after the pump had tripped out, the system had partially drained down via the bypass line which had been left open. Subatmospheric pressures, column separation and vapour cavity formation had occurred in the elevated section of pipe. A vent fitted in the return line may have contributed to the ease with which this happened.

On start-up of the standby pump, a column of water had moved at speed down the supply line, encountering bends and junctions, and collapsing vapour cavities. This generated high pressures and dynamic loads and gave rise to the pipe movements observed. It also emerged that the pipe supports were designed to support gravitational loads only. Provision had not been made for side loads, and hence the pipes were able to move off their supports.

A computer model of the system under part-load supported these conclusions and provided a basis for investigating remedial action. It was also used to investigate other part load configurations of the system, the results of which indicated that these tended to produce more severe transient effects than when operating from full flow. Among recommendations for suppression and control that emerged from these studies were:

- install two combined vacuum-relief/release valves and an air valve in and near elevated sections of the lines;
- modify the start-up procedures, including fitting a bypass line round the main pump discharge valves, so that a controlled initial flow could be used to purge the lines of air.

Lessons that may be learned from this case include:

- *All* operating scenarios should be adequately risk assessed prior to being instigated.
- Even under nominally low-flow conditions in complex networks in which sections are closed off, unacceptable pressures and changes in pressure can occur. The velocity and pressure distribution in the network can be significantly different from that expected under normal operation, leading to lower safety margins and higher Joukowsky pressure changes.
- When branch lines are not in service, they should be shut off where they connect to the main pipeline and not at remote ends that amplify any pressure changes.

2.4.10 Lessons to be Learned

The incidents recounted above represent a selection of those known to the author. Many others have occurred, and some will be found in the literature [see, for example, Chaudhry (1987), Hamilton and Taylor (1996), Jaeger (1963), Kletz (1998), Locher *et al.* (2000), Pulling (1976), Serkiz (1983) and the proceedings of the more recent Pressure Surge Conferences].

Irrespective of how much effort is made to anticipate situations involving unsteady and transient flows, to imagine perverse sequences of events that may conspire to come together at the right (or wrong?) time, the potential for an accident, however remote, is always there.

Every system must be considered on its merits. If there is the slightest risk that some event, or sequence of events, will happen, in due time it will. Some of the incidents described happened during commissioning, but others happened as much as 50 years afterwards, and then despite years of apparently safe operation. The designer, operator and owner must always be alert to threats from transient events for as long as the safe and reliable operation of pipeline systems remains their responsibility.

Among the general lessons that may be learned from these various cases are:

- Involve all key people in developing the fluid transient risk assessment. These include those who:
  - can contribute to defining what operating conditions may be encountered,
  - can provide relevant and complete data on the system and the fluid it is to convey.
- Although peak flows often create the most severe fluid transients, there are still

many other situations where unacceptable conditions can occur, especially during commissioning.

- Education of maintenance and operational staff in the causes and symptoms of fluid transients can play a useful part in anticipating problems that may occur and action to take should they, unfortunately, do so.
- It is essential to keep up to date with inspection, maintenance and service schedules.

SECTION 2.5

Transients: current status – future developments

The topic of 'fluid transients' has been described as a mature subject. This carries an implication that the fundamentals of transient behaviour are well known and understood, and that, by using appropriate computer codes, engineers are able, with confidence, to design and operate systems free from potential hazards due to unsteady and transient flows.

There is a vestige of truth in this, but it carries with it the essential proviso that the designer must be conversant with the subject. The present state of knowledge still requires the exercise of engineering judgement based on experience and understanding for there to be confidence in the integrity and security of piping systems. Much progress has been made since the pioneering work of Allievi (1903, 1913 and 1925), Joukowsky (1900), Korteweg (1878), Lamb (1898) and others, but there is still a large gap between current knowledge and modelling capability on the one hand and the situation implied in the previous paragraph on the other.

2.5.1 Summary of Fluid Transient Modelling Capability in 2003

This section is a broad-brush review of the current position to provide a platform from which to suggest the direction of, and topics for, future research. It is not intended as a fully referenced critique of work over the previous century.

Despite all the advances made in the last century, computer codes for fluid transient analysis are still limited to 'analysis' – analysis of preconceived systems – they do not *design* either systems or strategies for their protection. There is still much about fluid transients, and their suppression and control, that is more an art than a fully understood science. Also, while it is possible to undertake risk assessments of piping systems with a greater degree of confidence than before, much fluid transient analysis is still contracted out to 'experts'. The continuing programme of conferences on the subject, with the inclusion of sessions on accidents and incidents, is also testimony to the need for further work. Current capabilities may be summarized as follows:

- Software is available for purchase that can analyse transient flows:
  - of some liquids and slightly compressible homogeneous fluids,
  - with constant, or nearly constant, wave propagation speeds,
  - passing through routine boundary conditions,
  - using steady state friction and component characteristics,
  - in a number of common piping systems and networks.
- Resources are available through software developers and some specialist consulting companies that can provide guidance about transient flows in a limited range of more complex situations. These include:
  - combined free surface and full-bore flows, e.g. power station condenser cooling water systems, sewer and drainage systems, the filling of pipelines, etc.,
  - some multiphase and multicomponent flows,
  - some two- and three-dimensional flows,
  - special boundary conditions, e.g. surge shafts, hydroelectric turbines,
  - fluid–structure interaction (FSI) for relatively simple systems,
  - gas transmission lines,
  - trains moving through tunnels, i.e. some flows with moving boundaries.

The phrase 'provide guidance' is used deliberately since various assumptions will be incorporated to describe the behaviour of system components and the fluids involved. For example, in the simple pump trip scenario used at the beginning of Part 1 to introduce the basic concepts of fluid transient behaviour, the pipeline remained pressurized and full of liquid at all times. Figures 1.4 and 1.5 illustrate predictions from one of the standard computer codes for the pressure fluctuations downstream of the pump following loss of power. Effects due to friction are evident in these predictions (the additional changes on top of the Joukowsky pressure changes and the damping with time), but these do not, in fact, represent the full story.

If real data from an instrumented test on the actual system were to be obtained, it would be found that, although there would be quite close agreement during the initial fluctuations in pressure, cumulative discrepancies would develop. These can be attributed in part to the fluid and in part to the behaviour of the pipes and the system.

Also, when the pipeline route was changed to a higher elevation, it was shown that lower pressures would occur in the system. This would lead to air and gas release, bubble generation and separation of the liquid into two or more separate columns which subsequently collapse, generating sharp pressure rises and shock loads (see Fig. 1.8).

The predictions illustrated in these three figures may be regarded as 'fit for purpose' in the sense that they indicate that unacceptable conditions will occur. The analysis has served its purpose and need proceed no further since it has demonstrated that control and suppression methods must be introduced.

However, the fact that discrepancies between prediction and reality occur points to the need for a better understanding of the mechanisms at work.

Many man-months of research have gone into several of these issues (see, for example, the proceedings of the various conferences and symposia on this topic over the past 15 years or so sponsored by BHR Group, the Fluid Engineering Division of ASME, etc.). Much of this has been in response to local or specific needs, not surprisingly since it has to be financed, but much of it has consequently been rather uncoordinated and haphazard. One notable exception was IAHR-sponsored work on power station cooling water systems [IAHR (2000)].

Suppose it were possible to stand back and orchestrate the R&D. How might some focus be given to it? One approach is to look far into the future and keep in mind the desire to be able to *design* systems using computer codes. Such codes will ultimately be part of a broader integrated design approach that also incorporates functional design and optimization, both of which have evolved significantly in parallel with fluid transient work. Some might say that they have evolved further since some optimization programs are quite sophisticated and make use, for example, of knowledge-based expert systems and similar deductive and learning algorithms.

This may be anathema to some traditional engineers, but it is really only a matter of time before it happens. Irrespective of whether or not this view is supported, it is nevertheless a means by which to focus on what issues can usefully be addressed.

2.5.2 Knowledge Engineering and Fluid Transients

Increasingly, engineering design groups in many organizations are unable to retain the breadth of experience and expertise necessary to undertake complex engineering designs. Hence, the question may be posed: can the necessary body of purely technical knowledge and data be combined with the judgement and discretion of the 'human expert' to yield a Knowledge-Based Expert System (KBES) that incorporates both the engineering judgement of the experts and the ability to continue to learn?

With respect to fluid transients, the evaluation of risks is often quite complex and usually requires a significant amount of engineering judgement. This arises for two reasons. Firstly, no two systems are really quite the same, so problems that can occur, and the practical steps to overcome them, differ from one system to another. Secondly, important items of data that are needed for an analysis are frequently not available at the time the assessments are required and have to be assumed or estimated. It becomes a matter of judgement as to which particular numerical values are adopted and the effects they may have on the accuracy of the predicted response of the system in question to intended operational changes or the various fault conditions that may occur. This is a typical situation for which Knowledge-Based Expert Systems are designed.

The evolution of 'Knowledge Engineering' has provided scope to develop programs that will 'design' rather than merely 'analyse' predetermined systems. The early developments in Artificial Intelligence moved computers from functioning simply as data processors to functioning as knowledge processors. They can now incorporate the knowledge of human experts to solve difficult problems. While a significant amount of knowledge can be stored directly in a knowledge-based system, the real power comes from its ability to reason beyond the knowledge or data directly stored. This ability is often known as *inferencing*.

The major application areas of knowledge-based systems include:

- *Design and configuration.*
  Using criteria for an acceptable solution, knowledge-based systems can yield designs to fit the given criteria.
- *Planning and scheduling.*
  Knowledge-based systems can plan a sequence of actions or events that meet the constraints and needs of the end-user or process.
- *Diagnosis and classification.*
  Using symptoms, knowledge-based systems can reason about, identify and classify underlying problems and propose ways to overcome them.
- *Simulation, control and training.*
  Knowledge-based systems can simulate complex processes and indicate the consequences of changes in control variables.

Early knowledge-based systems were usually called *expert systems* because they were designed to produce solutions to problems that only experts could solve. These systems generally used *if–then* rules to store or *represent* the knowledge of the human experts. In this context, *rules* are pieces of knowledge that can be combined in various ways to solve a variety of problems.

Such rules are useful summaries of expert knowledge, but they express only the *surface* of that knowledge. A better approach is to build a *model* or *simulation* of the system in question. Such a model should capture the relationships between symptoms and their underlying causes. Reasoning that incorporates a model or simulation of this kind is called *model-based reasoning*. Although such models can often be created using only rules, in practice it is more convenient to combine rule-based and method-based reasoning within a domain of structured objects.

The *domain* of a knowledge-based system is that part of the world with which the KBES is concerned. In the present case, the 'world' is *pipeline design*. The problem that the system solves in that domain is defined as the *task*. The task of interest here is to assess risks due to fluid transients and to design strategies to control and suppress unacceptable conditions.

The rule-based approach focuses on the task, not on the domain, whereas the model-based approach focuses on the domain. The same model can be used to accomplish multiple tasks and is not restricted to using rules alone. Other problem-solving strategies, such as object-oriented programming can also be incorporated in the models.

The advantages of model-based systems include flexibility, ease of modification and the ability to handle novel problems. For example, a classic symptom-based system for car maintenance can make diagnoses only if it has explicit knowledge or a direct link between symptom and fault. On the other hand, a model-based system, programmed with a model of how a car works, can make a diagnosis by introducing certain faults and then run simulations to see whether they give rise to the symptoms in question. Similar developmental approaches have also been made with KBESs applied to medical diagnosis and the treatment of certain illnesses.

Model-based systems are not restricted to one task. The model is a general representation of the domain and can be used for more than one application or task. The core of the system, i.e. the model, is the same for each case but the behaviour of the system will be different. For this reason, the goal for KBESs applied to pipe system design is to develop models that can be used for functional design, optimization, risk assessment, simulation and the training of operators.

With years of experience in a problem domain, human experts come to understand in detail the components and interactions involved in their specialist field. The objective of knowledge engineering is to capture this experience, judgement and discretion in a computer system.

To achieve this it is necessary to break complex systems down into their most basic elements. The position in the system of each element is defined, and its relationship to other elements, the way it functions, the characteristics it possesses, etc., are all spelt out. All the necessary information can be stored in a model organized on a hierarchical basis just as, for example, companies are organized in which effective lines of communication must be set up between the various elements to enable them to function as a coherent system (see Fig. 2.45).

A model for pipe systems may be developed along similar lines as shown in Fig. 2.46. Some of the basic elements are clearly recognizable, others, such as the fluid and the descriptors for many of the various components, have been omitted owing to space constraints.

Much data that can be used in the descriptive elements already exist, such as material and many fluid properties and the characteristics of system components. Gaps in the knowledge form the basis for future research programmes as discussed below. Where much attention needs to be focused is in capturing the discretion and judgement of the various 'experts' and building on their experience while they

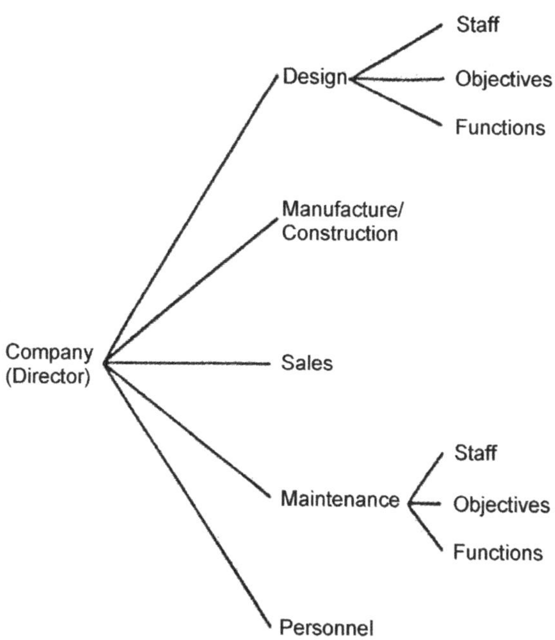

Fig. 2.45 *Symbolic representation of a hierarchical management structure*

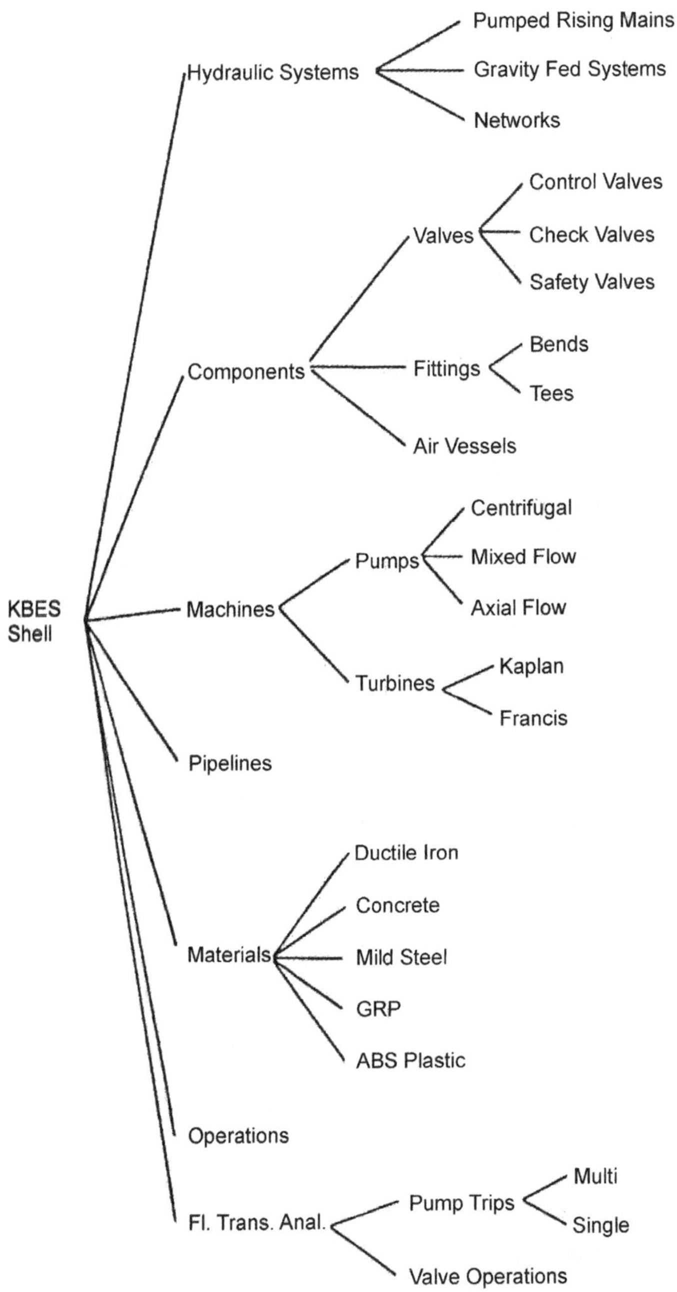

Fig. 2.46 *Hierarchical management structure for a representative hydraulic system*

are still available. This will also help ensure that accidents and incidents such as those described in Section 2.4 are less likely to happen

To summarize, Knowledge-Based Expert Systems are interactive computer programs that:

- incorporate the experience of human experts in computer models;
- focus on the activity domain;

- identify the structural and functional components;
- represent them by objects, functions, rules and methods.

They can be very efficient where:

- the solutions are not exclusively algorithmic;
- recognized experts exist;
- the problem presents several solution paths;
- the knowledge for making decisions evolves continuously.

To repeat a comment from the beginning of this subsection, by acknowledging that the ultimate achievement *will* be the development of Knowledge-Based Expert Systems later, if not sooner, future research and development can be coordinated and focused rather better than it might perhaps have been in the past. With this in mind, the present discussion continues under the following headings:

- Behaviour and response of the fluid.
- Dynamic behaviour of components and devices.
- Fluid–Structure Interaction (FSI).

2.5.3 Behaviour and Response of the Fluid

Energy loss and damping of transient pressure waves can be attributed to the behaviour of both the fluid and the mechanical system. With respect to the fluid, viscous effects and the structure of the fluid body must be considered. Various models have been explored to represent 'dynamic' or frequency-dependent friction to allow for the rate of change in velocity, and the velocity itself, in the vicinity of pipe and tunnel walls. This work continues.

Another flow scenario that has attracted much interest is when phase changes occur and/or the flow is multicomponent. With liquid flows, one of the first consequences of the pressure falling towards vapour pressure is that any dissolved gas present will come out of solution. Agitation of the flow, as through valves and pumps, will increase the rate at which gases are given up. It is generally assumed that gas comes out of solution in a time-dependent fashion and, if the pressure rises again, any reabsorption is at such a low rate as to be negligible.

In many analyses the gas, having come out of solution, is usually assumed to be uniformly distributed through the fluid or to collect at nodal points. Unless the flow is highly turbulent, the behaviour of the bubbles will be influenced more noticeably by gravitational effects. They will tend to be located as long strings along the top of pipes that are close to horizontal and have axial velocities that are different to the liquid velocities, especially in near-vertical pipes, and to congregate at high points.

Flow visualization studies have shown that, when vapour cavities and column separation occur rapidly, the fluid in the vapour/gaseous region is initially very frothy as rapid boiling takes place. However, as time progresses, it seldom remains like that and the flow will tend to become free surface and/or slug flow. The exact time depends on the fluid, degree of turbulence, inclination of the pipe, etc.

In general, studies of multiphase and multicomponent flows have concentrated on the two ends of the flow spectrum, i.e. bubbly flow and mist flow. In these two regions, much progress has been made in developing an understanding of the complex phenomena involved and how they can be modelled numerically. Much more work remains to be done, especially in the much more difficult region between these extremes, e.g. dynamic slug flow, partial free surface flow, etc.

Topics within the general field of multiphase and multicomponent flows that merit further study include the following:

- Interfacial friction in bubbly and stratified flows, including added mass effects.
- Thermal effects, heat and mass transfer and variable specific heats.
- Transitions between quasi-homogeneous and non-homogeneous flows, including stratified and slug flows – first in simple systems and then in networks.
- Scale effects – transients in large-bore systems, including storm sewers, very large-scale gravity-fed lines, etc.
- Gas release and reabsorption; bubble growth and movement. The vapour–liquid interface.
- Transients in liquids other than 'clean' water and in solid–liquid–gas systems; sewage, process fluids and volatile liquids, including self-ignition potential when combustible vapours are present.
- Numerical methods to mirror the physical behaviour adequately for design purposes.

While it might be argued that some of the topics would only lead to refinements of what is already known, the study of dynamic and complex flows is still a long way from completion.

2.5.4 Dynamic Behaviour of Components and Devices

Historically, the class of components that has attracted most interest is turbomachinery. For pumping plants it is now generally recognized that the Suter representation of pump characteristics is the most useful one available, especially when abnormal conditions of reverse flow and/or rotation occur. The advantage of this method is that, in flow regimes of particular interest in transient analysis, i.e. in the vicinity of flow reversal and speed reversal, the characteristics are still real and finite. This is in contrast to the normal steady state parameters of H/N^2 and Q/N that would have to pass through infinity.

However, even the Suter characteristics are still based only on steady flow data. The originally very limited set of curves available has recently been extended, but there is a need to extend the database further. Specific topics to pursue include:

- comparisons between predicted and experimental behaviour under rapidly changing flow conditions;
- scale effects and the use of non-dimensional parameters such as shape number to broaden the application of the curves to other pumps in a homologous series.

That such work has not yet been pursued with any degree of urgency is perhaps due to the fact that, in most installations, the pumps are normally protected by check valves. As the pump characteristic for the normal zone of operation is more likely to be available, a little extrapolation at each end is often accepted as good enough, especially against the background of uncertainties in all the other data upon which the analysis will be based.

This is less true in the case of turbines where a much greater effort has been expended, not only with regard to the machine itself but also to the control system and guide vanes. The capital investment in hydroelectric schemes is generally far greater than most pumping systems, and hence higher confidence levels in the risk assessments of unsteady conditions are demanded. Virtually all schemes are designed with the aid of both physical and numerical model studies. Much of the physical testing is still on the basis of steady state situations, even though dynamic conditions are also of concern. Linked to transient behaviour is the question of system stability, where the 'system' comprises the pipelines, machines, control system and the electrical system, and the interactions between them all.

After hydraulic machinery, probably the next major class of components is 'valves', with the wide range of types and the various functions they are intended to fulfil. Again, it tends to be only the steady state characteristics that are available to the transient flow analyst – a situation that needs to be remedied.

One notable exception for which dynamic performance characteristics have been identified is the undamped check valve. The principles have been defined and a few manufacturers are now willing to provide them. However, dynamic data for many check valve types are still not available, and more manufacturers need to be encouraged to provide it. Some work has recently been undertaken for a few types of damped check valve and a start has been made for vacuum relief and air release valves. This is still in its infancy, but it is a start and is to be strongly encouraged.

Other valve types for which very little dynamic performance data are available include pressure relief valves, pressure regulating valves, pressure sustaining valves and control valves. As numerical techniques advance, enabling better computer codes to be developed, the quality of the basic data available to describe the dynamic performance of components such as these is lagging behind – a situation that needs to be rectified.

Although various excursions into alternative numerical techniques have been made, the method of characteristics may be described as the 'industry standard' among transient analysts. It also has the benefit that it can be coupled to techniques for describing structural motion. Furthermore, even when other numerical techniques are tried, recourse to the method of characteristics is usually found desirable, if not essential, for handling the boundary conditions. Insofar as there needs to be further enhancements of the numerical modelling of transients these relate to devising suitable strategies, e.g. interpolation schemes on time and spatial lines, etc., for incorporating second-order effects associated with wave dispersion and energy dissipation.

2.5.5 Fluid–Structure Interaction (FSI)

Another consequence of the improving modelling techniques, especially the ability to couple dynamic events in the fluid and in the system, is the need for a better understanding of the system response. During the past two years or so, several papers have appeared in the various conferences and symposia addressing FSI. Some have reported models of quite simple systems and have provided a basis for devising practical ways of incorporating FSI in the models for real and more complex systems, especially in nuclear installations and petrochemical plant.

Further experimental and numerical work is required and should embrace traditional relatively rigid systems, plastically deforming tubes and the more flexible systems that make use of viscoelastic pipes and fittings.

As pointed out earlier, actual pressure–time histories depart cumulatively from those predicted by computer codes. Various mechanisms are at work, causing distortion, dispersion and attenuation of the pressure waves. Quite often, different sections of the pressure wave travel at slightly different speeds. For example, in a pressure wave that is dispersing, the leading edge travels faster than the trailing edge, and so this begs the question – which particular speed should be used in the Joukowsky equation to estimate the associated pressure change?

Attenuation mechanisms are associated with the radial and axial movement of pipe walls as a result of the internal changes in pressure. When an increase in pressure is transmitted through the fluid in a pipe, the pipe itself expands. This generates an axial strain wave that is propagated along the pipe wall at a speed similar to the speed of sound in the wall material. This then produces a precursor wave back into the fluid, being evident as a small increase in pressure.

For many systems, the changes are 'small' when compared with the other approximations involved in typical designs, and so workable solutions, using conservative factors of safety, can be produced on a 'fit for purpose' basis. However, several instances have been reported involving serious incidents and attributed to the inability to model adequately the response of the complete system. Pressures have exceeded normal Joukowsky changes owing to additive effects associated with physical movement of pipes and components, pipes have ruptured, moved off their supports and physical injury to personnel has occurred. This is unacceptable and must be remedied.

2.5.6 Concluding Remarks

'Fluid transients is a mature subject' – 'Fluid transients is a black art'. Both these views have been expressed by engineers involved in risk assessments of pipe systems. This section has proposed a means of bridging the gap between these two opposing statements, namely to focus on the development of Knowledge-Based Expert Systems for risk assessments of fluid transients in pipe systems. It also provides a basis for identifying further technical R&D.

It will take several years to develop such interactive computer codes, but this will, nevertheless, provide a focus and direction to the work necessary to do so. Initially, the focus could be on KBES models for a limited class of systems, such as single-pipe rising mains, or gravity-fed lines. Networks will follow as program developers gain experience with the interactive programming techniques involved and establish ways of incorporating engineering judgement and discretion. It is an ambitious target, but it will be achieved in time!

PART 3

This section of the book is intended as a self-contained database and reference section. Some elementary theory is introduced, but only sufficient to provide a basis for explaining the derivation of Joukowsky's equation for the head change across transient pressure waves, and the various equations for estimating the speed with which they travel.

Also included are some theoretical developments relevant to various 'approximate methods' that can often be used for preliminary risk assessments of pipeline systems and the initial design of surge protection strategies. These include being able to make reasonable estimates of air vessel capacities, valve closure times, pump and motor inertias, etc. prior to initiating computer-aided analyses. Numerous charts and tables are included to assist in this.

It may be observed that, in developing the various techniques, some of the symbols, or the suffixes, denoting various parameters may change from one technique to another. The reason for this is that, in most cases, the same symbols that were used in the source reference have been used here so that, should readers wish to refer to the original text, they may do so with ease.

SECTION 3.1

Some basic theory

Two of the most basic parameters used in transient flow studies are the speed, c, with which pressure waves travel and the difference, Δp, in pressure across them. The full development of the equations for predicting pressure and flow changes in a pipe system can be quite complex, and would be required as a basis for any computer model. However, by restricting interest to a single transient in a simple pipe, the development of the equations for the two parameters needed can be illustrated more easily.

3.1.1 Change in Pressure across a Transient

Consider a section of a simple straight pipe in which there has been a sudden change in the flow further downstream and, as a result, a transient pressure wave is travelling upstream, i.e. to the left in Fig. 3.1a, at a speed c relative to the fluid. The speed relative to the pipe wall is $(c - v)$. Initial conditions in the undisturbed flow are denoted by

$$\text{Pressure} = p, \quad \text{Density} = \rho, \quad \text{Velocity} = v$$

The unsteady flow situation shown in Fig. 3.1a can be converted to an equivalent steady flow situation by attaching a frame of reference to the pressure wave, with

Fig. 3.1 *In (a) the pressure wave is travelling upstream at speed (c–v) relative to the pipe. In (b) the unsteady flow is converted to steady flow with respect to a reference frame attached to the pressure wave*

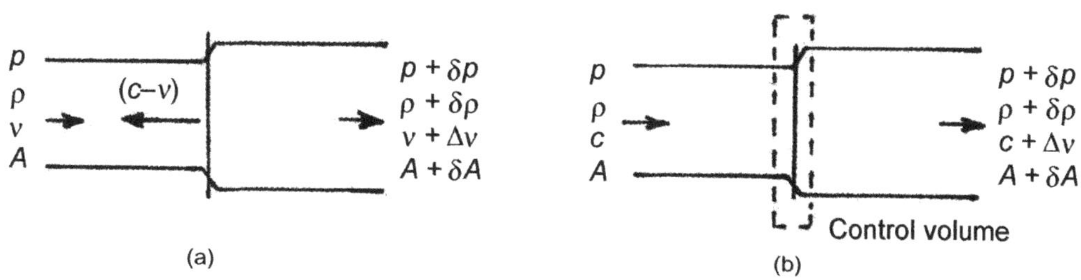

the flow velocities into and out of the little control volume being modified accordingly (see Fig. 3.1b).

Newton's Second Law, for the conservation of linear momentum, can now be applied to the flow through this control volume, i.e. the net force across the control volume is equal to the net loss in linear momentum from it

$$-\Delta pA = \text{mass flowrate} \times [(c + \Delta v) - c] \tag{3.1}$$

For the vast majority of liquid flows in pipelines, the change, Δv, in the flow velocity is negligible compared with the propagation speed, c, and so the mass flowrate through the control volume of Fig. 3.1b is given by ρAc. Substitute this into equation (3.1) and rearrange to obtain

$$\Delta p = -\rho c\Delta v \tag{3.2}$$

The physical significance of the minus sign is that, as the transient is moving upstream into the flow and reducing it, the associated pressure change is an increase. In the corresponding equation for a pressure wave moving downstream into a flow, and also reducing it, both sides of equation (3.2) would be positive. In general discussion, the negative sign is often dropped and reference is made only to the *magnitude* of pressure changes.

Equation (3.2) is one form of the *Joukowsky equation* [Joukowsky (1900)]. An alternative, and useful, form is to express the pressure change in terms of a change in pressure head, which is achieved by dividing both sides of equation (3.2) by ρg to give

$$\Delta H = -\frac{c\Delta v}{g} \tag{3.3}$$

In order to use equation (3.2) or (3.3) it is necessary to determine the propagation speed c of the transients, and the equation to do this is derived by applying the law of conservation of mass to the control volume.

3.1.2 The Wave Speed Equation

The Law of Conservation of Mass dictates that, for steady (or quasi-steady) flow, the flow leaving a control volume is equal to the flow into it. Referring to Fig. 3.1b and letting the small changes in density and cross-sectional area be $\delta\rho$ and δA respectively

$$(\rho + \delta\rho)(A + \delta A)(c + \Delta v) = \rho Ac$$

Dividing the left-hand side by the right-hand side, and substituting for $\Delta v = -\Delta p/\rho c$ from equation (3.2), gives

$$\left(1 + \frac{\delta\rho}{\rho}\right)\left(1 + \frac{\delta A}{A}\right)\left(1 - \frac{\Delta p}{\rho c^2}\right) = 1$$

Multiplying out and neglecting the products of very small quantities leads to

$$\frac{\delta\rho}{\rho}+\frac{\delta A}{A}-\frac{\Delta p}{\rho c^2}=0 \tag{3.4}$$

The density change $\delta\rho$ can be related to the bulk modulus, K, of the liquid by

$$\frac{\delta\rho}{\rho}=\frac{\Delta p}{K}$$

Substituting this into equation (3.4), and rearranging, leads to the general form

$$c=\sqrt{\frac{1}{\rho\left[\frac{1}{K}+\frac{1}{A}\frac{\partial A}{\partial p}\right]}} \tag{3.5}$$

This is the basic equation for determining the speed of propagation of transient pressure waves through liquids in pipes and tunnels. The bulk modulus represents the elasticity of the fluid, while the last term in the brackets in the denominator, the area strain per unit change in pressure, takes account of the elasticity of the conduit.

3.1.3 Equations for Calculating Wave Speeds

Further development of the above equation into a more usable form requires assumptions to be made concerning the type of conduit through which the transient pressure waves are being propagated. Several examples are fully derived in the major textbooks [Chaudhry (1987), Fox (1989), Wylie and Streeter (1993), etc.] and technical papers [e.g. Halliwell (1963), Hills (1999), Pearsall (1965) and Thorley and Twyman (1976)], but a summary is given in the following sections, with graphical examples provided in Figs 3.2 to 3.9.

3.1.3.1 Pipes of Circular Cross-section

For *thin-walled pipes* of internal diameter D and wall thickness e (i.e. for pipes of $D/e > 10$), propagation speeds can be calculated from

$$c=\sqrt{\frac{1}{\rho\left[\frac{1}{K}+\frac{D}{Ee}\phi\right]}} \tag{3.6}$$

where ϕ is a Restraint Factor and E is the Young's Modulus of Elasticity for the pipe wall material.

The value of ϕ depends upon the way the pipe is supported and the Poisson's ratio ν for the wall material. When the Young's modulus is large, as for metal pipes, the numerical value of ϕ may be taken as unity without significant error. Physically, this can be taken to imply frequent expansion joints along the line.

Other cases often quoted include

$$\phi = (1 - \nu^2)$$

if the pipes are completely restrained against axial movement, and

$$\phi = (1 - 0.5\nu)$$

if axial motion occurs owing to increased pressure on a closed end.

In practice, none of these three idealized situations will be fully realized. Hence, since the influence on the calculated wave speed will be less than that due to the uncertainty in the assumed numerical values for the elastic modulus E, the fluid density ρ and the bulk modulus K, ϕ is frequently set to 1, i.e. it is simply ignored (see Figs 3.2 to 3.4 for some examples).

For *thick-walled pipes* the values of the Restraint Factor ϕ corresponding to the three cases cited above are respectively:

Pipes with frequent expansion joints

$$\phi = \frac{D}{D+e} + \frac{2e}{D}(1 + \nu)$$

Fully restrained pipes

$$\phi = \frac{D}{D+e}(1 - \nu^2) + \frac{2e}{D}(1 + \nu)$$

Partially restrained pipes

$$\phi = \frac{D}{D+e}(1 - 0.5\nu) + \frac{2e}{D}(1 + \nu)$$

For *lined pipes* having a circular cross-section it is necessary to account for the influence of both the pipe itself and its lining. The area strain term in equation (3.5), namely the last term in the denominator, becomes more complex since it is necessary to account for the diameter, wall thickness, elastic modulus and Poisson's ratio for both pipe and lining.

This topic has been explored by Halliwell (1963) and Hills (1999), from which some results are presented here. The equations given below have been modified from their work to be consistent with symbols used in this text. It should be noted that, if reference is made to the original work, the paper by Halliwell (1963) unfortunately suffers from a number of typographical errors, so some caution is advised should his results be used directly.

Recalling that the general form of the wave speed equation (3.5) is

$$c = \sqrt{\frac{1}{\rho\left[\frac{1}{K} + \frac{1}{A}\frac{\partial A}{\partial p}\right]}}$$

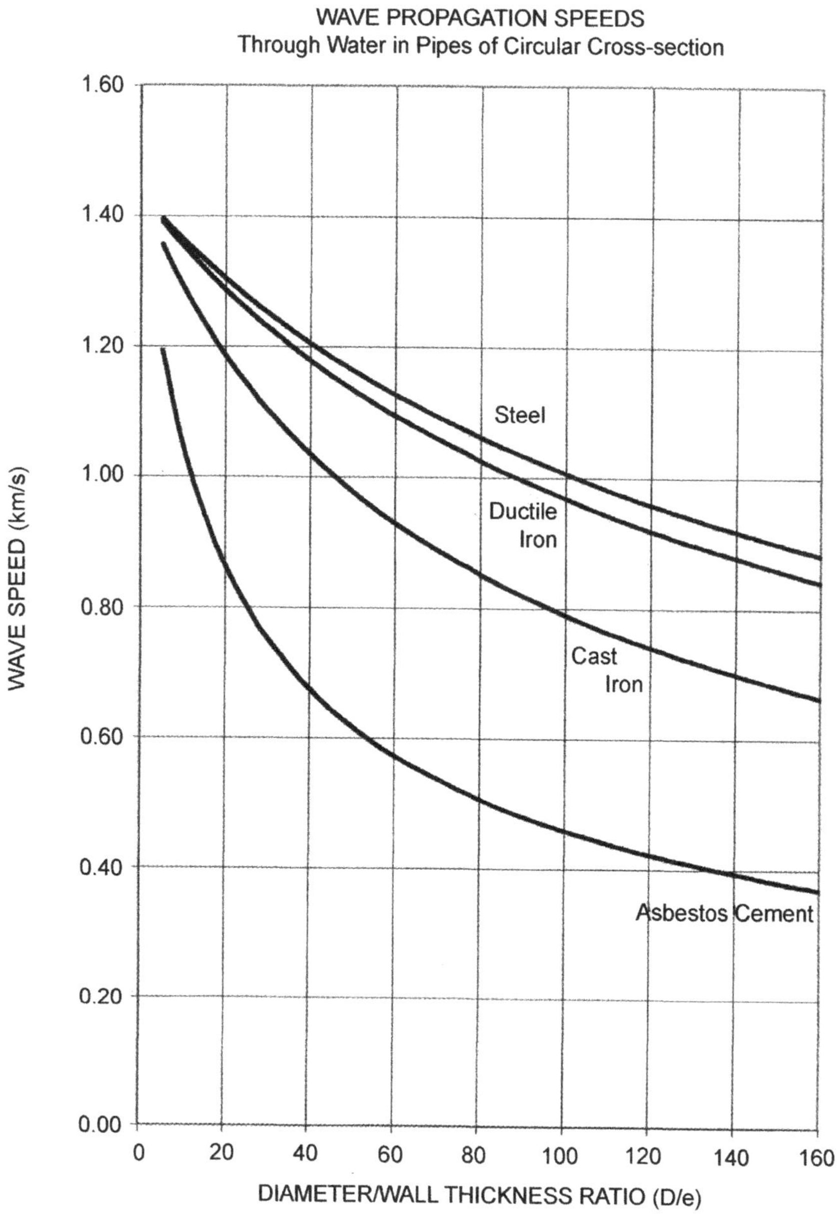

Fig. 3.2 *Wave propagation speeds through water contained in steel, ductile iron, cast iron and asbestos cement (transite) pipes*

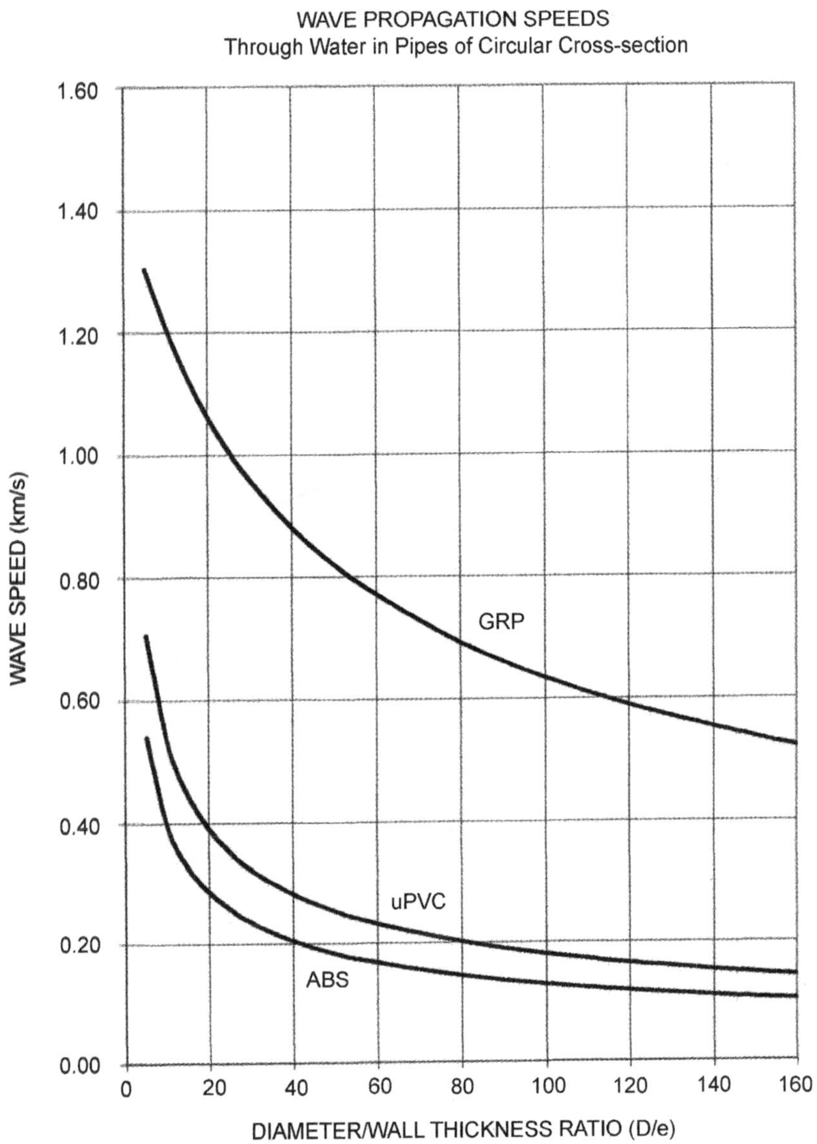

Fig. 3.3 *Wave propagation speeds through water contained in GRP, uPVC and ABS plastic pipes. These graphs are intended to provide only a guide as the actual values are affected by material composition, temperature and rate of strain*

Hills develops the area strain term such that equation (3.5) can be expressed in the form

$$c = \sqrt{\frac{1}{\rho\left[\frac{1}{K} + \frac{1}{E_1}\psi\right]}} \tag{3.7}$$

The term ψ is a composite term combining the effects due to the deformation of

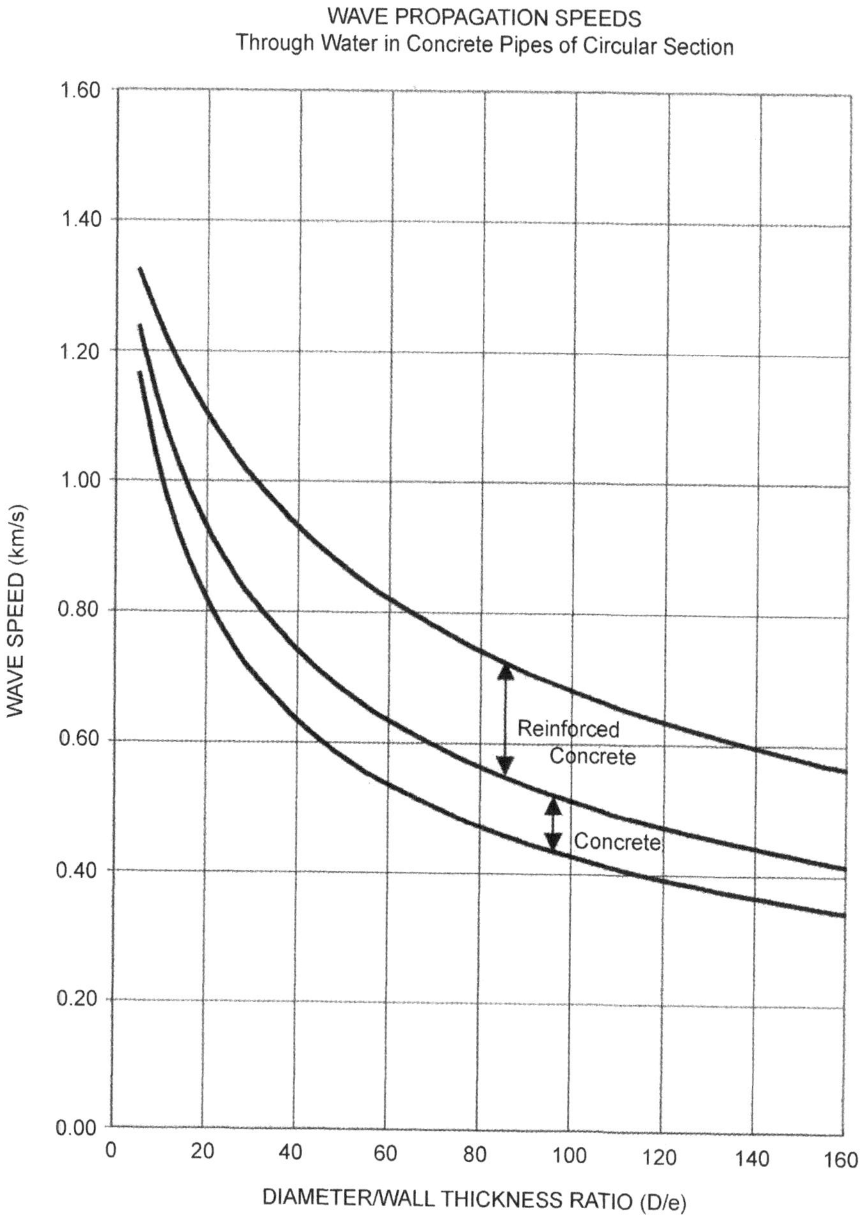

Fig. 3.4 *Wave propagation speeds through water contained in pipes made of concrete. The upper range is for pipes made of reinforced concrete; the lower range is based on material properties for unreinforced concrete pipes*

both the pipe and lining, and includes terms for the pipe sizes, shape and material properties. The other symbols have their usual meaning. It is assumed that the pipes are all of circular cross-section, that the lining and the pipe remain in physical contact and that they are anchored against axial motion. The fluid is also assumed to be homogeneous and Newtonian.

Thick-walled pipes are usually taken, as previously stated, to be those in which the diameter–wall thickness ratio is about 10 or less. To summarize the development of equation (3.7) for *Lined Thick-Walled Pipes* very briefly, the area strain term may be calculated from

$$\frac{1}{A}\frac{\partial A}{\partial p}=\frac{\psi}{E_\text{l}}=\frac{2(1+\nu_\text{l})D_1^2}{E_\text{l}(D_1^2-D_0^2)}\left[(1-\lambda)+(1-2\nu_\text{l})\left(\frac{D_0^2}{D_1^2}-\lambda\right)\right] \tag{3.8}$$

where E_l is the elastic modulus of the lining material and λ can be determined from

$$\lambda=\frac{2E_\text{p}(1-\nu_1^2)D_0^2(D_2^2-D_1^2)}{\begin{array}{c}E_\text{l}(1+\nu_\text{p})(D_1^2-D_0^2)[D_2^2+(1-2\nu_\text{p})D_1^2]\\ +E_\text{p}(1+\nu_\text{l})(D_2^2-D_1^2)[D_0^2+(1-2\nu_\text{l})D_1^2]\end{array}} \tag{3.9}$$

For many applications the pipelines may be regarded as thin walled, and equations (3.8) and (3.9) can then be simplified for convenience, since the added complexity can easily be shown to change the final result only marginally. Also, to put this into a practical context, it is likely that the accuracy of the basic data to be used in these equations will not be known with any degree of accuracy, and so only 'typical' or 'representative' values will be used. Hence, results that are usually perfectly adequate for design purposes for *Lined Thin-Walled Pipes* of circular section may be found from

$$\frac{1}{A}\frac{\partial A}{\partial p}=\frac{\psi}{E_\text{l}}=\frac{D_\text{l}}{E_\text{l}e_\text{l}}(1-\lambda)(1-\nu_1^2) \tag{3.10}$$

where D_l is the mean diameter of the lining and λ can now be calculated from

$$\lambda=\frac{2D_1^2E_\text{p}e_\text{p}(1-\nu_1^2)}{\left[D_\text{p}^2E_\text{l}e_\text{l}(1-\nu_\text{p}^2)+D_1^2E_\text{p}e_\text{p}(1-\nu_1^2)\right]} \tag{3.11}$$

In achieving these simplifications, e_l and e_p denote the wall thickness of the inner lining and outer pipe walls respectively and, likewise, D_l and D_p denote the corresponding mean diameters.

A yet further simplification can be achieved if the Poisson's ratio ν for both the pipe wall and its lining may be assumed to be the same and a mean pipe diameter is used, i.e.

$$\lambda\approx\frac{2D_1^2E_\text{p}e_\text{p}}{\left[D_\text{p}^2E_\text{l}e_\text{l}+D_1^2E_\text{p}e_\text{p}\right]}\approx\frac{E_\text{p}e_\text{p}}{(E_\text{l}e_\text{l}+E_\text{p}e_\text{p})} \tag{3.12}$$

Substituting these simplified results back into equation (3.5) gives

$$c=\sqrt{\frac{1}{\rho\left[\dfrac{1}{K}+\dfrac{D_\text{l}}{E_\text{l}e_\text{l}+E_\text{p}e_\text{p}}(1-\nu^2)\right]}} \tag{3.13}$$

If the lining material reduces to zero thickness, this is the same result as equation (3.6) for a pipe anchored against longitudinal movement.

For pipes made from a fairly rigid material such as steel, and lined with a material such as rubber or something similar, it may be found adequate to calculate the wave speed on the basis just of the steel pipe alone and round it up, as a first approximation.

3.1.3.2 Tunnels

For *circular section tunnels* the relevant equations can be developed very easily from those for thick-walled pipes. For a tunnel the wall thickness tends towards infinity, i.e. $(D+e) >>> D$, and all three of the restraint factors above for thick-walled pipes reduce to

$$\phi = \frac{2e}{D}(1+\nu)$$

Substituting this back into equation (3.6) leads to

$$c = \sqrt{\frac{1}{\rho\left[\frac{1}{K} + \frac{2(1+\nu)}{G}\right]}} \qquad (3.14)$$

where G is the Modulus of Rigidity of the rock through which the tunnel is bored.

For *lined* circular section tunnels several equations have been developed [Halliwell (1963) and Suo (1988)]. For a steel-lined tunnel a convenient expression for the wave speed is

$$c = \sqrt{\frac{1}{\rho\left[\frac{1}{K} + \frac{2D}{GD + 2Ee}\right]}} \qquad (3.15)$$

where e is the thickness of the circular liner for which E is the Young's Modulus, D is the diameter of the tunnel and G is again the Modulus of Rigidity of the rock (see Fig. 3.5).

For *tunnels lined with both steel and concrete* the equations become quite complex. However, the combined effect of the steel and concrete, supported by the surrounding rock, will be to present a fairly rigid conduit, and hence the wave propagation speeds will approximate to those in very rigid cylinders.

Equation (3.15) may also be used to obtain a reasonable approximation for the wave speed in *buried pipes* if it is confirmed that the backfill material around the pipe provides additional support. However, this may not always be the case, especially with the more elastic pipes such as those made from uPVC and similar plastics. In extreme conditions, where significant fluctuations in pressure occur, with correspondingly large pipe deformations, the expanding pipe may push the backfill material away and then, on relaxation of the internal pressure, the pipe becomes unsupported and lies on the bottom of a tunnel with which it is not in permanent contact. Also, it may not remain circular in shape.

Fig. 3.5 *Wave propagation speeds through water contained in unlined rock tunnels*

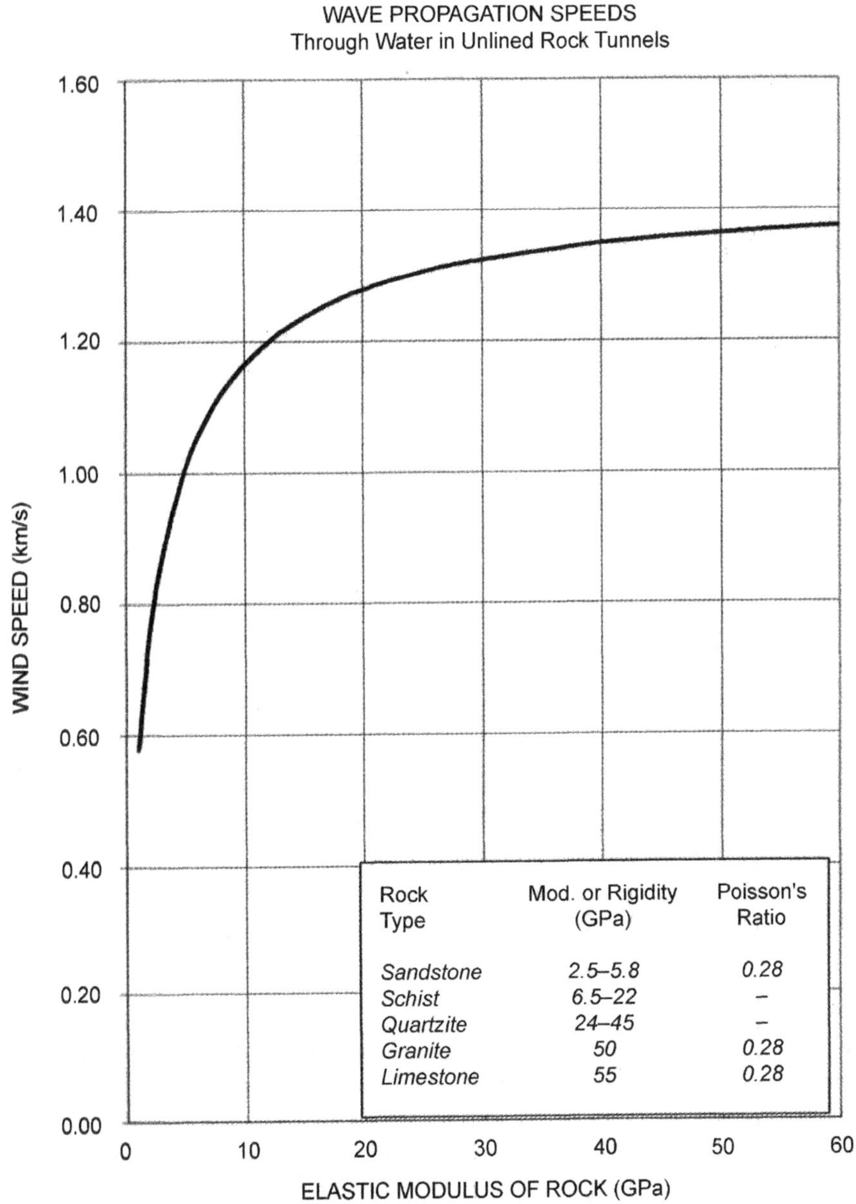

3.1.3.3 Plastic, uPVC and Glass-Reinforced Plastic Pipes

Wave propagation speeds in pipes manufactured from Glass-Reinforced Plastic (GRP), plastic, uPVC and similar non-metallic materials can be calculated with the aid of equation (3.6) above when the deformations are within the elastic range. Figure 3.3 provides an indication of wave speeds in these pipes. Care must be taken, however, to ensure that the appropriate, i.e. *dynamic*, values of the physical properties of the wall material are used.

The elastic moduli of plastic and uPVC pipes are affected by both temperature and rate of strain. Increasing the temperature reduces the elastic modulus and lowers the wave speeds. Increasing the rate of strain leads to higher (dynamic) moduli of elasticity that give higher speeds. Numerical values quoted by manufacturers should be checked to ensure they are the dynamic values appropriate for the temperature range expected. On the positive side, it may be noted that the viscoelastic behaviour of many plastics does mean that pressure waves are attenuated at a much faster rate than in metal and concrete pipes.

For pipes made from GRP and similar composite materials of a matrix construction, the elastic modulus depends on both the materials used and their relative proportions. The method of manufacture may also be significant. If an experimental value for the elastic modulus E is not available, an estimate may be made from

$$E = V_{\mathrm{f}}E_{\mathrm{f}} + (1 - V_{\mathrm{f}})E_{\mathrm{b}} \tag{3.16}$$

where

V_{f} = volume fraction of fibre
E_{f} = elastic modulus of the fibre
E_{b} = elastic modulus of the bonding material

3.1.3.4 Non-circular Ducts

In some aeronautical and nuclear applications, square and hexagonal ducts are sometimes used. For conduits having a regular polygonal cross-section, including square, hexagonal and octagonal pipes, the area strain term in equation (3.5) may be expressed [Thorley and Enever (1979)] as

$$\frac{1}{A}\frac{\partial A}{\partial p} = \frac{1}{E}\left[\frac{D}{e} + \frac{\tan^4\alpha}{15}\left(\frac{D}{e}\right)^3\right] + \frac{\tan^2\alpha}{2G}\left(\frac{D}{e}\right) \tag{3.17}$$

In this equation, D is the perpendicular distance across the flats of the regular polygon and $\alpha = 180^\circ$/number of flat sides. The complexity of this expression is because most of the deformation of the duct wall is in the form of a bending deflection. If this becomes excessive, plastic deformation of the corners occurs (the above equation then becomes invalid) and the wave speed drops dramatically – see below. If the pressures are high enough, the duct will tend towards a circular profile and equations (3.6) and (3.16) become more appropriate. For small-

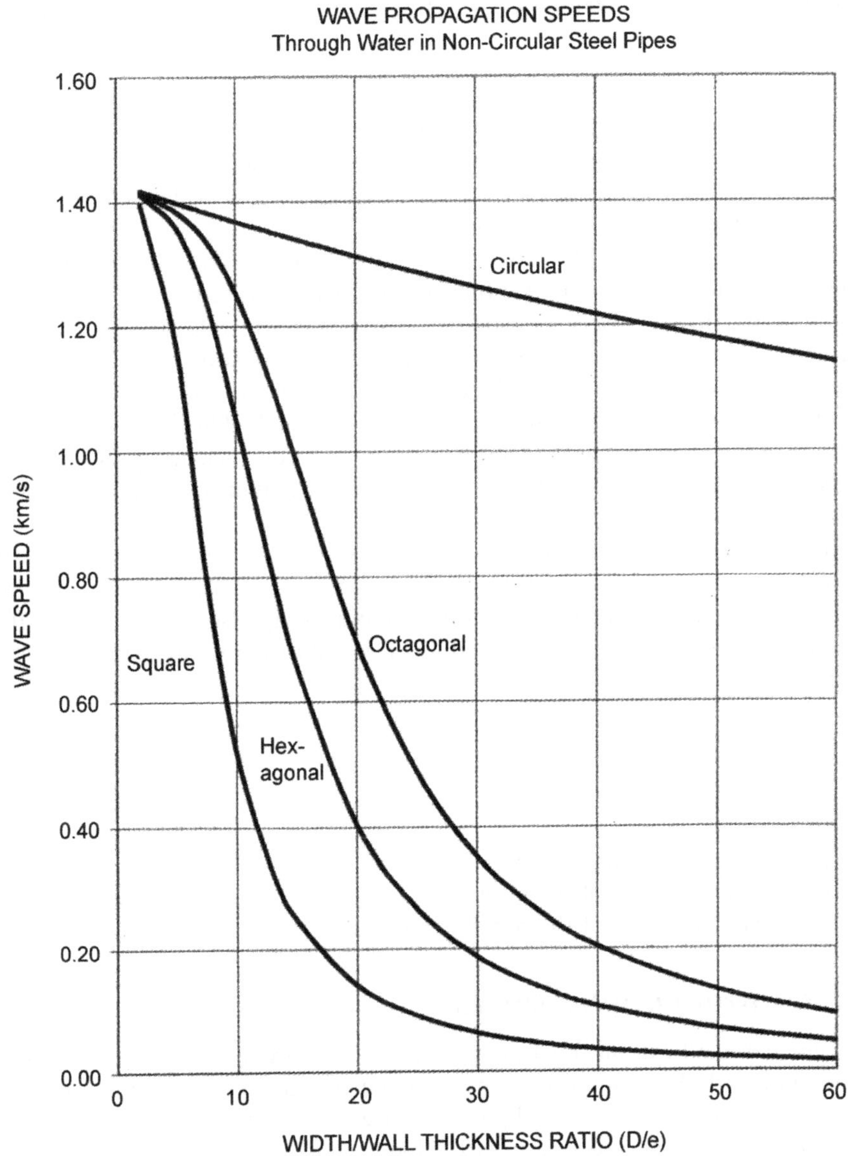

Fig. 3.6 *Wave propagation speeds for small-amplitude pressure waves through water in steel pipes of various cross-sectional shapes. 'Small amplitude' means that the pipe walls deflect only elastically, and, for non-circular section pipes, this is predominantly by bending deflection of the side walls*

amplitude pressure waves, i.e. with no plastic deformation at the corners, Fig. 3.6 illustrates the effect of duct shape on their speed of propagation.

3.1.3.5 Liquids Other than Water

For many liquids the speed of sound c_0 in the unbounded fluid is more likely to be available than the bulk modulus K. To accommodate this, equation (3.6) can be rearranged as

$$c = \sqrt{\frac{1}{\left[\frac{1}{c_o} + \frac{\rho D}{Ee}\phi\right]}} \tag{3.18}$$

Some typical values of the acoustic velocities of various liquids are given in Table 3.3, and a few examples of wave speeds through liquids contained in stainless steel pipes are illustrated in Fig. 3.7.

3.1.3.6 Multiphase and Multicomponent Fluids

The presence of entrained air and gases, or of solid particles, in an otherwise pure liquid can have a marked effect on the wave propagation speeds. Even small quantities of gas, distributed as small bubbles, can lead to wave speeds of a mere 10 per cent of those in the liquid phase alone, as shown in Fig. 3.8 [Zhou *et al.* (2002)]. The influence of solid particles, on the other hand, can be to raise or lower the wave speeds, depending on the physical properties of the particles [Thorley and Wiggert (1985)].

For fluids that may be regarded as homogeneous mixtures, comprising either finely dispersed bubbles or solid particles in a continuous liquid medium, and conveyed in a pipe, a general form of the wave speed equation is

$$c = \sqrt{\frac{1}{[\alpha\rho_d + (1-\alpha)\rho_c]\left[\frac{\alpha}{K_d} + \frac{(1-\alpha)}{K_c} + \frac{D}{Ee}\right]}} \tag{3.19}$$

in which the following new symbols are introduced

ρ_c, ρ_d = densities of the continuous and discrete phases respectively
K_c, K_d = bulk moduli of the continuous and discrete phases
α = local volumetric fraction of the discrete phase

For bubbly gas–liquid mixtures having a low volumetric gas content, the bulk modulus K_d is equal to the local pressure p. Assuming isothermal behaviour of the mixture and that the influence of the duct wall is negligible except at very high pressures, equation (3.19) may therefore be simplified to

$$c = \sqrt{\frac{1}{(1-\alpha)\rho_c\left[\frac{\alpha}{p} + \frac{(1-\alpha)}{K_c}\right]}} \tag{3.20}$$

An even simpler approximation may be made if the compressibility of the liquid can also be neglected

$$c = \sqrt{\frac{1}{(1-\alpha)\rho_c\frac{\alpha}{p}}} \tag{3.21}$$

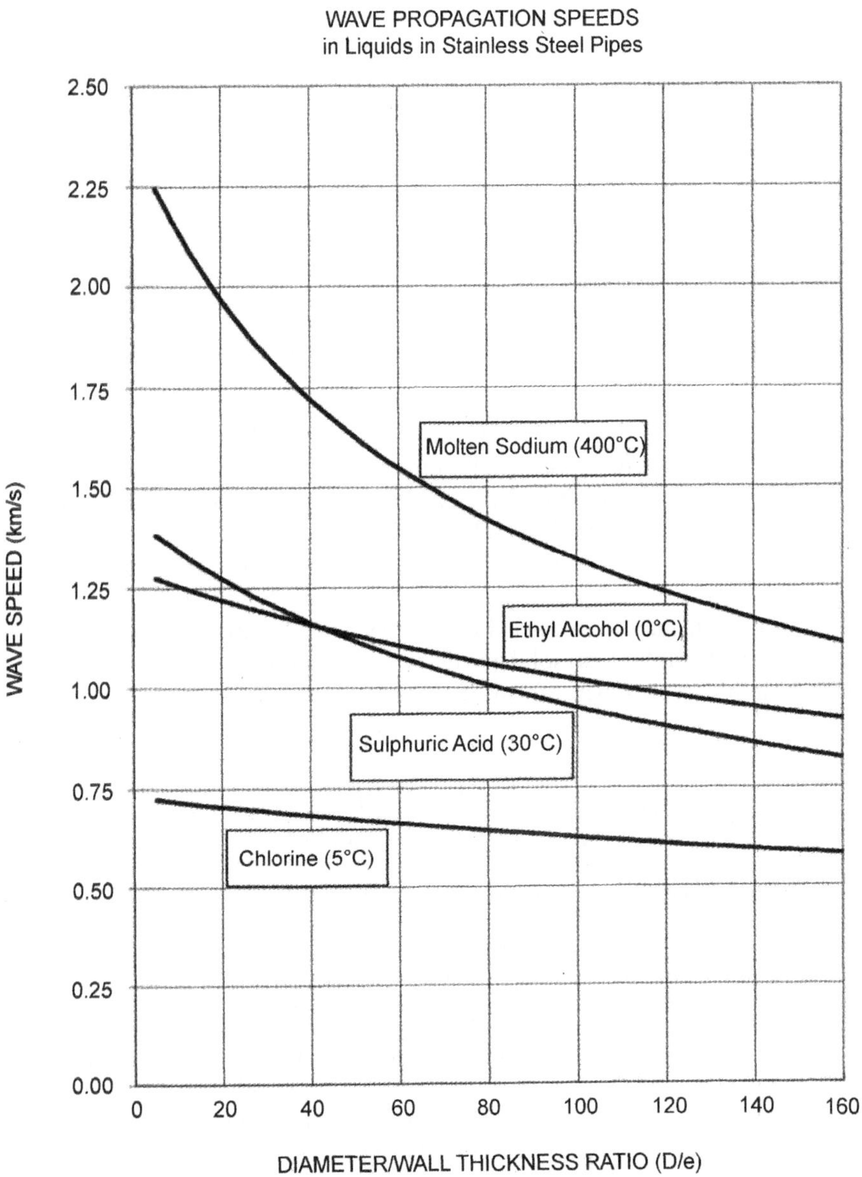

Fig. 3.7 *Wave propagation speeds through various liquids contained in stainless steel pipes of circular cross-section*

Note that, since the local pressure p appears in the equations for gas–liquid flows, the wave speed will vary throughout the system.

For homogeneous solid–liquid and slurry flows, the full version of equation (3.19) should be used, including the duct wall effects. The bulk modulus K_d for the solids component may, for approximately spherical particles, be estimated [Conway (1950)] from

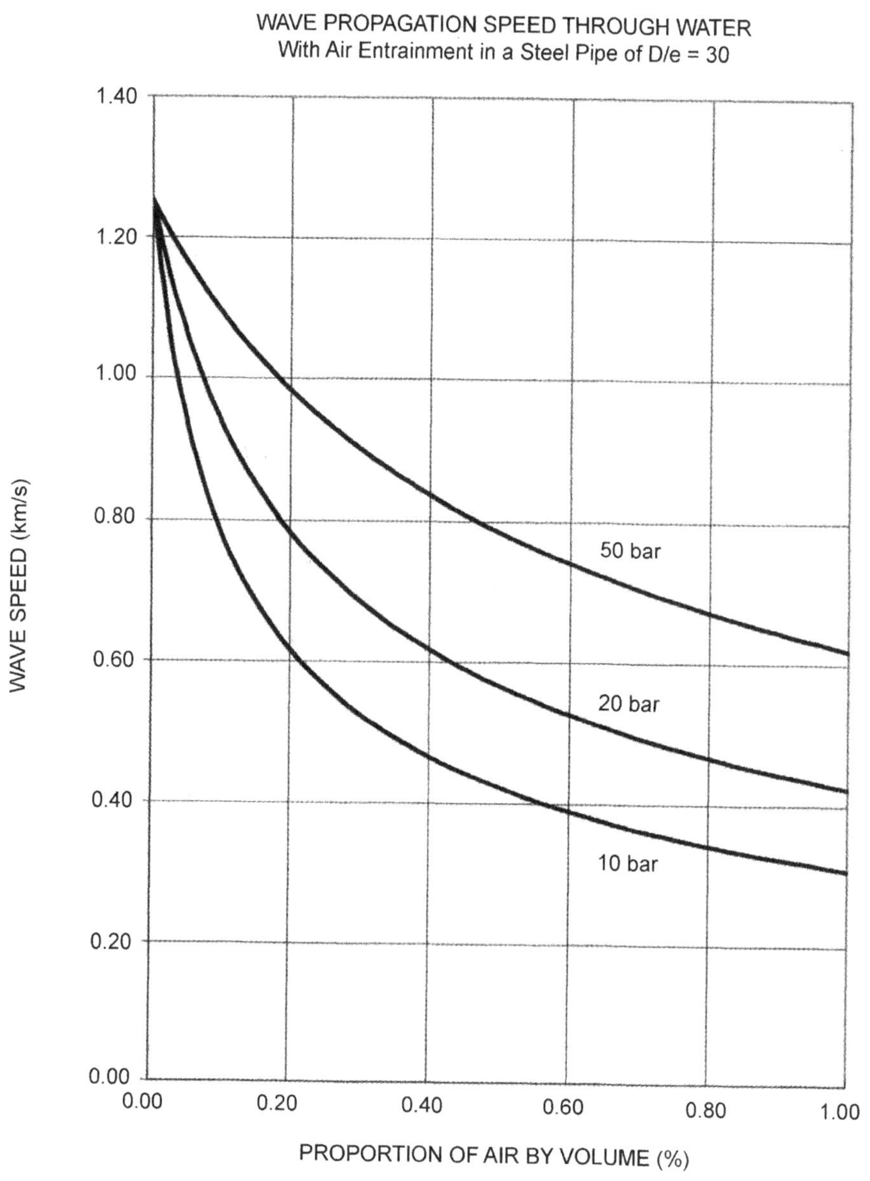

Fig. 3.8 *The presence of even small quantities of air can significantly reduce the wave propagation speeds through water in pipes and tunnels. These graphs show the effect of less than 1 per cent of dispersed air bubbles on the wave speed in a steel pipe having a D/e ratio of 30*

$$K_d = \frac{E}{3(1 - 2\nu)}$$

where

E = Young's modulus for the solids
ν = Poisson's ratio

When the flow can no longer be regarded as homogeneous and stratification begins to develop, the above equations are no longer valid. The transient flows that will then occur are considerably more complex and specialist advice will be required.

3.1.3.7 Plastically Deforming Tubes

A few situations can arise where the amplitude of the initial pressure wave is large enough to cause plastic yielding of the duct wall [e.g. see Twyman *et al.* (1980)]. The leading edge of the pressure rise will travel at a speed associated with the elastic behaviour of the pipe, and can still be obtained from equation (3.6). As the pressure increases above the yield pressure of the pipe wall and it stretches plastically, these 'excess pressures' are propagated at a much lower velocity and with quite rapid attenuation. The speed of propagation of the various pressure levels can be calculated from

$$c = \sqrt{\frac{1}{\rho\left[\frac{1}{K} + \frac{D}{e(E+2\sigma)}\right]}} \tag{3.22}$$

in which

D = instantaneous radius of the pipe
e = instantaneous wall thickness
E = tangent modulus of the duct wall material
σ = true local stress

Should plastic deformation occur, even without bursting the pipe, very rapid attenuation of the pressure wave usually takes place [Thorley and Twyman (1976)]. After travelling little more than a few metres, the pressure behind the leading edge of the transient is reduced to that associated with the elastic limit for the duct in question.

Data for a specific example are shown in Fig. 3.9 for illustrative purposes. In this case, the pipe was originally non-circular and the initial wave speed was influenced mainly by bending deflection of the side walls of the pipe. As the pressure increased, the pipe assumed a more circular shape and the wave speed increased. Once yielding of the pipe wall occurred, as the pressure increased above about 12 MPa, initially at the corners and then more generally, the wave speed dropped dramatically. Rupture of the pipe wall in the locality of the event initiating the transient, at a pressure of the order of 24 MPa, gave a zero wave speed for the highest pressure levels at this location. Further away from this site, a pressure wave having an amplitude associated with elastic deformation of the duct was propagated in the usual manner.

To determine the theoretical results, it is necessary to obtain accurate load–extension and stress–strain curves for the pipe wall material so that the tangent modulus, *true* stress and *true* strain characteristics of the material can be obtained for substitution into equation (3.22). Note that this equation reduces to equation (3.6) for pressures less than those that cause the material to yield plastically.

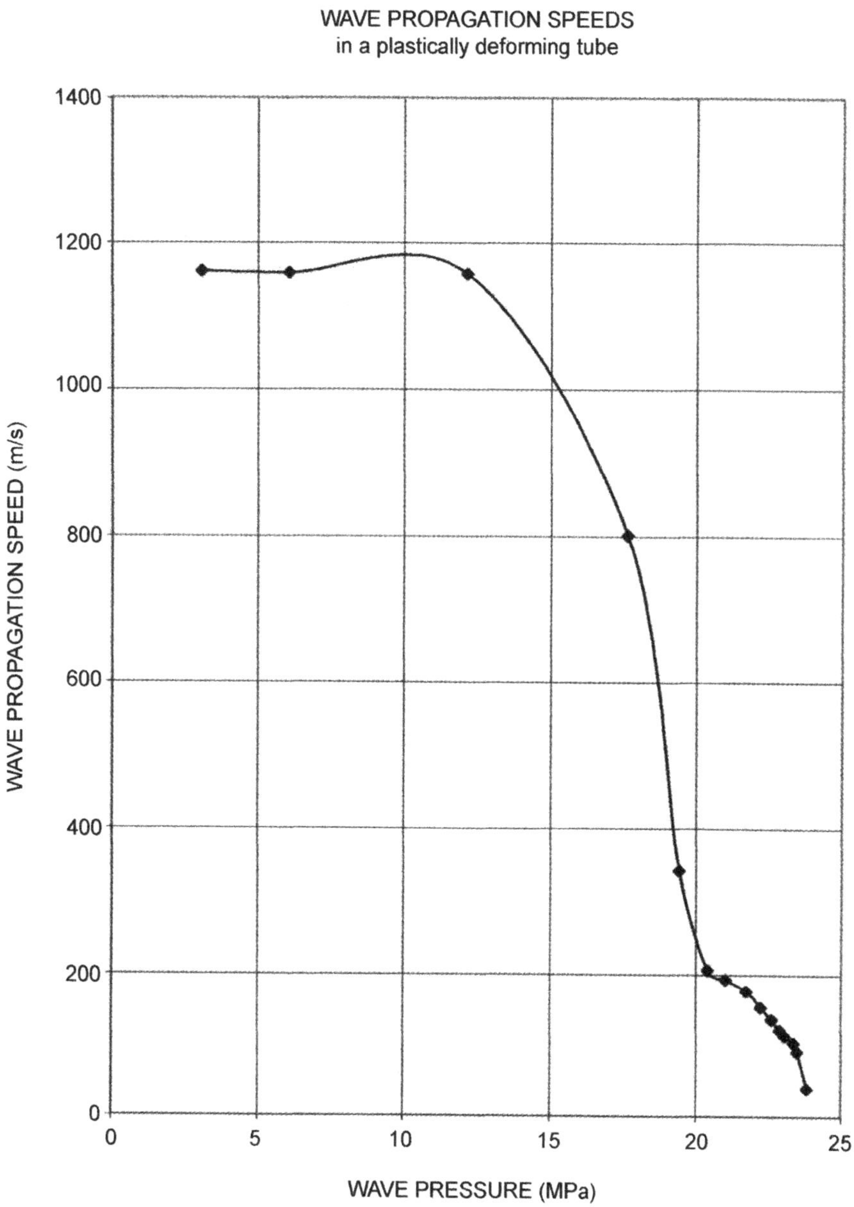

Fig. 3.9 *An example of how the speed of propagation of very large amplitude pressure waves through a liquid contained in an alloy steel pipe is affected by the onset of plasticity in the pipe wall*

3.1.3.8 Flexible Hoses

Wave speeds in flexible hoses, e.g. as used in offshore loading systems, are affected by system pressure, material properties and hose wall construction. This can be quite complex, involving various materials and steel reinforcing wire, etc. Tests [Thorley (1986)] on pipes of 500 mm ND yielded wave propagation speeds through water in such hoses in the range 340–440 m/s. Note that *dynamic* material properties are required for theoretical estimates, and that the flexibility of the pipes is likely to lead to increased uncertainties in predictions.

3.1.3.9 Data for Wave Speed Estimates

Tables 3.1 to 3.3, and the diagrams that accompany them, provide data for the estimation of wave propagation speeds. The information has been collated

Table 3.1 Physical Properties of Some Common Liquids at Atmospheric Pressure

Liquid	Temperature (°C)	Bulk modulus K (GN/m^2)	Density ρ (kg/m^3)
Benzene	20	1.1	879
Ethyl alcohol	20	1.32	789
Kerosene	20	1.32	805
Methanol	20	1.0	791
Mineral oils	25	1.5–1.9	860–888
Phosphate esters	25	0.9–0.95	1265–1315
Sea water	15	2.27	1025
Sodium	400	4.8	856
Sulphuric acid	30	2.7	1330
Water	20	2.19	998

Table 3.2 Physical Properties of Some Common Pipe Wall Materials

Material	Young's modulus E (GN/m^2)	Shear modulus G (GN/m^2)	Poisson's ratio ν
Aluminium	69	27.6	0.33
Aluminium alloys	69–75	26–28	0.33
Asbestos cement	24	–	–
Cast iron	90–160	40–50	0.25
Concrete	20–30	8.5–13	0.15
Ductile iron	172	83	0.30
Glass	7–8	3	0.24
GRP*	50	3.9	0.35
Mild steel	200–210	80–84	0.28
Phosphor bronze	120	43	0.38
Plastics*			
ABS	1.7	0.65	0.33
Nylon	1.4–2.8	0.45–0.90	0.50
Perspex	6.2	2.3	0.33
uPVC (at 20 °C)	3.3	1.1	0.5
Reinforced concrete	30–60	–	–
Stainless steel	200–215	80–85	0.28
Titanium	103	45	0.34

* Manufacturers' data should be obtained for the intended application. The values quoted here are for general guidance only. For GRP it has been assumed that the volume fraction, V_f, of fibre, having a Young's modulus of 20 GN/m^2, is 0.7. The bonding material, with a Young's modulus of 3.5 GN/m^2, occupies the remaining volume fraction of 0.3.

Table 3.3 Speed of Sound in Various Liquids

Liquid	Speed of Sound (m/s)	Temperature (°C)
Acetic acid	1581	50
Acetone	1190	20
Acetylene tetrachloride	1155	28
Ammonia	{1715	{−33
	{2000	{−77
n-Amyl alcohol	1224	29
Aniline	1660	20
Benzene	1326	20
n-Butanol	1265	20
i-Butanol	1220	20
n-Butyl alcohol	1260	20
Chlorine	850	20
Chlorobenzene	1300	25
o-Cresol	1540	20
m-Cresol	1500	20
Cyclohexane	1280	20
Cyclohexanol	1620	30
Diethylene glycol	1535	30
Ethanol	1156	20
Ethyl Alcohol	1162	20
Ethyl benzene	1340	20
Ethylene glycol	1670	20
n-Heptane	1130	20
Heptene	1080	30
Hexane	1200	20
Methane	1320	−160
Methanol	1123	20
Methyl alcohol	1120	20
Naphthalene	1250	100
Nitrobenzene	1460	24
Nonanol	1390	20
Octanol	1360	20
n-Octane	1190	20
Oleic acid	1440	20
Pentane	1010	20
Phenol	1275	100
n-Propanol	1220	20
n-Propyl alcohol	1225	20
Sodium	2500	150
Sodium chloride	1990	850
Toluene	1330	20
Trichloroethylene	1050	20
Water – distilled	1483	20
Water – sea	1490	10
o-Xylene	1360	20

from several sources, among which are Chaudhry (1987), Kaye and Laby (1966, 1966), Tennant (1979) and various consultancy projects undertaken by the author.

More often than not, 'typical' values such as these, as well as nominal dimensions for the conduits, are all that are available. Assumptions about gas content may also be required for some liquids, especially water. It is often prudent to assume that the wave propagation speeds that can be determined with the various equations given above, or read from the charts, are unlikely to be estimated to an accuracy better than about ±8 per cent, even for 'pure' liquids.

SECTION 3.2

Rigid column approximations

For accurate results, theoretical models of unsteady flow should incorporate the elastic behaviour of the liquid. However, a few occasions do arise where acceptable approximations can be developed without the complications this brings. In 'Rigid Column' analyses the liquid is assumed to be incompressible and to behave as a solid, although it can move around bends.

For such an approach to have any validity, the time over which flow changes occur should be considerably longer than the Periodic Time $T = 2L/c$ of the liquid column. Note that the length of the liquid column may not, in some cases, be the same as the length of the pipeline. Examples include when liquid column separation occurs and when a deep-well pump is started up with an air/water interface in the riser pipe.

Situations where a Rigid Column model may be adopted – at least as a first approximation for some of the assessments – include:

(a) A valve closure/opening at a uniform rate, or a pump trip/run-up that extends over more than, say, 50 times the Periodic Time.
(b) Owing to falling pressure in a system, e.g. following a pump trip, a vapour cavity opens at a high point, splitting the liquid column into two. The downstream column will then often decelerate relatively slowly under the action of a modest net static head and friction. Estimates can be made, with a rigid column approximation, of the size of the air or vapour cavity, the reverse velocity of the liquid column on impact as the cavity closes and the associated pressure rise.

(c) A Feed Tank or Vacuum-Breaking Valve is possibly being considered for the suppression of transient pressure changes in a system. When such devices operate, the liquid column downstream decelerates relatively slowly. A rigid column approximation can enable a first estimate to be made of the required capacity of the Feed Tank or Vacuum-Breaking Valve(s) for use in a computer-aided analysis to confirm or refine the design.

(d) The influence of surge shafts and large air vessels is to convert a transient event into a mass oscillation, and the time-scale of events can be lengthened considerably. Rigid column models can then give quite good descriptions of system behaviour.

3.2.1 Equation of Motion

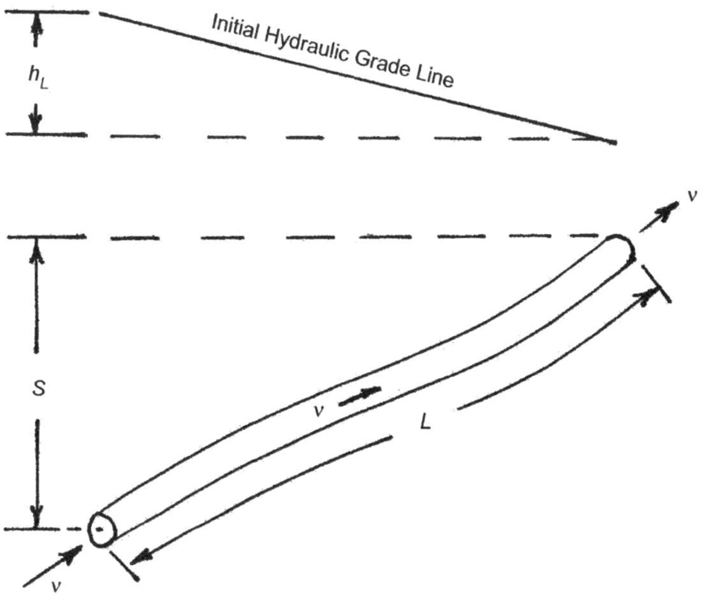

Fig. 3.10 *A representative 'Rigid Column' of liquid being decelerated under the action of friction and an effective static head, or net back pressure head, S*

An outline of a rigid column model follows. It is intended to lead to some simple equations that can be used with caution in the situations described in (a) to (c) above. Case (d), and surge shaft design, is outside the scope of this text and will not be discussed further.

For an incompressible column of liquid of length L, being decelerated under the action of an effective static head S and frictional pressure head h_L, in a pipeline of diameter D, Newton's second law of motion takes the form

$$S + h_L = -\frac{L}{g}\frac{\mathrm{d}v}{\mathrm{d}t} \tag{3.23}$$

For turbulent flows, the frictional head may be expressed as

$$h_L = C_f v^2 \tag{3.24}$$

where

$$C_f = \frac{fL}{2gD}$$

and f is the Darcy–Weisbach friction factor, which will be assumed constant.

Combining equations (3.23) and (3.24) leads to

$$\mathrm{d}t = -\frac{L}{g}\frac{\mathrm{d}v}{(S + C_f v^2)} \tag{3.25}$$

Integrating over a time interval t for a change in velocity from v_1 to v_2 yields

$$t = \frac{L}{g\sqrt{C_f S}}\left(\tan^{-1} v_1\sqrt{\frac{C_f}{S}} - \tan^{-1} v_2\sqrt{\frac{C_f}{S}}\right) \tag{3.26}$$

Further integration and manipulation of equation (3.26), after the manner of Kephart and Davis (1961), enables the distance travelled X_s by the liquid column in coming to rest from an initial velocity v_o to be approximated by

$$X_s = \frac{Lv_o^2}{2gh_L} \times \ell n\left(1 + \frac{h_L}{S}\right) \tag{3.27}$$

3.2.2 Cavity Formation and Collapse in a Rising Main

If a rising main pipeline has a distinct knee, or high point, such that separation of the liquid column may occur, owing either to the formation of a vapour cavity or the admission of air, two independent water columns will exist. Figures 3.11 and 3.12 illustrate these situations and define symbols to be used in the following equations. By considering the motion of the two liquid columns, some useful results can be obtained.

For the system illustrated in Fig. 3.11, assume that, following a pump trip, a vapour cavity forms at the location shown. The two liquid columns are of length L_1 and L_2 and the *effective* static heads, allowing for the vapour head in the cavity, are S_1 and S_2 respectively. The upstream column of liquid, L_1, comes to rest and is prevented from reversing by a non-return valve at the pumps. The downstream column, L_2, also comes to rest, but then reverses until the cavity collapses with an instantaneous rise in pressure.

The length of the gap, X_E, that opens up between the two columns is

$$X_E = X_D - X_U \tag{3.28}$$

where X_D and X_U are the distances moved by the Downstream and Upstream liquid columns in coming to rest. Evaluating X_D and X_U from equation (3.27)

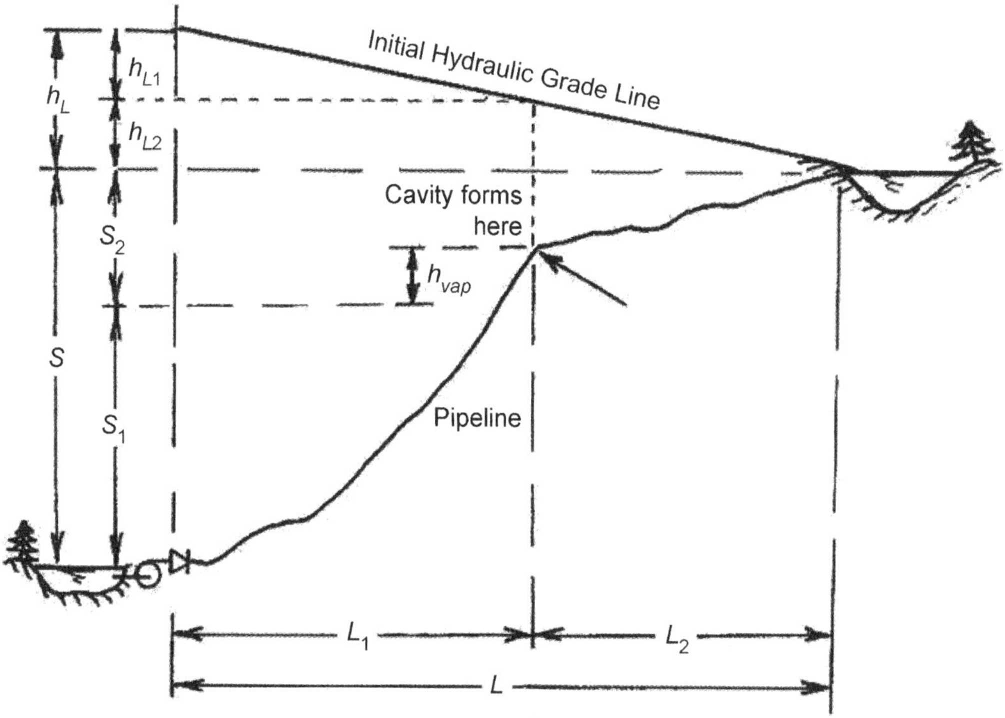

Fig. 3.11 *A typical profile for a rising main pipeline that would be exposed to vapour cavity formation following loss of power to the pumping station*

leads to

$$X_E = B\frac{v_o^2}{2g} \tag{3.29}$$

in which

$$B = \left[\frac{L_2}{h_{L2}}\ell n\left(1 + \frac{h_{L2}}{S_2}\right) - \frac{L_1}{h_{L1}}\ell n\left(1 + \frac{h_{L1}}{S_1}\right)\right] \tag{3.30}$$

If, when calculating X_E (i.e. B) for a particular case, the result is negative, the physical significance is that the liquid column does not, in fact, separate.

Following reversal, the downstream liquid column reaches a maximum velocity, v_E, just prior to impact of

$$v_E = v_o\sqrt{\frac{S_2}{h_{L2}}\left(1 - e^{-(Bh_{L2})/L_2}\right)} \tag{3.31}$$

and the associated rise ΔH in the pressure head is

$$\Delta H = \frac{cv_E}{2g} \tag{3.32}$$

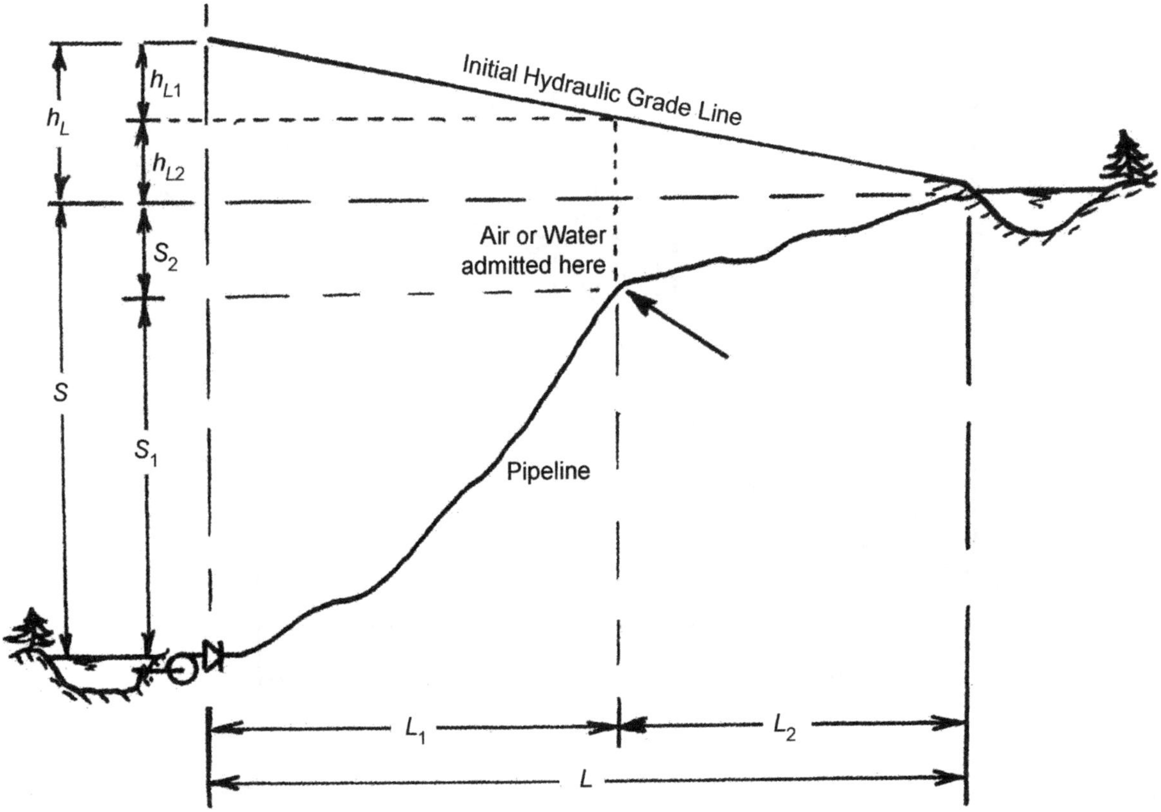

Fig. 3.12 *A pipeline profile similar to Fig. 3.11, but air valves or a feed tank are located at the high point where subatmospheric pressures may first tend to occur*

Note that, after impact, the resultant velocity of the reunited liquid column is one-half of the impact velocity, since the kinetic energy of the moving and stationary columns is shared between them as the velocity at the site of the impact is reduced from v_E to $v_E/2$. The resulting pressure change is propagated both ways from the point where the liquid columns recombine. At the downstream (open) end, the pressure wave is reflected as a reduction in pressure. At the upstream end, however, the consequence of the pressure wave meeting the closed non-return valve on the pump discharge is to double the pressure change. The local pressure at the valve then, for a while, becomes

$$H = S - M + \frac{cv_E}{g} \tag{3.33}$$

in which M is the elevation of the non-return valve above the upstream reservoir. The pressure at the pump discharge could also rise a little further owing to the addition of the frictional head associated with the reversed flow.

Should the rising main have a relatively uniform slope, and the pressure drop following a pump trip be sufficient to cause a vapour cavity to form at the pump discharge, equation (3.31) reduces to

$$v_E = v_o \sqrt{\frac{S}{S + h_L}} \tag{3.34}$$

Note that here S is the static head on the system plus the vapour pressure head. The Joukowsky pressure change at the pump discharge will be

$$\Delta H = \frac{c v_E}{g} \tag{3.35}$$

which will be increased by the addition of the static head as a pressure wave propagates downstream, plus some friction head.

3.2.3 Air or Water Admission at a Low-Pressure Point

To avoid the problem of excessive subatmospheric pressures, the admission of air or water may be acceptable in some lines at a location such as indicated in Fig. 3.12. In the case of air admission, this might be done at several points but depends heavily on the pipeline profile. The size and number of air admission valves can often be determined from manufacturer's charts and is discussed in more detail in Section 3.6.

Feed tanks are not generally produced as standard components and have to be sized individually. As a first estimate, and to provide the initial data for a computer-aided study, the approximate capacity required can be calculated using results from the previous section.

The principal assumptions in the preceding section are adopted, but note that the feed tank is assumed to be sited so that it supplies water into the line to maintain the local pressure at atmospheric. This means that S_1 and S_2 are now the actual static heads on the two liquid columns.

The maximum distance, X_E, that separates the two water columns is again given by equations (3.29) and (3.30), but with the revised values for S_1 and S_2. If this cavity is to be filled with water from a feed tank, the required volume, V_{FT}, may be estimated from

$$V_{FT} = A X_E \tag{3.36}$$

where A is the internal cross-sectional area of the pipe. To avoid dewatering the feed tank, and acknowledge the idealizations made in the theory, this value should be increased by, say 15–20 per cent.

SECTION 3.3

Estimation of air vessel capacities

The important role of air vessels and air cushion surge chambers in suppressing transient pressures has been described in Subsection 1.3.3.1. They are among the more common devices used for this purpose, and perhaps the major problems that system designers face are questions of how large the initial air volume should be, and what the total capacity of the air vessel (or vessels) should be in order to provide a given degree of protection.

Accurate theoretical descriptions of the dynamic behaviour of air vessels are extremely complex, especially if any allowance is made for heat transfer effects. Design charts have, however, been devised that enable reasonable estimates to be made of the capacities required to provide specified levels of protection. Nevertheless, it should be emphasized that they are *only* estimates, and computer-based analyses should be undertaken for confirmation and optimization.

3.3.1 Rising Mains

The usual location for an air vessel in a rising main is downstream of the check valves on the pump discharge. Small and medium-sized vessels may be installed inside the pumping station, whereas larger ones are often situated externally. However, they should be located as close as possible to the pumping station, since they really only protect the pipeline downstream. Figure 3.13 shows a typical rising main application.

Several 'approximate methods' have been developed, over the past 60 years or so, to assist with the sizing of air vessels. A comprehensive set of charts, created by Graze and Horlacher (1989) and Graze (1989), are reproduced here, with the agreement of the authors. Advantages they provide over earlier design charts are:

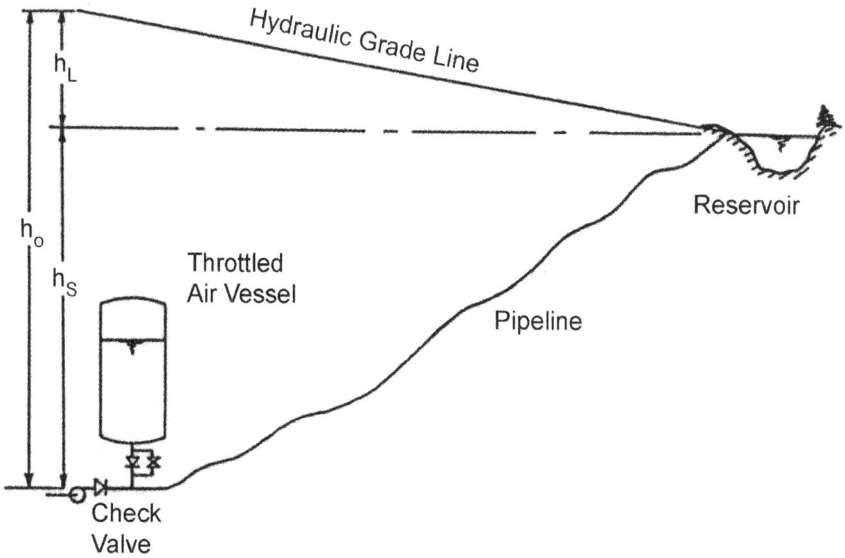

Fig. 3.13 *Typical rising main installation protected by an air vessel from the consequences of a pump trip*

(a) They enable the sensitivity of the design to the assumed polytropic index to be checked.
(b) They permit optimization of the inflow orifice losses. This has the advantage that, by suitable throttling, the overall size of the vessel can be reduced when the need to limit excess pressures in the system is a critical design case.

In addition to charts for sizing air vessels, other charts from Graze and Horlacher provide an indication of the extreme pressure envelopes along the longitudinal profile of the pipeline.

3.3.1.1 Unthrottled Air Vessels

The Graze and Horlacher Design Charts are presented as Figs 3.14 to 3.23. They are in a non-dimensional form and make use of the following parameters:

Friction parameter

$$\bar{h}_{\mathrm{f}} = \frac{h_L}{h_{\mathrm{j}}} \tag{3.37}$$

Joukowsky head

$$h_{\mathrm{j}} = \frac{cv_{\mathrm{o}}}{g} \tag{3.38}$$

$$\bar{h}_{\mathrm{o}} = \frac{h_{\mathrm{o}}^{*}}{h_{\mathrm{j}}} = \frac{h_{\mathrm{o}} + h_{\mathrm{atm}}}{h_{\mathrm{j}}} = \frac{h_{\mathrm{s}} + h_L + h_{\mathrm{atm}}}{h_{\mathrm{j}}} \tag{3.39}$$

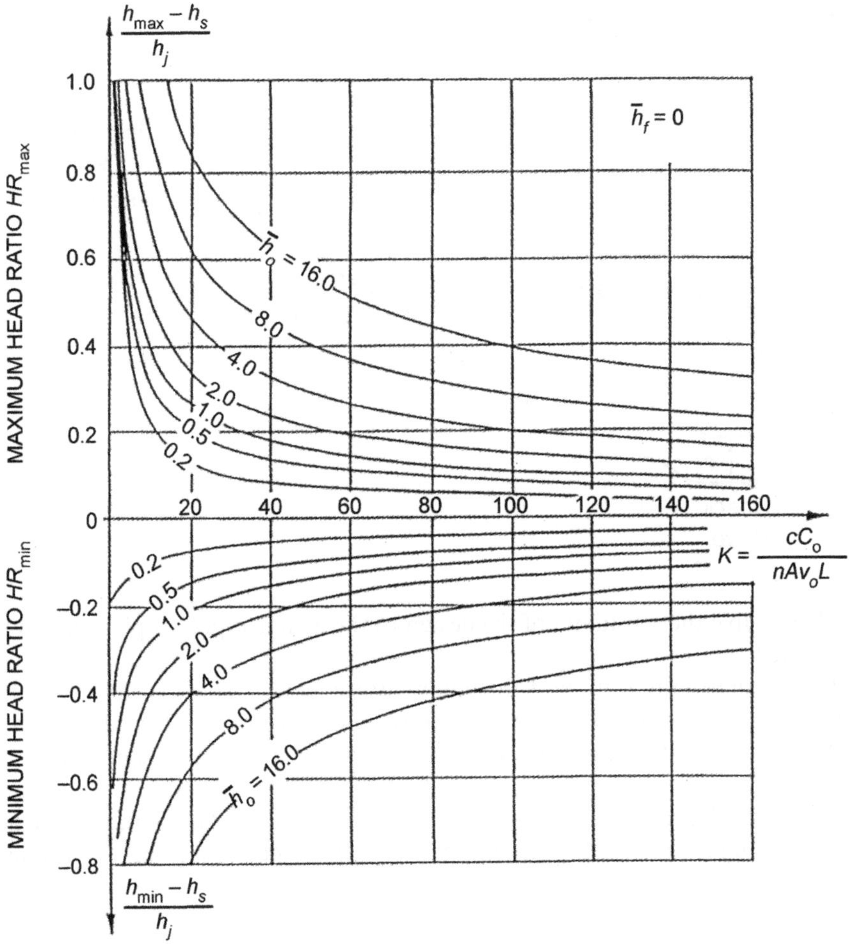

Fig. 3.14 *Maximum and minimum pressure head ratios when the friction parameter $h_f = 0$ and without throttling*

Maximum head ratio

$$\mathrm{HR}_{max} = \frac{h_{max} - h_s}{h_j} \tag{3.40}$$

Minimum head ratio

$$\mathrm{HR}_{min} = \frac{h_{min} - h_s}{h_j} \tag{3.41}$$

A composite air vessel and pipeline parameter, K, defined as

$$K = \frac{cC_o}{nAv_oL} \tag{3.42}$$

in which

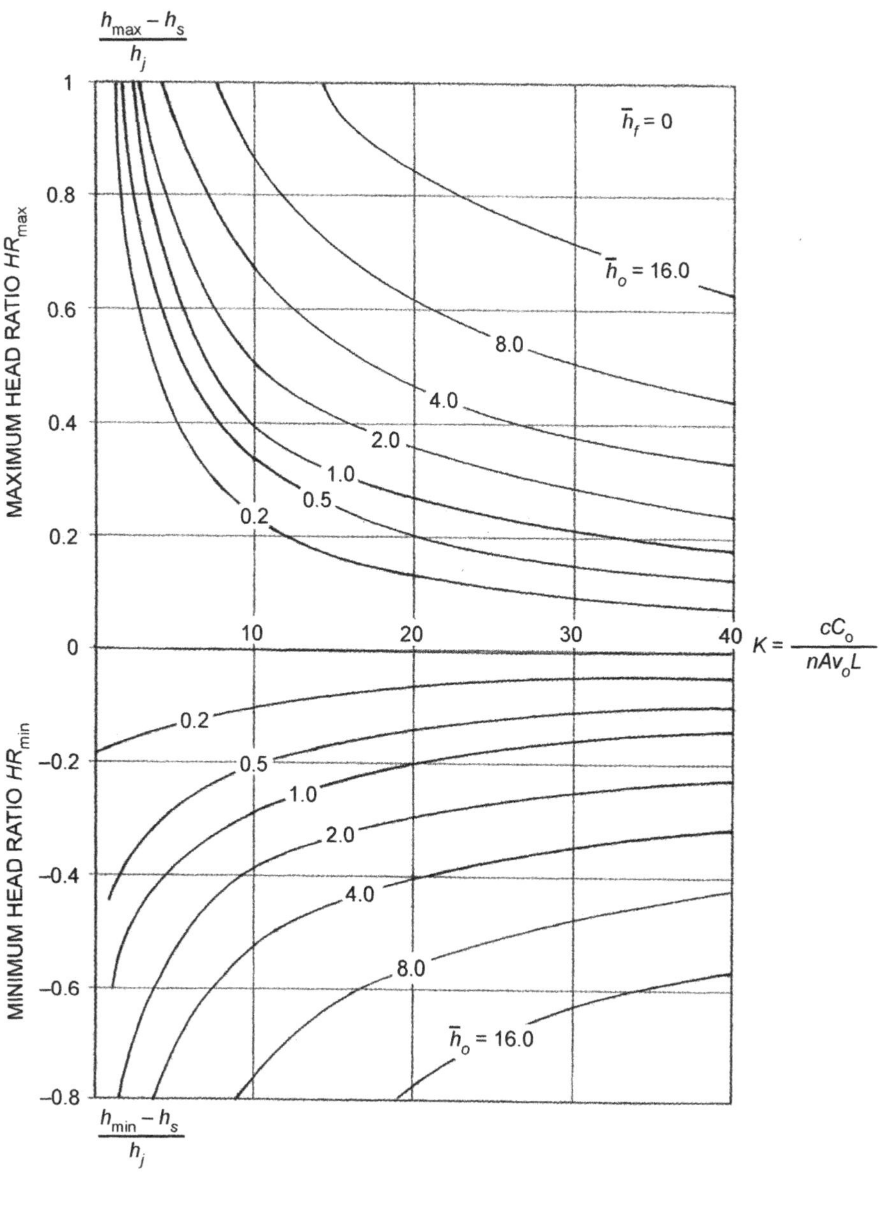

Fig. 3.14a *Figure 3.14 repeated, but with the x axis expanded near the origin*

A = pipeline cross-sectional area
c = transient propagation speed
C_o = initial volume of air in the air vessel (i.e. the unknown to be found)
g = acceleration due to gravity
h_{atm} = local atmospheric head
h_L = frictional pressure head drop in the pipeline
h_o = sum of static and friction heads, often approximating to the pump delivery head
h_s = static head on the pump discharge

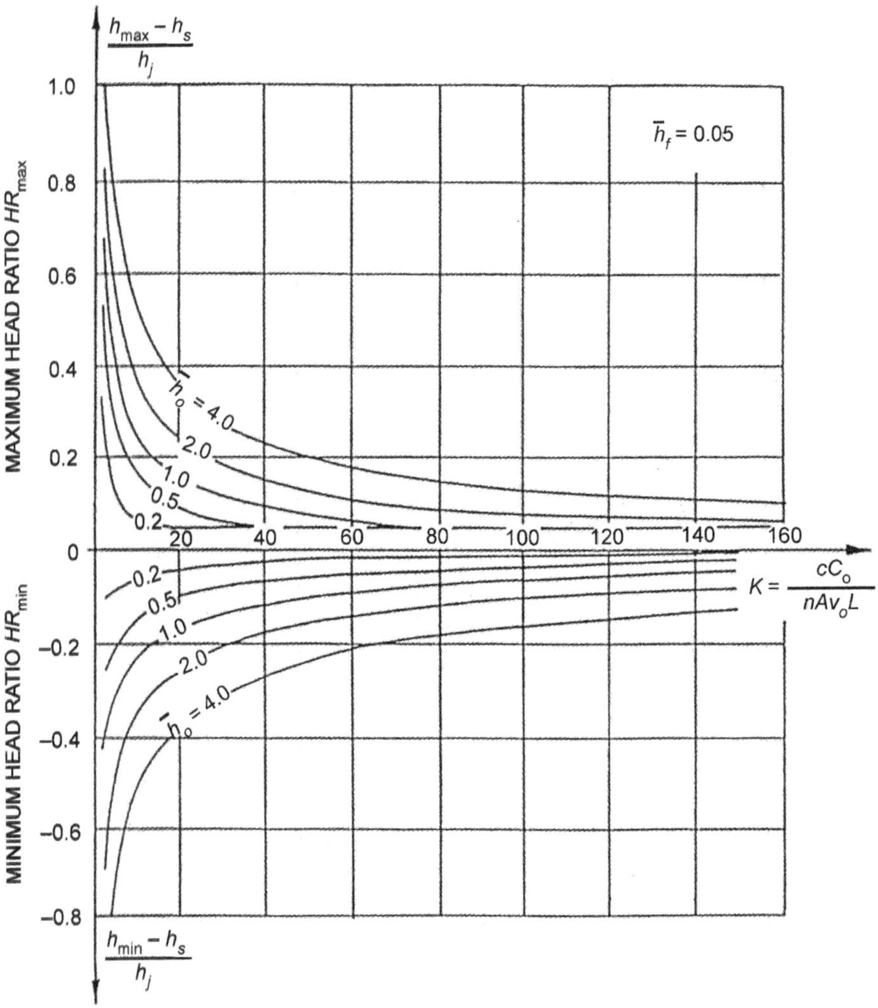

Fig. 3.15 *As Fig. 3.14, but with $\bar{h}_f = 0.05$*

L = pipe length
n = mean polytropic index
v_o = initial steady flow velocity

Parameters h_{max} and h_{min} are the permitted maximum and minimum pressure heads respectively in the pipeline, in the vicinity of the connection to the air vessel. They are usually specified by the system designer. The asterisk, as used in equation (3.39), denotes an absolute pressure.

The mean polytropic index, n, is usually taken to fall in the range 1–1.4. It is referred to here as the 'mean' because under dynamic conditions in the air vessel it is unlikely to be a constant. In one typical installation [Graze and Horlacher (1989)] it has appeared to fluctuate within an apparent range of approximately 0.6–2.0, with associated fluctuations in pressure and temperature being from 4 to 13 bar and from −45 to +50 °C respectively.

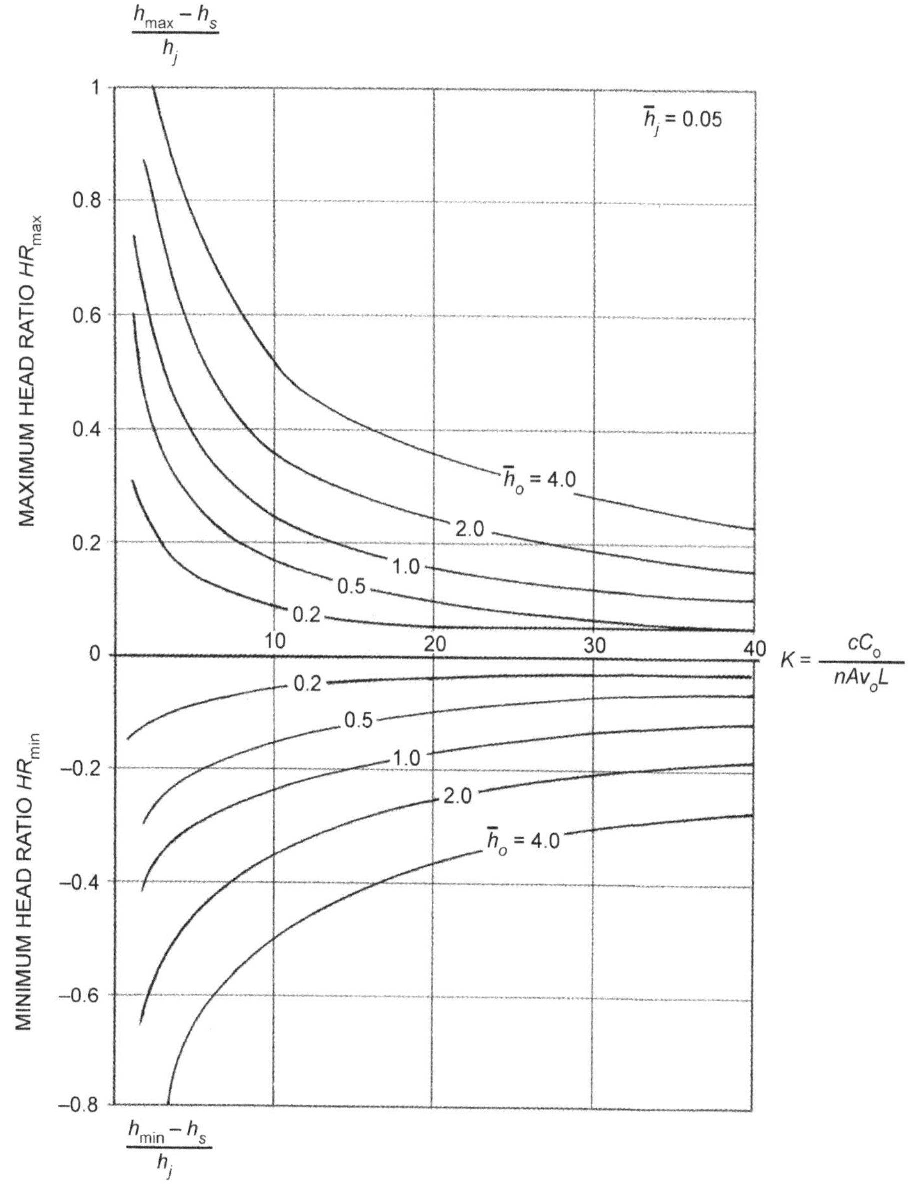

Fig. 3.15a *Figure 3.15 repeated, but with the x axis expanded near the origin*

For present purposes, it is deemed appropriate to assume a value of n in the region of 1.3–1.4 when a little heat transfer is expected to occur, whereas a value of between 1.05 and 1.1 should be used if significant heat transfer, i.e. almost isothermal behaviour, is expected. An initial compromise might be to use a value in the region of 1.175–1.2.

The first five charts, Figs 3.14 to 3.18, are for the extreme pressure heads in the pipeline, in the vicinity of the air vessel, and assume no throttling. For a given application, the appropriate chart to use is governed by the friction parameter,

Fig. 3.16 *As Fig. 3.14, but with $\bar{h}_f = 0.1$*

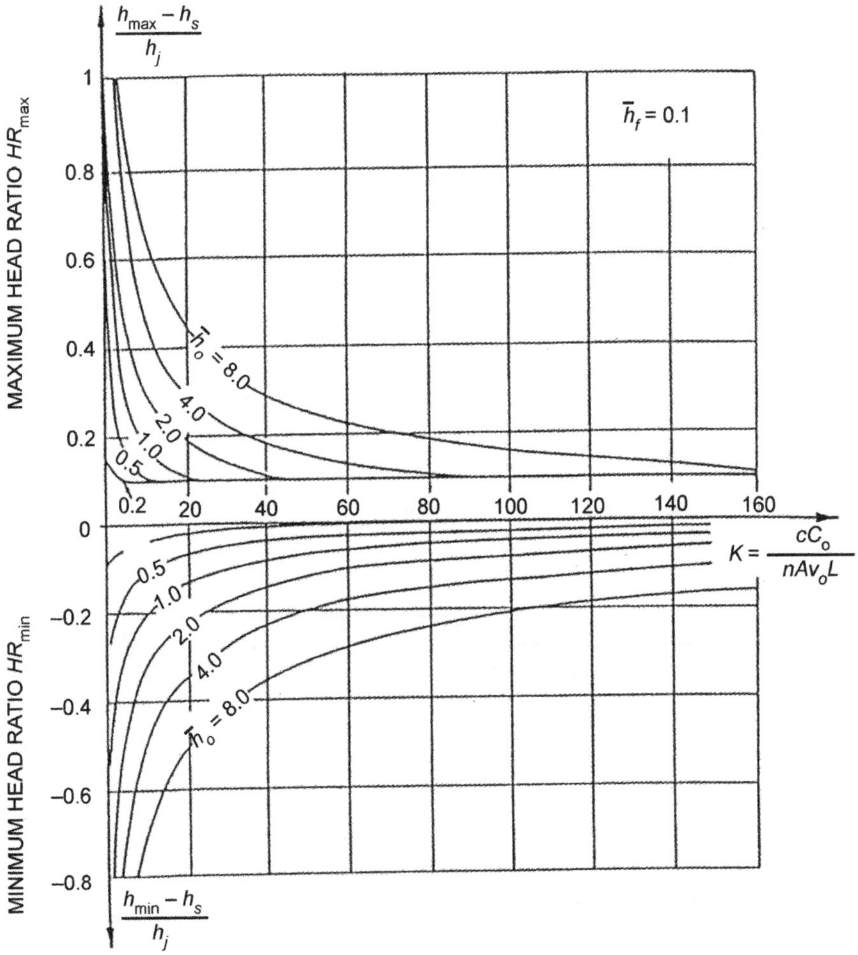

equation (3.37). These charts are used to obtain an estimate of the required air volume in the air vessel under initial steady flow conditions.

Figures 3.19 to 3.21 present a selection of maximum and minimum head envelopes along the pipeline, based on the non-dimensional parameters defined above. These charts may be used to check that high or low points along the line do not experience unacceptably low or high pressures respectively.

As above, the appropriate chart to use is the one for which the friction parameter, equation (3.37), is closest to that of the pipeline of interest.

Having determined the initial air volume, C_o, the total volume, V_T, of the air vessel can be evaluated by calculating the maximum volume, C', to which the air expands on the downsurge. A safety margin should then be added to allow for assumptions made in creating the charts and the quality of the basic system data available, to prevent dewatering of the vessel, etc.

Assuming a polytropic behaviour of the air in the vessel, the volume, C', to which it expands is given by

$$C' = C_o\left(\frac{h_o^*}{h_{min}^*}\right)^{1/n} \tag{3.43}$$

where

$$h_o^* = h_s + h_L + h_{atm}$$

and

$$h_{min}^* = h_{min} + h_{atm}$$

The total air vessel volume, V_T, is therefore

$$V_T = C' \times \text{factor of safety} \tag{3.44}$$

A typical factor of safety would be 1.2 or 1.25, giving a 20–25 per cent increase over the estimated maximum air volume, which occurs when the pressure inside the vessel is a minimum.

It is emphasized that the purpose in obtaining these estimates is to provide the basic data for computer programs that model transient behaviour much more accurately. When undertaking a computer analysis, it is important that the analyses are run long enough to capture *both* the maximum and minimum values of the pressure head envelopes over the whole system.

The time-scale of the oscillations in pressure in a system protected by an air vessel is very much longer than without it. Figures 3.22 and 3.23 enable the times at which the maximum and minimum heads occur, in systems having modest head losses, to be estimated. If there is no friction in the system, the minimum and maximum pressure heads occur, respectively, at

$$t_{h_{min}} = \frac{T}{4} \quad \text{and} \quad t_{h_{max}} = \frac{3T}{4} \tag{3.45}$$

in which the period, T, of the oscillation is given by

$$T = \frac{2\pi L}{c}\sqrt{\frac{K}{\bar{h}_o}} \tag{3.46}$$

and K is defined by equation (3.42) above.

Fig. 3.16a *Figure 3.16 repeated, but with the x axis expanded near the origin*

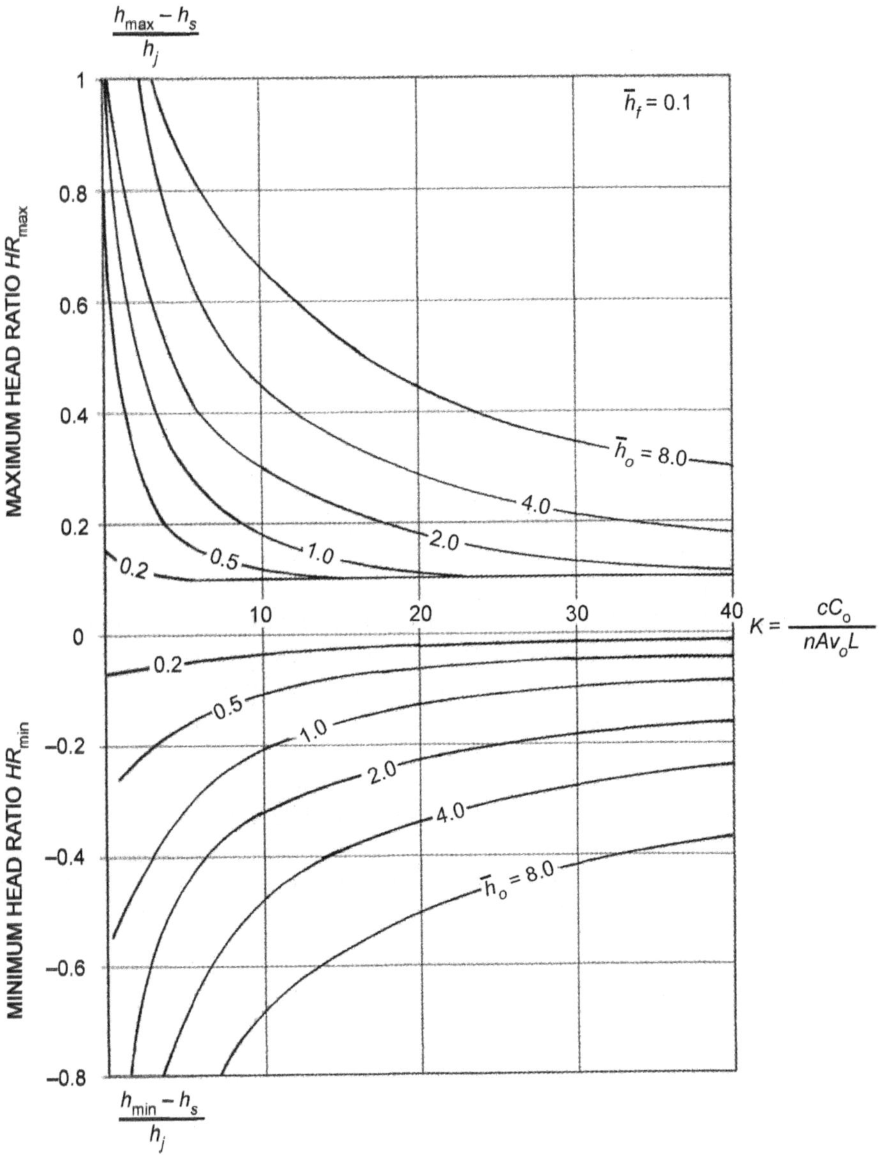

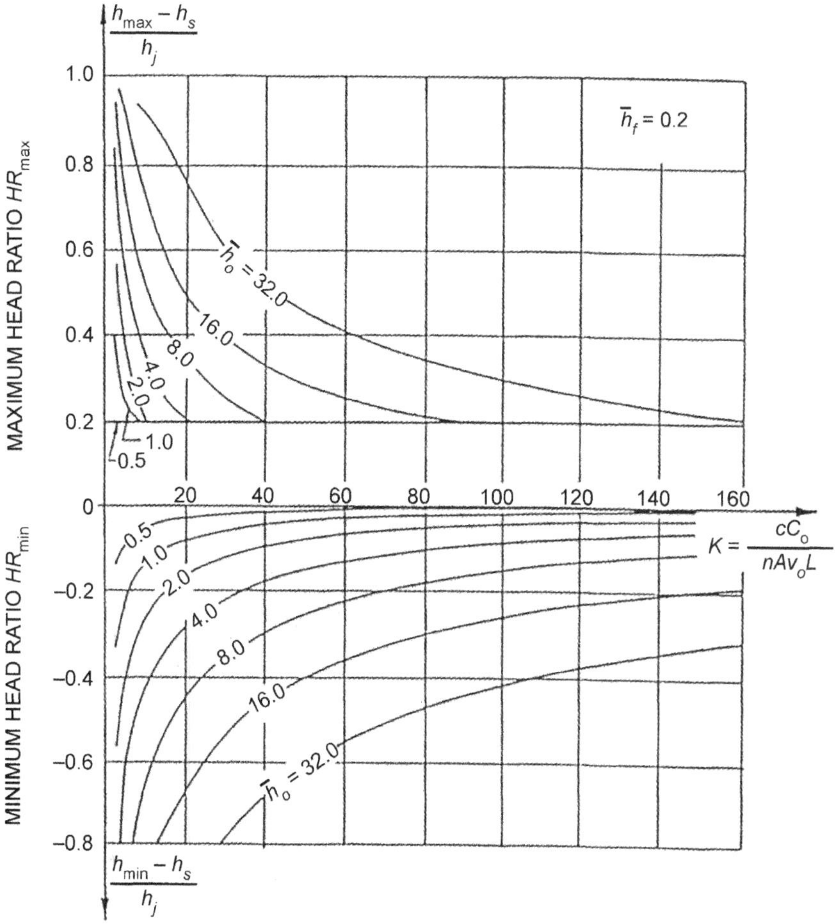

Fig. 3.17 *As Fig. 3.14, but with* $\bar{h}_f = 0.2$

Fig. 3.17a *Figure 3.17 repeated, but with the x axis expanded near the origin*

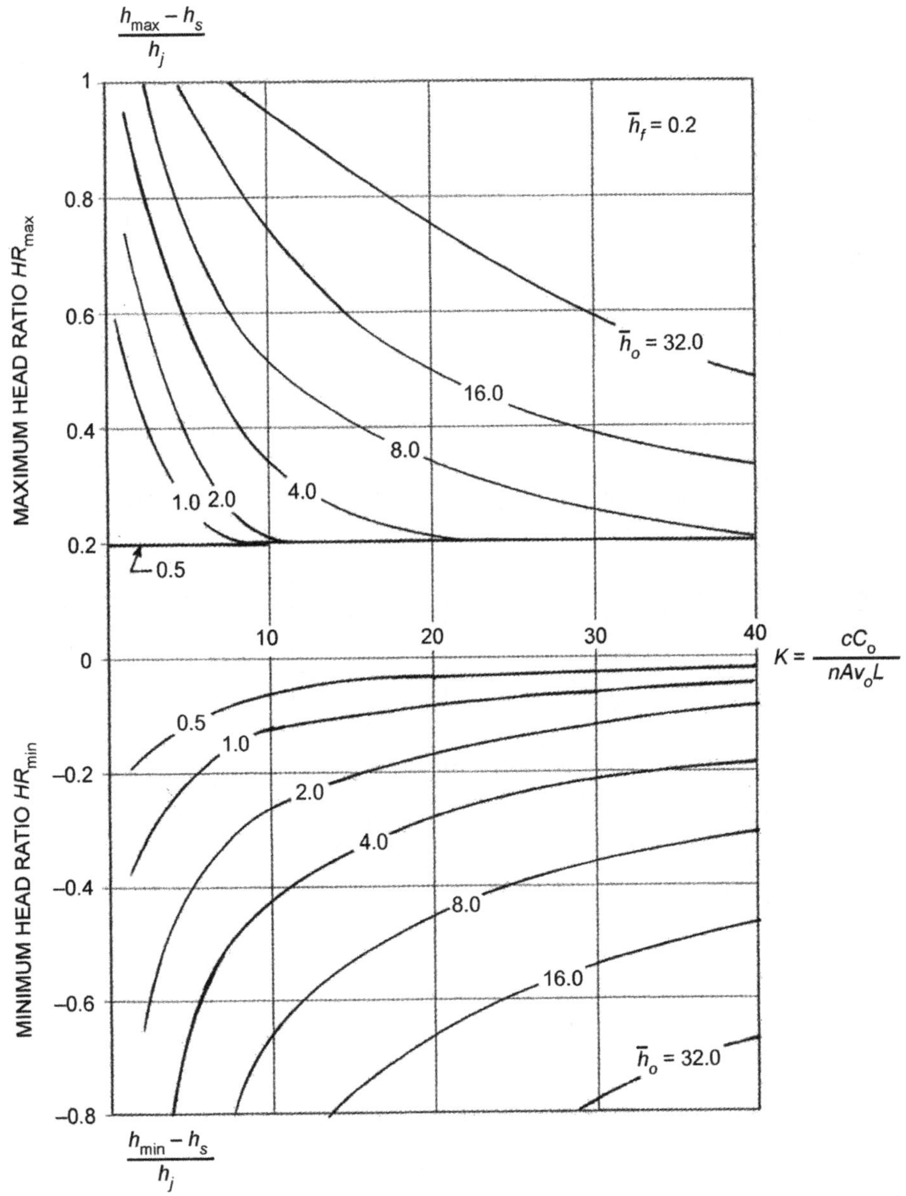

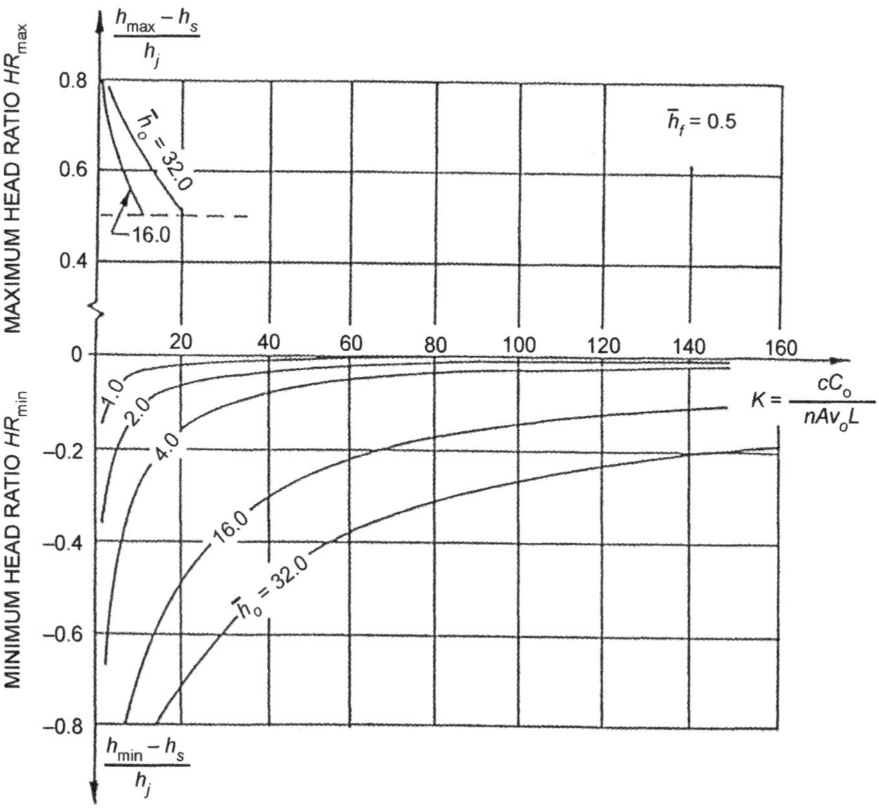

Fig. 3.18 *As Fig. 3.14, but with $\bar{h}_f = 0.5$*

Fig. 3.18a *Figure 3.18 repeated, but with the x axis expanded near the origin*

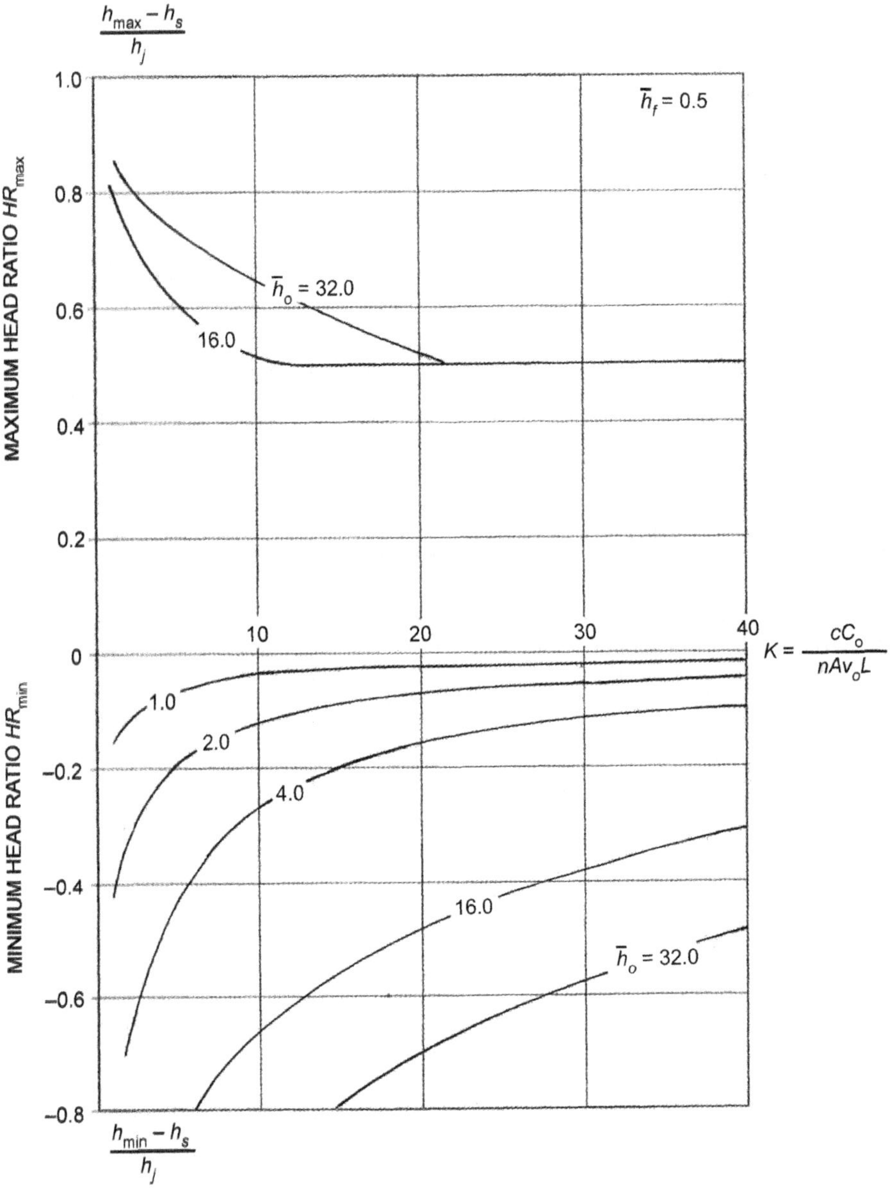

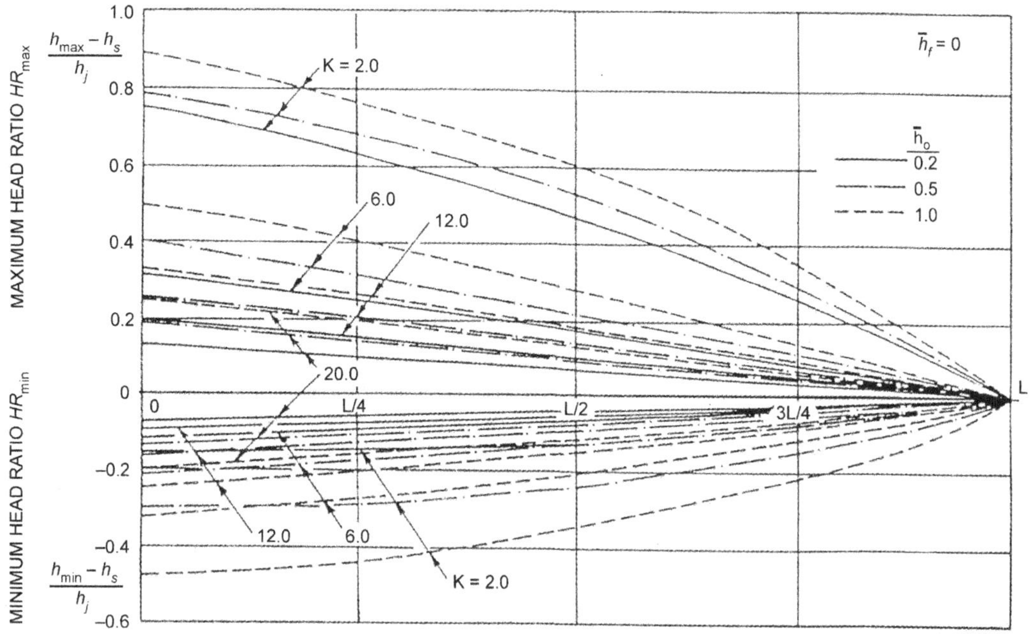

Fig. 3.19 *Envelopes of maximum and minimum pressure head ratios against pipe length, when the friction parameter* $\bar{h}_f = 0$

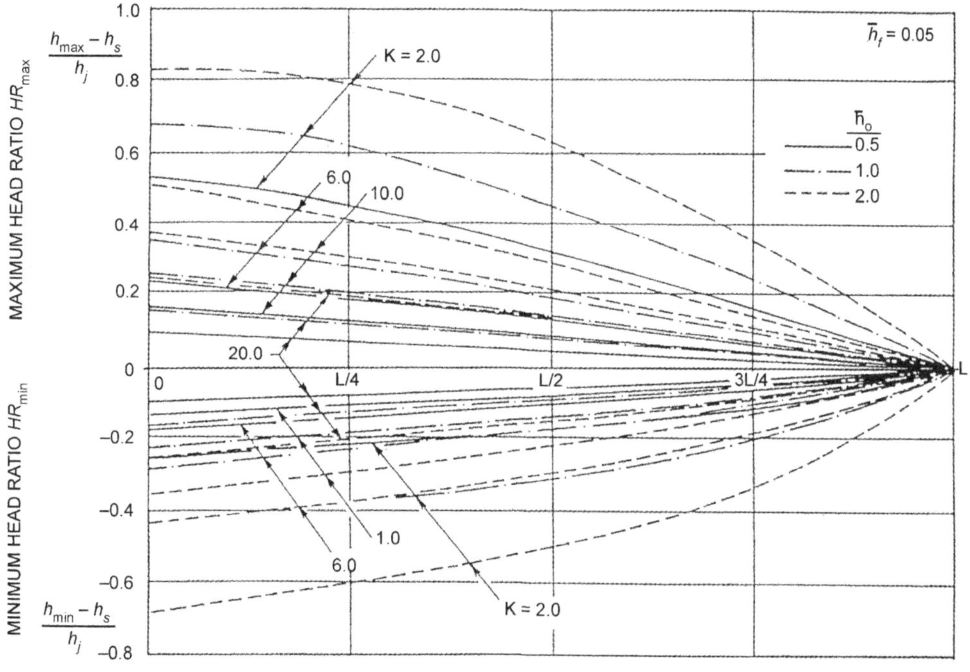

Fig. 3.20 *Envelopes of maximum and minimum pressure head ratios against pipe length, when the friction parameter* $\bar{h}_f = 0.05$

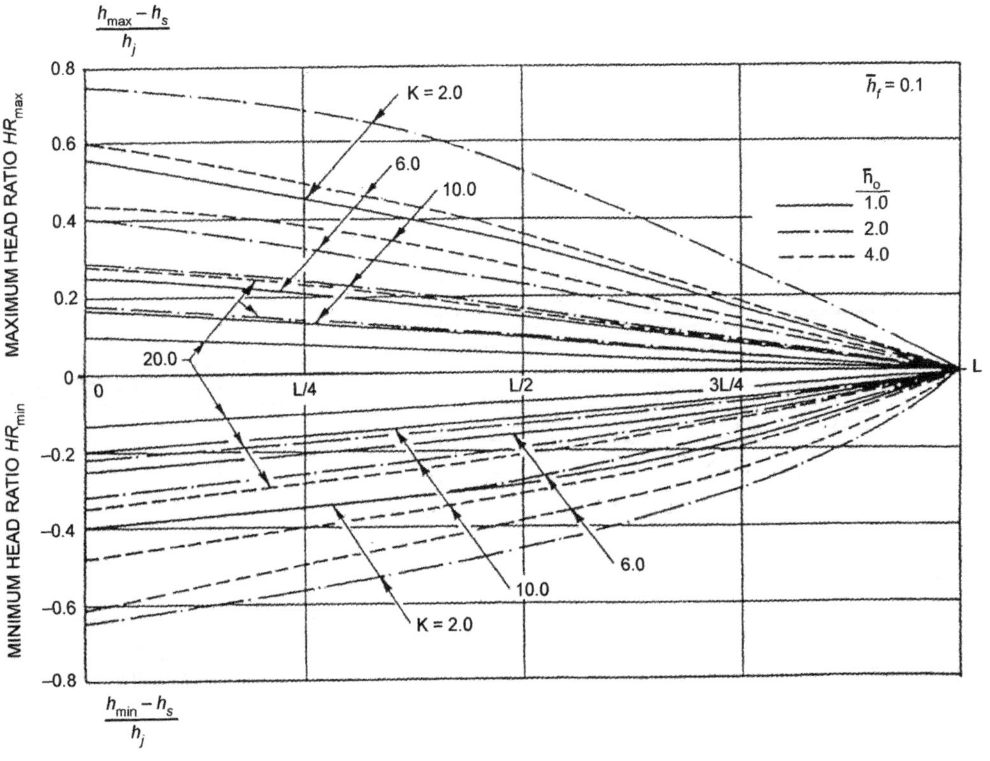

Fig. 3.21 *Envelopes of maximum and minimum pressure head ratios against pipe length, when the friction parameter* $\bar{h}_f = 0.1$

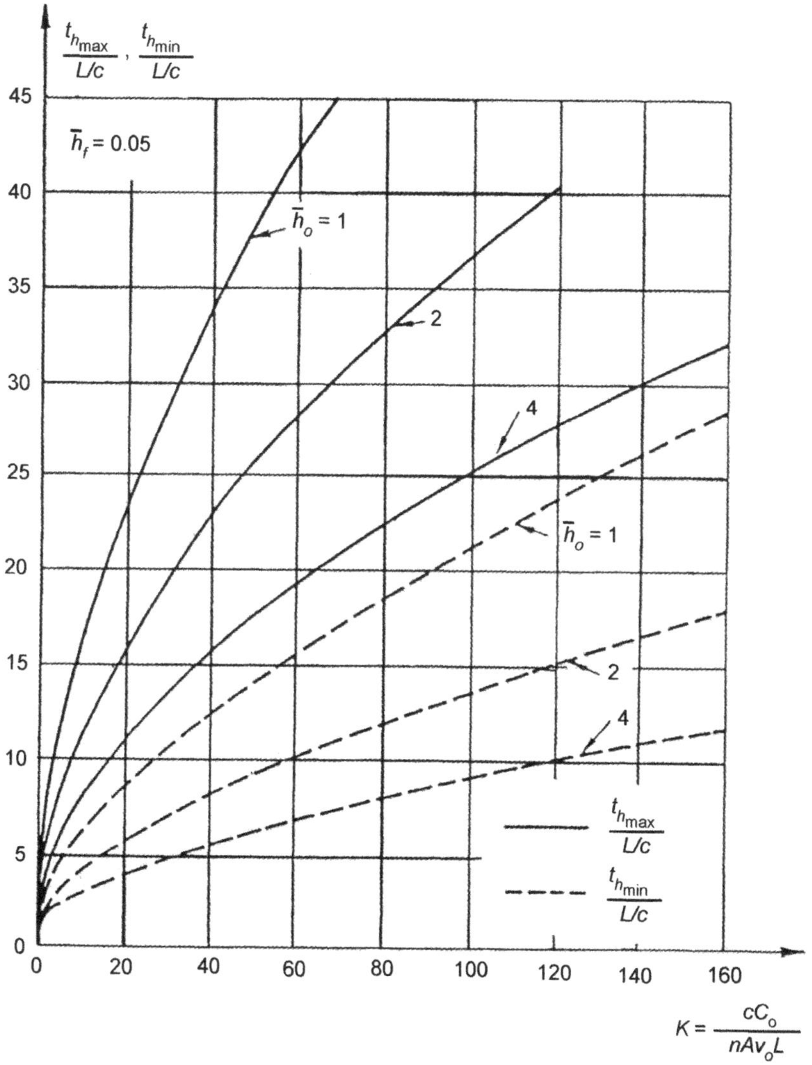

Fig. 3.22 *Time (non-dimensionally) to the occurrence of the maximum and minimum pressure heads at the pump discharge following a pump trip. Friction parameter* $\bar{h}_f = 0.05$

Fig. 3.23 *Time (non-dimensionally) to the occurrence of the maximum and minimum pressure heads at the pump discharge following a pump trip. Friction parameter $\bar{h}_f = 0.1$*

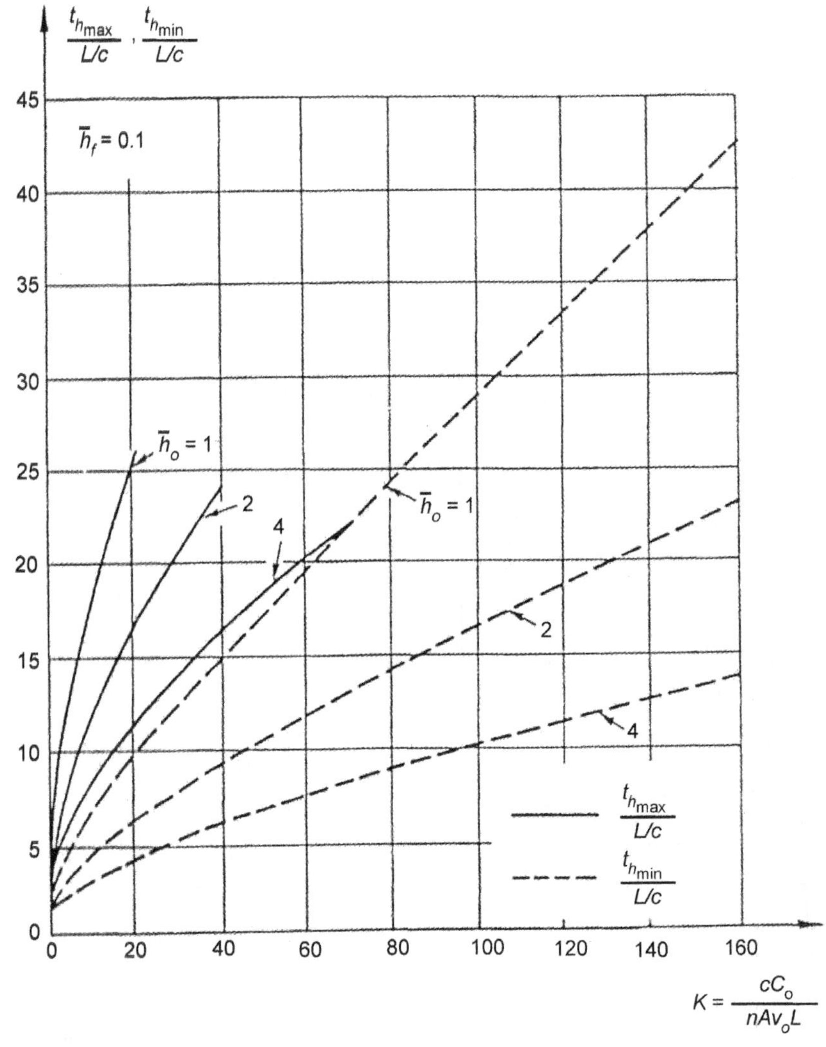

3.3.1.2 Throttled (Bypass) Air Vessels

The flow out of an air vessel should be as free and unrestricted as possible, to minimize the risk of cavitation and subatmospheric pressures developing. However, there can often be advantages in restricting the inflow by means of a throttle.

The physical arrangement is shown in Figs 1.14 and 3.13. The main potential benefits are that the overall time-scale of the pressure fluctuations in the system and the size (and hence cost) of the vessel required may in some circumstances be reduced. This is particularly useful when very large vessels might otherwise be needed.

A further potential benefit is that the maximum pressure in a system can be reduced, as shown by Fig. 3.24, from Graze and Horlacher (1989).

The head loss across the throttle is given by an equation of the form

$$h_{\mathrm{or}} = \xi \frac{v^2}{2g} \tag{3.47}$$

in which ξ is a non-dimensional loss coefficient.

Following an extensive programme of computer simulations, design charts based on optimum loss coefficients were devised [Graze and Horlacher (1989)]. These are reproduced here as Figs 3.25 and 3.26, in which

$$\bar{h}_{\mathrm{or_{in}}} = \frac{h_{\mathrm{or_{in}}}\,\mathrm{optimum}}{hj} \tag{3.48}$$

and

$$h_{\mathrm{or_{in}}}\,\mathrm{optimum} = \xi \frac{v_{\mathrm{o}}^2}{2g} \tag{3.49}$$

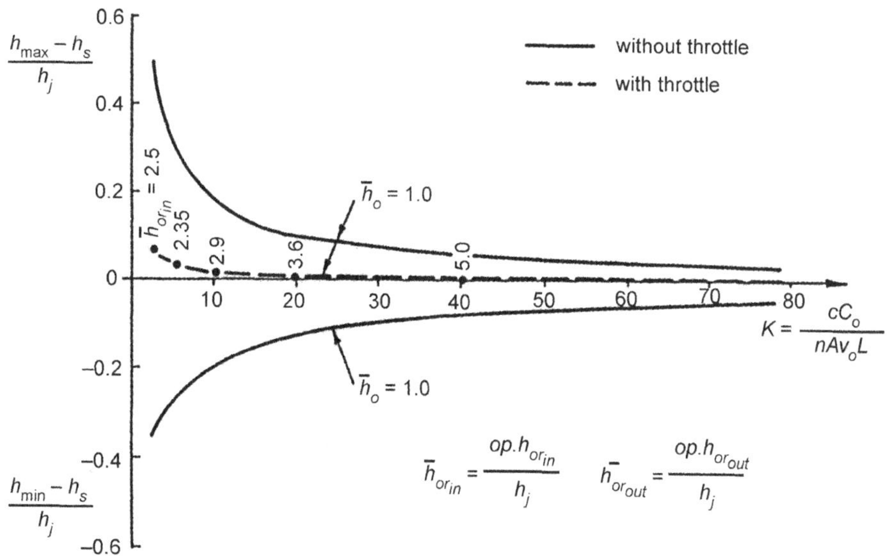

Fig. 3.24 *Comparison of extreme pressure heads close to an air vessel with and without throttling*

Fig. 3.25 *Generalized diagram for the sizing of air vessels with optimum throttling of the inflow, in pumping installations – for large maximum pressure head ratios*

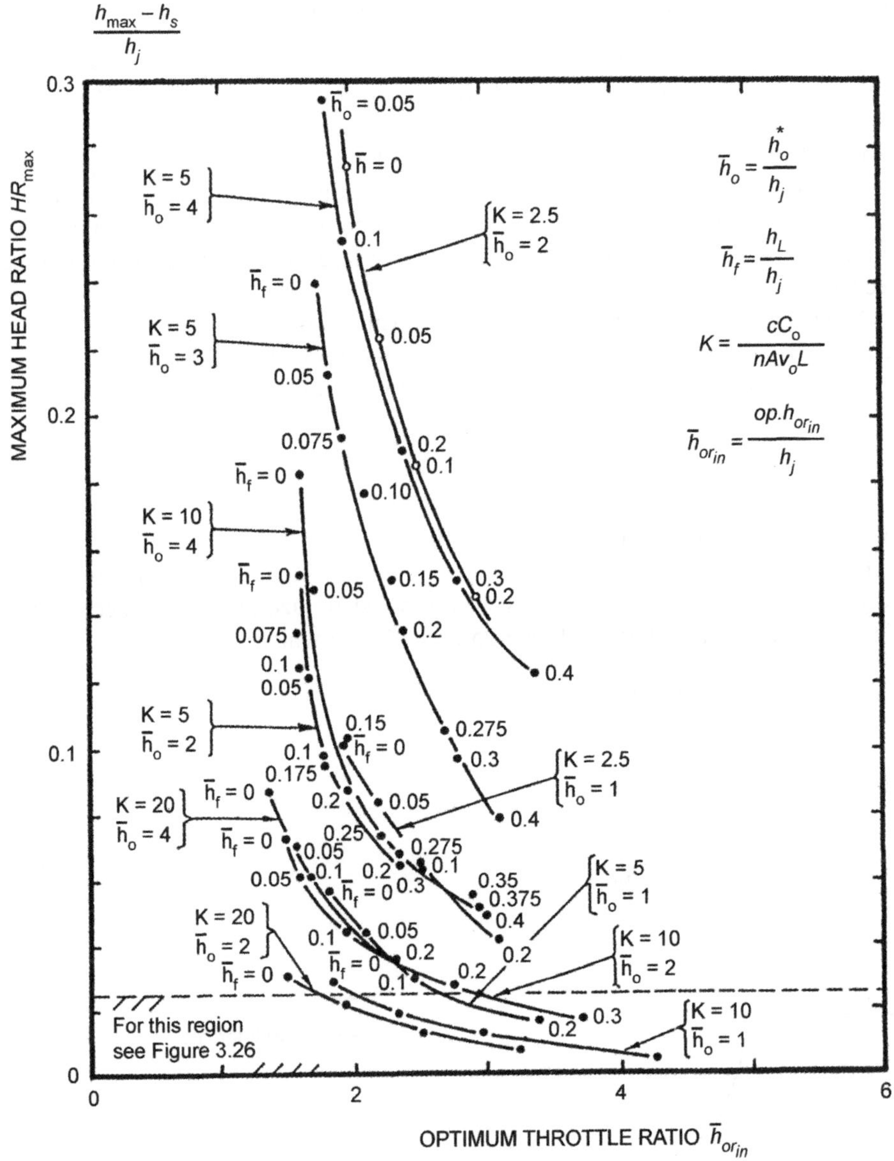

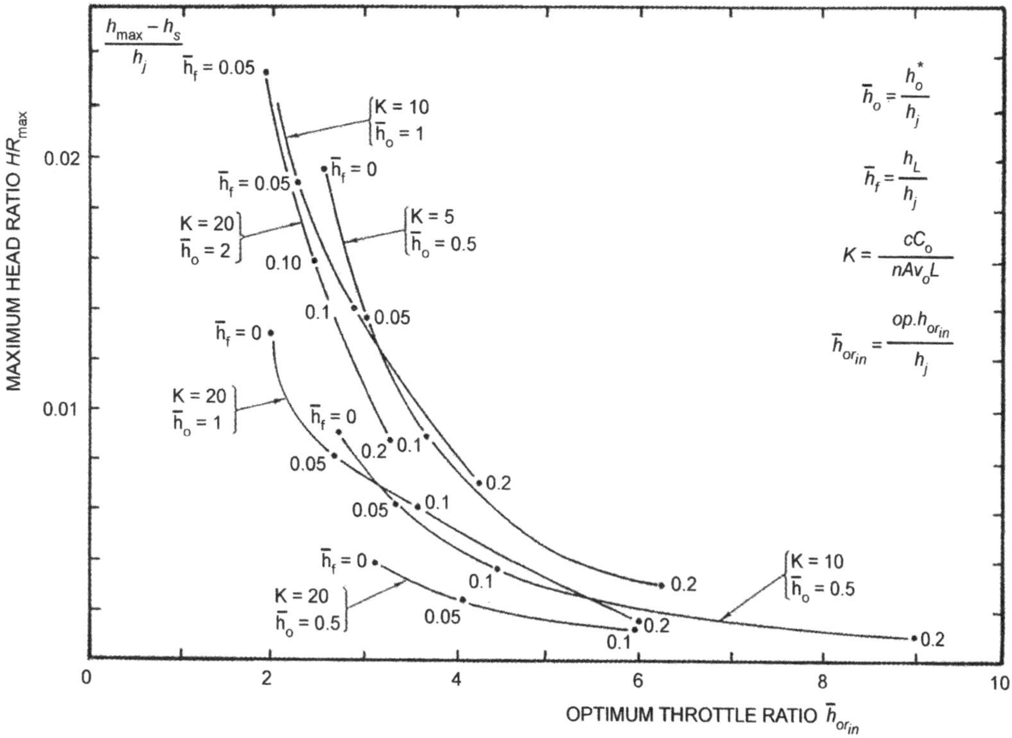

Fig. 3.26 *Generalized diagram for the sizing of air vessels with optimum throttling of the inflow, in pumping installations – for large maximum pressure head ratios*

3.3.1.3 Worked Example and Outline Procedure

The following data refer to a rising main pipeline conveying water, for which protection by an air vessel, against excessive pressure fluctuations following a pump trip, is being considered:

Pipe length L	3800 m
Cross-sectional area A	0.16 m^2
Static head h_s	130 m
Frictional head drop h_L	8 m
Initial steady flow velocity v_o	0.6 m/s
Local atmospheric pressure h_{atm}	10 m head of water
Transient wave speed c	1200 m/s

The maximum and minimum pressure heads to be permitted at the pump discharge, to avoid both excess pressures and subatmospheric pressures in the system, are $h_{max} = 162$ m head and $h_{min} = 103$ m head respectively.

First perform the following preliminary calculations:

1. The Joukowsky head from equation (3.38) is $h_j = 73.4$ m.
2. The friction parameter from equation (3.37) is

$$\bar{h}_f = \frac{8}{73.4} = 0.109 \text{ (say 0.1)}$$

3. From equation (3.39)

$$\bar{h}_o = \frac{130 + 8 + 10}{73.4} = 2.016 \text{ (say 2.0)}$$

4. The maximum head ratio from equation (3.40) is

$$\text{HR}_{\text{max}} = \frac{162 - 130}{73.4} = 0.436 \text{ (say 0.44)}$$

5. The minimum head ratio from equation (3.41) is

$$\text{HR}_{\text{min}} = \frac{103 - 130}{73.4} = -0.368 \text{ (say } -0.37)$$

6. Partial evaluation of the composite air vessel and pipeline parameter, K, from equation (3.42) is

$$K = \frac{1200 \times C_o}{n \times 0.16 \times 0.6 \times 3800} = 3.29\frac{C_o}{n}$$

7. Now select the chart that is most appropriate for a head loss parameter of 0.1, from step 1 above. This is Fig. 3.16 or, better still, Fig. 3.16a.
8. Refer to the curve for $\bar{h}_o = 2$ on both the top and bottom halves of the chart.
9. On the top half, read across from the Maximum Head Ratio of 0.44 (step 4 above) to intersect with the curve for $\bar{h}_o = 2$, then down to the horizontal axis to read off a value for K. In this example, a value in the region of $K = 5$ will be found.
10. Similarly, on the lower half, read across from the Minimum Head Ratio of –0.37 to the curve for $\bar{h}_o$, then up to the horizontal axis. This produces a value for K in the region of 7.5.
11. The larger of the two values from Steps 9 and 10 is used to calculate the required initial air volume. In the present example this is 7.5, which should be set equal to the term calculated in step 6, i.e.

$$K = 7.5 = 3.29\frac{C_o}{n}$$

12. The required air volume, C_o, is therefore

$$C_o = \frac{7.5}{3.29} \times n = 2.28 \times n$$

Because it is a modest size of air vessel, it might be assumed that there is some heat transfer and that $n = 1.175$, say, leading to

$$C_o = 2.28 \times 1.175 = 2.68\,\text{m}^3$$

13. 'From equation (3.43), the maximum volume C' to which this initial air volume expands is

$$C' = 2.68 \times \left(\frac{130 + 8 + 10}{103 + 10}\right)^{1/1.175} = 3.37\,\text{m}^3$$

14. Increasing this by, say, 20 per cent leads to a final volume, V_T, for the vessel of $4.0\,m^3$.

Examination of Fig. 3.21 indicates that, even without throttling, the maximum pressure head ratio on the upsurge on this basis would be within the permissible level of 0.44 at the pump.

In this example, the low pressure condition was the critical design case for sizing the air vessel. If, however, the maximum permissible pressure head at the pump discharge had been designated as, say, only 142 m head, this would have become the critical case. For an unthrottled vessel

$$\mathrm{HR}_{max} = \frac{142 - 130}{73.4} = 0.1635$$

and from Fig. 3.16a, $K = 24$, leading to

$$C_o = \frac{24}{3.29} \times 1.175 = 8.57\,m^3$$

To obtain the minimum pressure (corresponding to the maximum air volume C') read down through the K value (of 24) in Fig. 3.16a to intersect with the curve for $\bar{h}_o = 2.0$, then across left to get a value of around –0.2 for the Minimum Head Ratio HR_{min}. Substitute in equation (3.41) to obtain

$$h_{min} = h_j \times \mathrm{HR}_{min} + h_s = 73.4 \times (-0.2) + 103 = 88.32\,m$$

$$h^*_{min} = 88.32 + 10 = 98.32\,m\ abs$$

Then

$$C' = 8.57 \times \left(\frac{130 + 8 + 10}{98.32}\right)^{1/1.175} = 12.14\,m^3$$

leading to a total volume V_T, after incorporating a margin of safety, of around $15\,m^3$.

Since this is significantly larger than the requirement to limit the minimum pressure head, consideration may be given to installing a throttled inflow. The procedure involves the use of Figs 3.25 and 3.26 and is as follows.

For each of the four K values (2.5, 5, 10 and 20) for which the data are plotted on the charts, select (or interpolate) the curves for the appropriate value of h_o.

On each curve, interpolate the point corresponding to the value of the friction parameter $\bar{h}_f$ (0.1 here), and read off the x and y axes the values of $h_{or_{in}}$ and HR_{max} respectively. It is useful to tabulate them, as shown below for the current example.

Table 3.4 Data Extracted from Figs 3.25 and 3.26 for the Demonstration Example

K	$\bar{h}_f$	$\bar{h}_o$	HR_{max}	$\bar{h}_{or_{in}}$
2.5	0.1	2	0.18	2.5
5	0.1	2	0.11	1.85
10	0.1	2	0.45	2.0
20	0.1	2	2.50	2.5

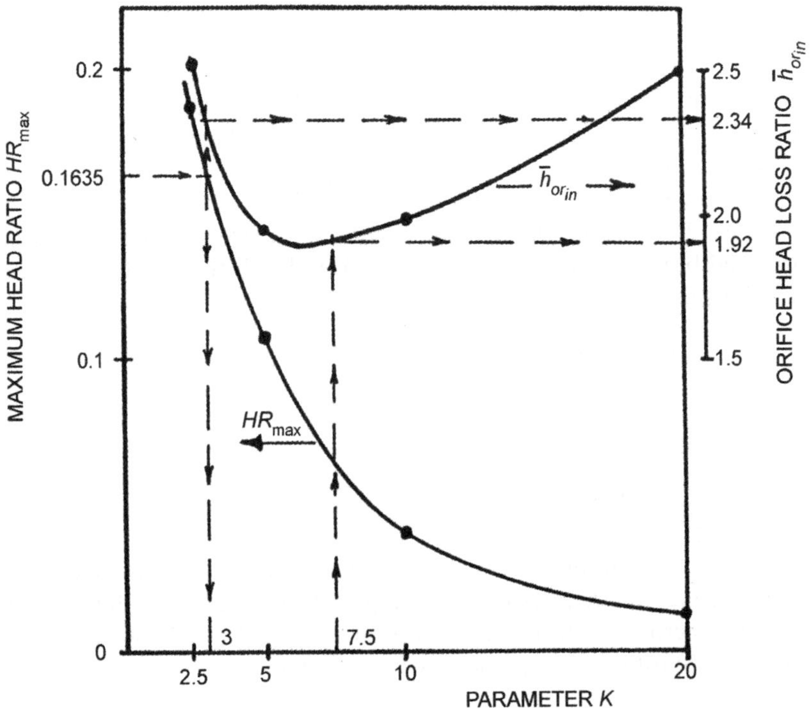

Fig. 3.27 *Curves for the Maximum Head Ratio and non-dimensional Optimum Throttle Ratio*

These data can then be used to plot Fig. 3.27 for the maximum head ratio and the optimum throttle ratio.

The maximum permissible head ratio for this demonstration example was specified as 0.1635. Reading across to the head ratio curve, and down, in Fig. 3.27 gives a value of $K = 3$, with an associated optimum orifice head ratio of 2.34.

For a value of $K = 3$, the corresponding initial air volume is

$$C_o = \frac{3}{3.29} \times 1.175 = 1.07\,\text{m}^3$$

This is less than the volume required to protect the system against the initial downsurge, and hence the latter ($K = 7.5$, $C_o = 2.68\,\text{m}^3$, $V_T = 4\,\text{m}^3$) should be used as the critical design case.

The optimum orifice head ratio, $h_{or_{in}}$, for this vessel, read off from Fig. 3.27 for $K = 7.5$, is around 1.92.

Substituting this into equations (3.48) and (3.49) gives optimum $h_{or_{in}} = 141\,\text{m}$ head, with the throttle loss coefficient $\xi = 7680$.

The maximum inflow will be much less than this, and the corresponding head loss between the pipeline and the air vessel will also be considerably smaller.

3.3.2 Start-up of Deep-Well Pumps

Many water recovery systems utilize submersible deep-well pumps. Figure 3.28 illustrates a typical scenario in which a pump (or pumps), mounted below ground level, pump water up to a treatment works or a holding reservoir. When the pump

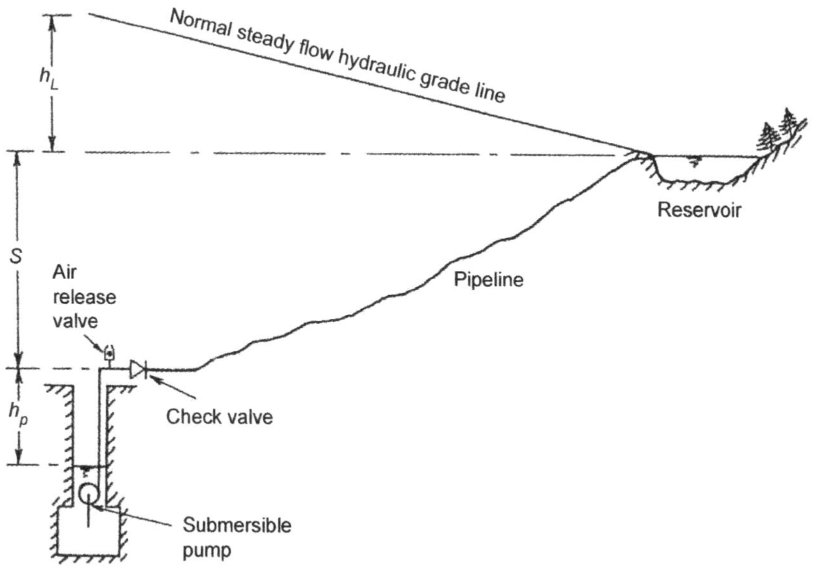

Fig. 3.28 *Schematic arrangement of a deep-well installation*

is shut down, reverse flow is prevented by a check valve at the top of the riser in the well shaft.

Even if a foot valve is fitted on the pump suction, the water column in the riser will usually fall to some extent, with a vapour cavity forming at the top. There is the risk of air ingress as the system ages, and so in a safety audit it is prudent to assume that the pump can be started up without any water in the riser and that some form of protection involving air release is required.

This subsection deals with the use of air vessels for this purpose, but other strategies are available, such as pressure regulating valves (see Subsection 1.3.3.5) and hydraulically controlled pressure relief valves (see Section 3.7). These operate on the principle of allowing full flow out of the system initially, then close gradually to force up the pressure in the riser until the check valve opens gradually. Whatever protection strategy is adopted, it is important that all the air is expelled from the riser before the check valve opens to admit flow to the main pipeline. The operational challenge is to achieve this rapidly and at the same time initiate the flow of water smoothly.

On start-up, the pump is working initially against zero head, and will jump to an operating point on the H–Q curve where it crosses the zero head axis (see Fig. 3.29). As the riser fills, and assuming that the air is rapidly and completely expelled to prevent any being pushed into the main pipeline, the system curve rises towards the lower one shown in the figure. The instant the riser is just full, i.e. the static head is only h_P, the flowrate is Q_S.

A standard large-orifice air release valve would then close, and this high-speed water column would slam into the liquid column being held back by the check valve in the main pipeline. The ensuing rapid deceleration would lead to a Joukowsky-type pressure rise, which would be propagated both ways from the point of impact, i.e. from the check valve.

The magnitude, ΔH, of the pressure rise propagated upstream and downstream from the check valve when the liquid column slams into it may be estimated from

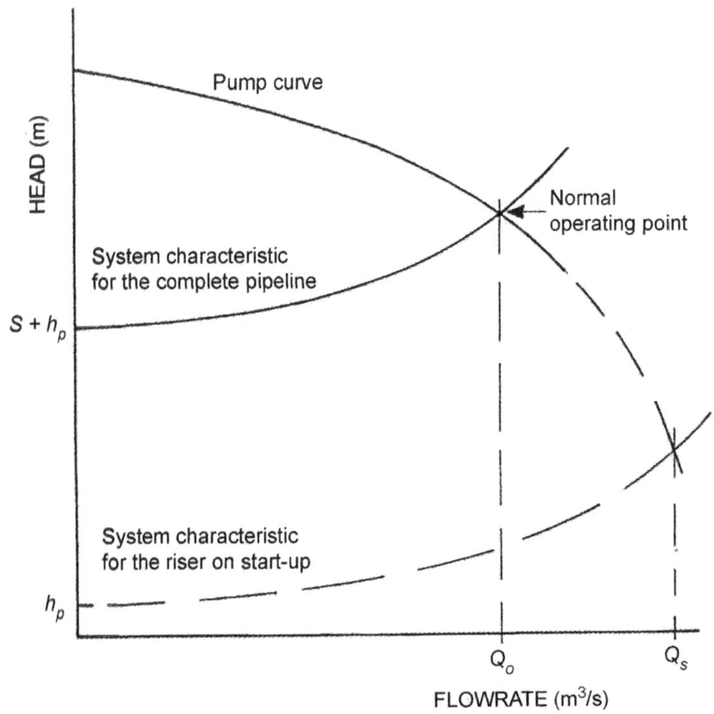

Fig. 3.29 *Pump and system characteristics for a deep-well installation*

Joukowsky's equation applied to this situation, i.e.

$$\Delta H = \frac{c}{g}\frac{Q_S}{2A} \tag{3.50}$$

The peak pressure that occurs at the check valve is $S + \Delta H$. Note that, since the Joukowsky pressure change is propagated both ways, the local flow velocity is only halved, not reduced to zero.

If either the peak pressure or the shock loading to the pipe supports is unacceptable, a control and suppression strategy must be devised. One option is to incorporate an air vessel as indicated in Fig. 3.30.

To be effective, the air vessel must be of an appropriate size and contain a suitable volume of air. To generate estimates of this, some simple design charts developed by Tucker and Young (1960) may be used. Among the assumptions used in creating these charts were:

(a) All the air in the riser has been completely exhausted when the check valve opens.
(b) The flow remains constant at Q_o when the check valve opens. This implies that the H–Q characteristic is very steep in the low-head region.
(c) It also implies that there is no frictional head drop in the entry to the air vessel.

Figure 3.31, and the enlarged section shown in Fig. 3.32, enable the air volume required to limit the pressure rise to be estimated. Figure 3.33 allows the minimum pressure, and hence the maximum volume, to be estimated so that the total volume of the vessel required can be determined.

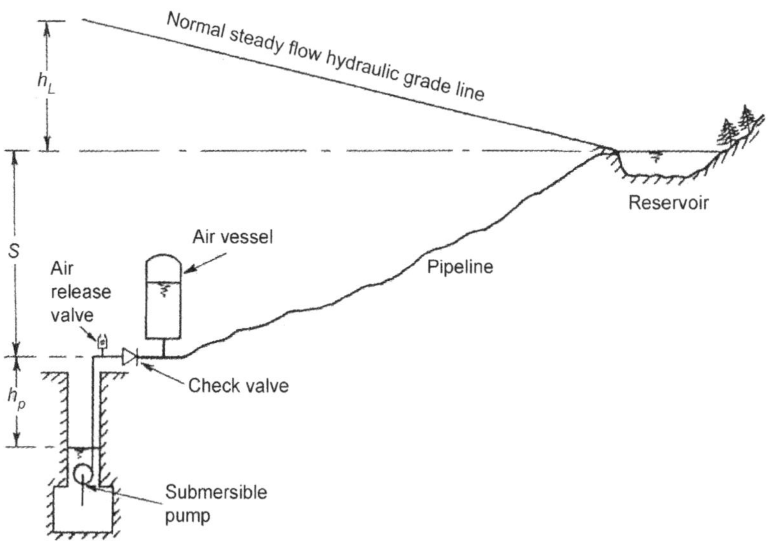

Fig. 3.30 *Schematic layout of a deep-well installation protected by an air vessel and air release valve*

The parameters on the charts shown in Figs 3.31 to 3.33 are defined as follows:

Air vessel parameter

$$\frac{1}{\sigma_{\mathrm{TY}}} = \frac{LQ_{\mathrm{o}}^2}{2AgS^* C_{\mathrm{o}}} \tag{3.51}$$

Maximum head ratio

$$\mathrm{HR}_{\max} = \frac{H^*_{\max}}{S^* + h_L} \tag{3.52}$$

Minimum head ratio

$$\mathrm{HR}_{\min} = \frac{H^*_{\min}}{S^* + h_L} \tag{3.53}$$

Friction parameter

$$n_{\mathrm{f}} = \frac{h_L}{S^*} \tag{3.54}$$

Maximum air volume

$$C' = C_{\mathrm{o}}\left(\frac{S^*}{H^*_{\min}}\right)^{1/1.2} \tag{3.55}$$

Several symbols are defined in Fig. 3.30. Of the remainder:

L = length of the main pipeline
A = pipe cross-sectional area
C_{o} = required air volume at zero flow prior to start-up

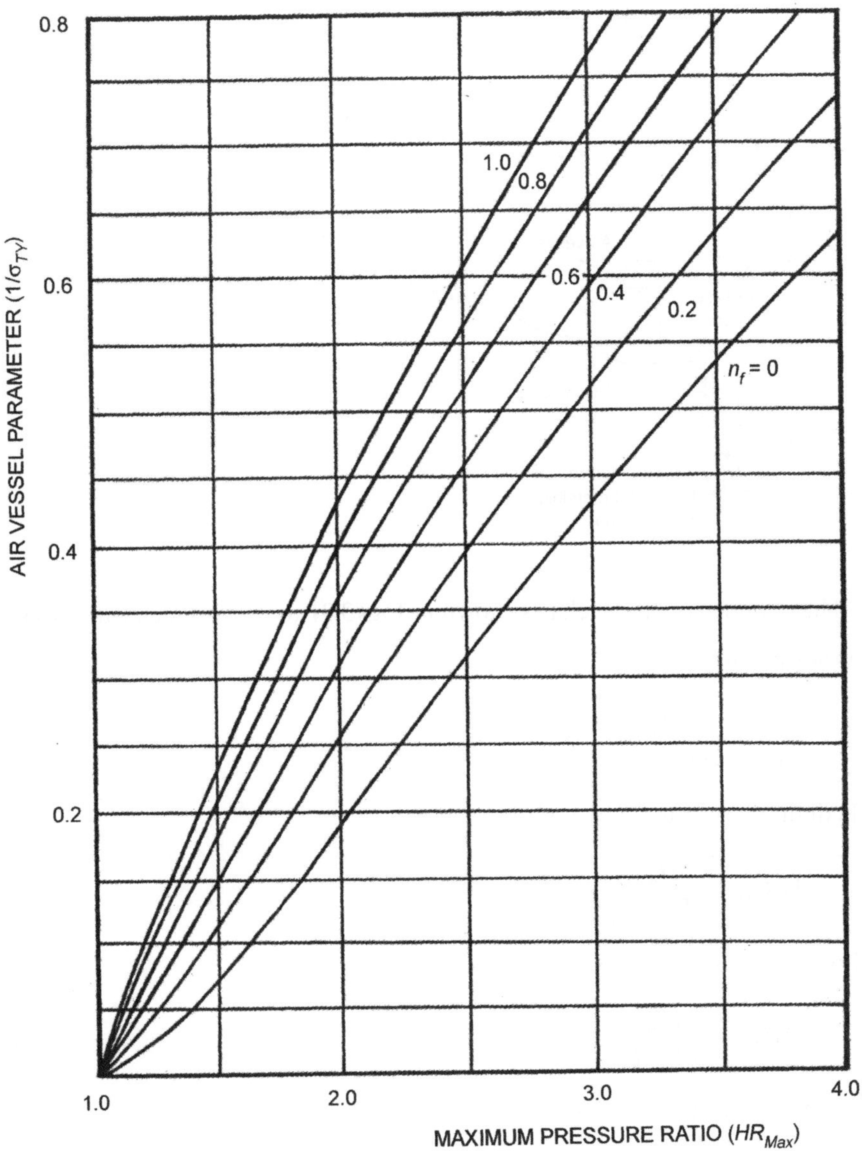

Fig. 3.31 *Air vessel parameter plotted against the maximum pressure head ratio – for large ratios*

The asterisk denotes that absolute pressures (i.e. gauge pressure plus local atmospheric pressure) must be used because of making use of the gas laws relating changes in pressures and volumes.

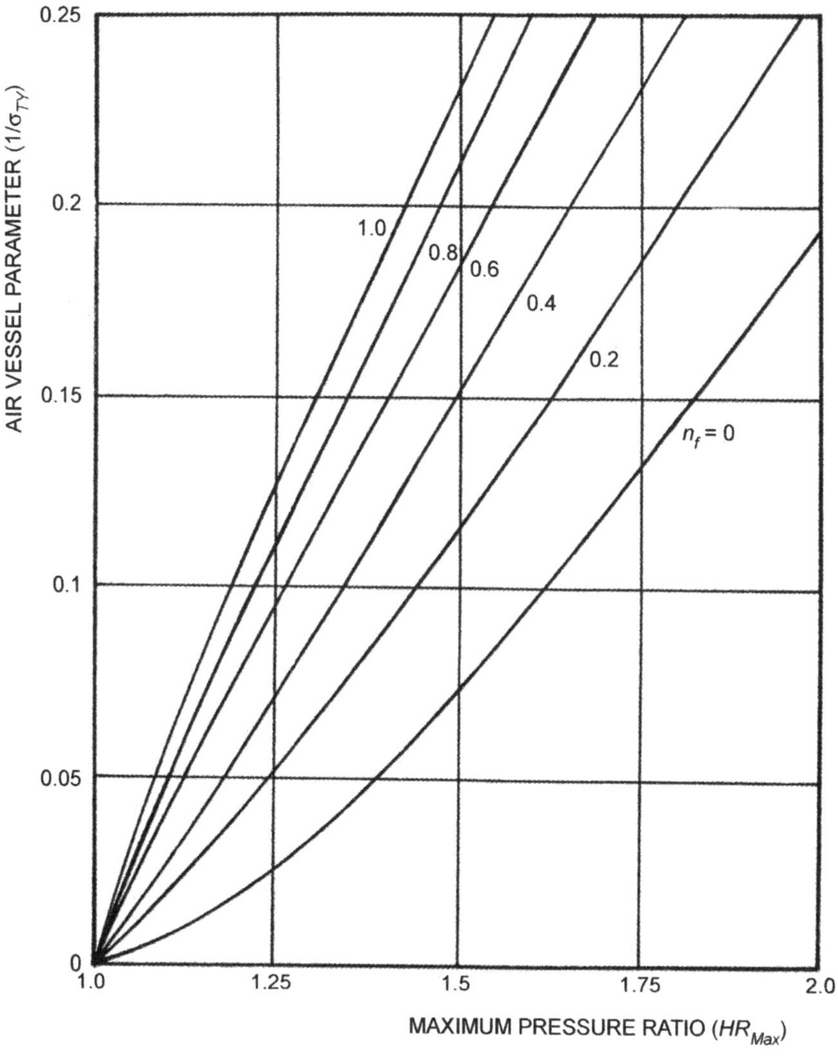

Fig. 3.32 *Air vessel parameter plotted against the maximum pressure head ratio – for small ratios*

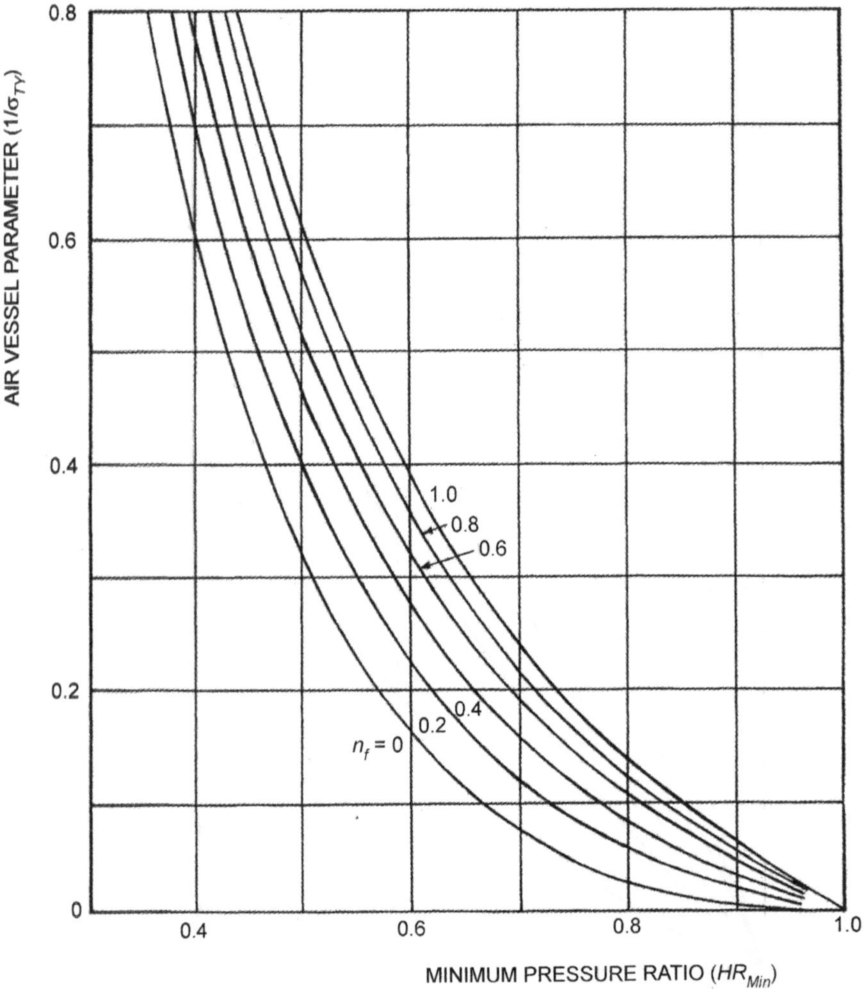

Fig. 3.33 *Air vessel parameter plotted against the minimum pressure head ratio*

3.3.2.1 Outline Procedure

The procedure for using the charts is as follows:

1. Determine the static lift S and convert to absolute pressure head S^*.
2. Evaluate the frictional head loss h_L associated with the normal operating flowrate Q_o, and calculate the friction parameter n_f from equation (3.54).
3. Determine the maximum permissible head (absolute) H^*_{max} at the check valve.
4. Calculate the maximum head ratio HR_{max} from equation (3.52).
5. Interpolate a curve for the friction parameter n_f on either Fig. 3.31 or Fig. 3.32, as most appropriate.
6. Read up from the maximum head ratio calculated above to the intersection with the friction parameter, then across to the air vessel parameter $1/\sigma_{TY}$.
7. Using equation (3.51), rearranged, calculate the required initial air volume C_o.
8. Interpolate a curve for n_f (as above) in Fig. 3.33 for the minimum head ratio.
9. Using the air vessel parameter, found in Step 6, read across to the intersection with the friction parameter, then down to get the Minimum Head Ratio

HR$_{min}$, and then calculate the minimum absolute pressure head, H^*_{min}, from equation (3.53).

10. Calculate the maximum volume to which the air expands on the downsurge from equation (3.55).
11. Decide on the total volume required for the air vessel, using a suitable safety factor (add, say, 20–25 per cent) to avoid the risk of the vessel dewatering on the downsurge.

In view of the assumptions made in developing the charts and the behaviour of the pump, this procedure, naturally, gives only an approximate value for the required size of the air vessel. It will, however, be on the conservative side in most cases, and provide a good starting point for a computer-aided analysis.

Note that this procedure applies only to the start-up of the pump. Surge protection for a pump trip must be assessed separately using, for example, the procedures outlined in Subsection 3.3.1, so that the worst-case scenario can be established and used as the critical design case.

3.3.2.2 Demonstration Example

For a system such as that shown in Fig. 3.28, determine the peak pressure that would occur following a pump start and estimate the size of air vessel that would be required to ensure that the pressure remains within the rated pressure of 16 bar. System details are as follows:

Pipe length L	5000 m
Pipe diameter D	400 mm
Darcy friction factor f	0.0124
Static lift S	40 m
Length of dry riser h_P	10 m
Design flowrate Q_o	0.2 m^3/s
Wave speed c	1150 m/s

The pump performance characteristic, as provided by the manufacturer, is shown by the solid line in Fig. 3.34. It is necessary to extrapolate this in the fashion shown by the dashed line. This is typical of the way the H–Q curves for centrifugal pumps usually fall away very steeply, to cross the zero head line at a flowrate of about 30–50 per cent greater than the rated flow.

Two other curves have been added to this diagram, the main system characteristic, starting at $S + h_P$, for normal steady flow and the system characteristic for the riser alone, starting from h_P on the vertical axis.

The frictional head drop in the riser is quite small and the flowrate Q_s, when the riser is just full, can be read off the graph as 0.28 m^3/s.

The maximum head at the check valve as the water column in the riser is suddenly decelerated is

$$S + \frac{cQ_s}{2gA} = 40 + \frac{1150 \times 0.28}{2 \times 9.81 \times 0.1257} = 170.6\,\text{m}$$

To estimate the initial air volume required in the vessel, and its total capacity, to limit the pressure at the top of the riser to the rated head of 16 bar, i.e. 160 m head,

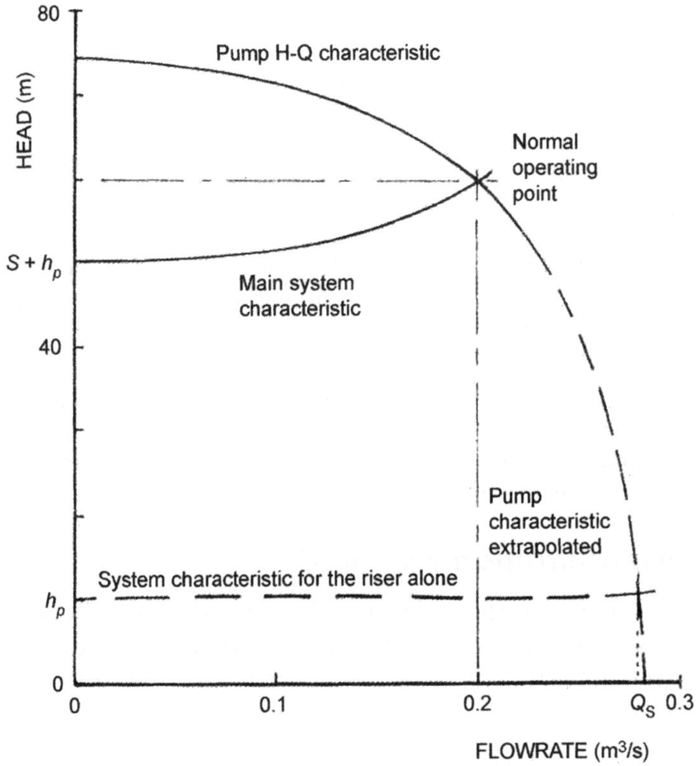

Fig. 3.34 *Pump and system characteristics for the specimen example*

follow the steps outlined above. Assuming local atmospheric pressure is 10 m head water gauge:

Absolute static head

$$S^* = 40 + 10 = 50\,\text{m}$$

Frictional head drop

$$h_L = \frac{8fLQ_o^2}{g\pi^2 D^5} = 20\,\text{m}$$

Friction parameter

$$n_f = \frac{20}{50} = 0.4$$

From equation (3.52) the maximum head ratio is

$$\text{HR}_{\text{max}} = \frac{160 + 10}{50 + 20} = 2.429 \text{ (say, 2.43)}$$

In Fig. 3.31, read up from $\text{HR}_{\text{max}} = 2.43$ to the curve for $n_f = 0.4$, then across left to the vertical axis (see Fig. 3.35) to obtain a value of 0.436 for the air vessel parameter $(1/\sigma_{\text{TY}})$.

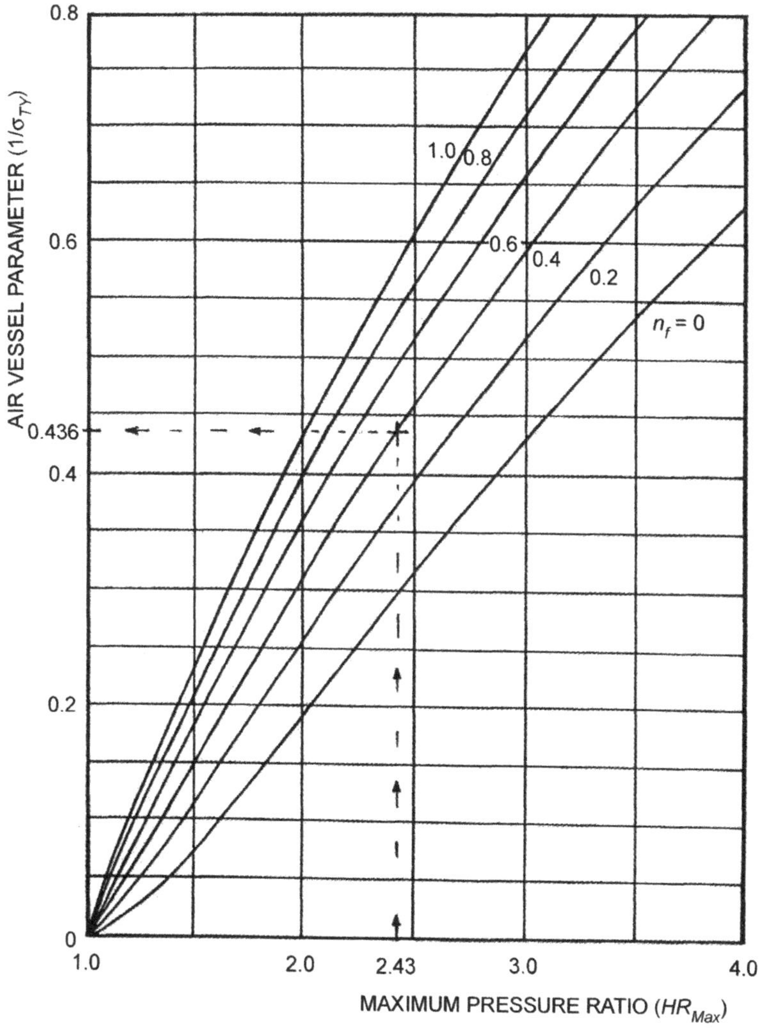

Fig. 3.35 *Determination of the air vessel parameter for the demonstration example*

From the definition of the air vessel parameter given in equation (3.51), calculate the initial air volume, C_o, of the trapped air

$$C_o = \frac{LQ_o^2}{(1/\sigma_{TY})2AgS^*} = \frac{5000 \times 0.2^2}{0.436 \times 2 \times 0.1257 \times 9.81 \times 50} = 3.72\,\text{m}^3$$

Use Fig. 3.33 to obtain the minimum pressure head in the air vessel (see Fig. 3.36). Read across from $(1/\sigma_{TY}) = 0.436$ to the appropriate line for $n_f = 0.4$, then down to the baseline to get a value of 0.514 for the Minimum Pressure Ratio HR_{min}.

From equation (3.53)

$$H^*_{min} = 0.514 \times (S^* + h_L) = 0.514 \times 70 = 35.98\,\text{m}$$

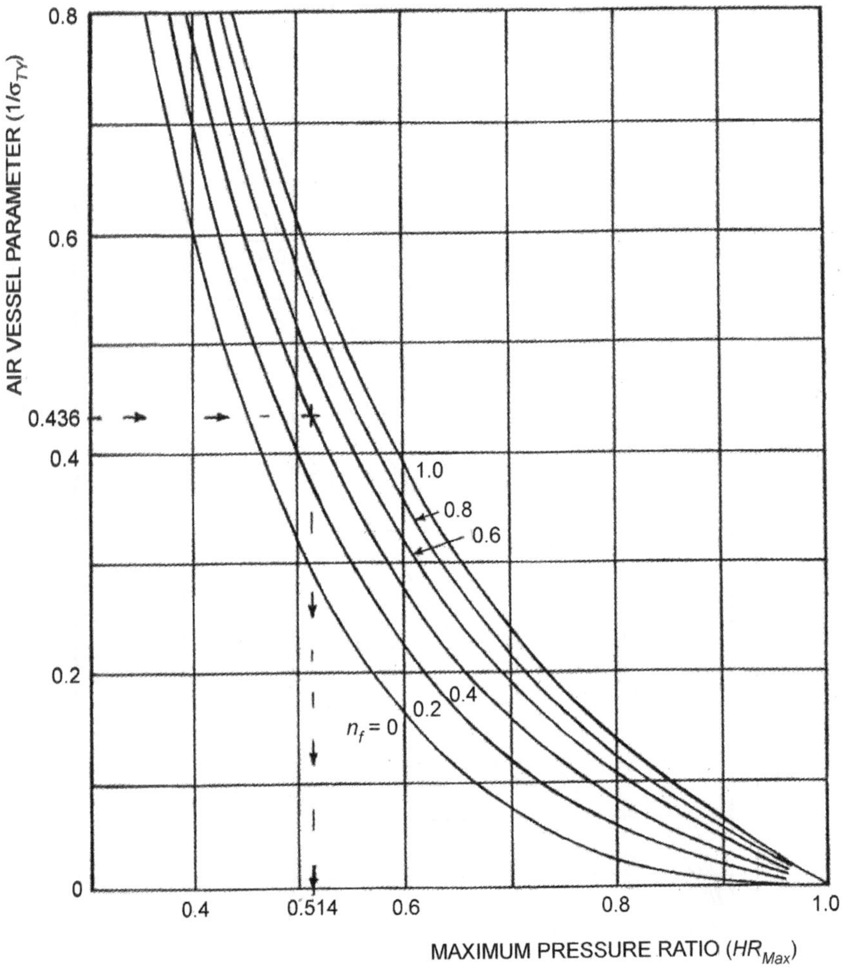

Fig. 3.36 *Determination of the minimum pressure ratio for the demonstration example*

and so the maximum air volume C' from equation (3.55) is

$$C' = 3.72 \times \left(\frac{50}{35.98}\right)^{0.833} = 4.89\,\text{m}^3$$

Allowing approximately 20 per cent safety margin, the final estimated volume for the air vessel is

$$V_T = 5.87\ (\text{say, } 6)\,\text{m}^3$$

If it is assumed that the air in the vessel expands isothermally, the theoretical air volume C' would be larger, by approximately $0.28\,\text{m}^3$.

SECTION 3.4

Pump data

Most piping systems for conveying liquids, including mixtures of liquids and either solids or gases, incorporate pumps. Their behaviour under transient flow conditions must be recognized in any assessment of the response of the system to these changing conditions – even if the pump (or pumps) is/are not the direct cause of the changing pressures and flows. For example, in a system with an emergency shutdown valve that has been activated well downstream of the pumps, the changing flow will cause a change in delivery pressure as the operating point of the pump moves up its performance characteristic.

At the time of the preliminary assessment of the fluid transient hazards to which a given system may be exposed, it is often the case that little more than the intended operating point of the proposed pump(s) is known. While this may be adequate for an initial assessment of the system, and the design changes and/or operational strategies that may be proposed for alleviating them, such limited data are generally insufficient for the full computer-based analyses that will be undertaken to optimize the suppression and control strategies being recommended.

The pump data that are most needed, especially at the time of specifying computer studies, are the pump performance characteristics and the inertia of the rotating parts. It will also be noted that the head, flow and speed changes that occur may go well beyond the normal range of the available data. This section aims to provide a data source to help remedy this need. The first part deals with the performance characteristics, and the later subsection provides a means of estimating the inertia of the rotating parts of both the pumps and their driving motors.

3.4.1 Pump Performance Characteristics

For situations where pumps continue to run at constant speed, while transient events are initiated elsewhere, the pressure heads and flows will usually remain within the normal zone of operation. One approach that may be sufficient and is sometimes used to represent the head–flow characteristic when it is expected to remain within the normal zone is in the form

$$H = a^2 + bQ + cQ^2 \tag{3.56}$$

where a, b, and c are constants determined by the shape of the H–Q curve.

More often, however, it is the transient events associated with pump start-up and shutdown (both intended and through fault conditions) that are to be analysed. In such cases, the computer modelling must provide for changes in speed and flow, including the reversal of both, and rapidly changing pressure heads between the upstream and downstream sides.

Of the many methods that have been developed and refined for this purpose, the procedure that has become the one most favoured by transient analysts for computer codes is that devised by Suter [Marchel *et al.* (1965) and Suter (1966)].

The main difficulty, however, is being able to provide the necessary data covering all four zones of pump operation, i.e. ± speed and ± flow. For many years only four sets of complete data were in the general literature, but these have more recently been extended to 14 [Thorley and Chaudry (1996)] covering a range of pumps of various Shape Numbers. These data are presented in this section in both graphical and tabular form.

The key features of the method attributed to Suter are summarized below as background to help explain the format in which the data are presented.

The usual form of the non-dimensional flow, head and power coefficients for rotodynamic pumps are

$$\frac{Q}{ND^3} = \text{constant}, \qquad \frac{gH}{N^2D^2} = \text{constant}, \qquad \frac{P}{\rho_f N^3 D^5} = \text{constant} \tag{3.57}$$

Additional terms are diameter and rotational speed of the impeller, D and N respectively, power P, gravitational acceleration g and fluid density ρ_f. For present purposes, since power is the multiple of torque T and the rotational speed N, the power coefficient can be rewritten as

$$\frac{P}{\rho_f N^3 D^5} = \frac{TN}{\rho_f N^3 D^5} = \frac{T}{\rho_f N^2 D^5} = \text{constant} \tag{3.58}$$

The reason for replacing 'power' with the product of torque and speed is that it is more useful in solving the full set of equations for the pump boundary condition in computer codes since they contain the equation of motion for the pump which relates the deceleration of the pump impeller to its inertia and the applied torque.

Simplified forms of these flow, head and power coefficients for machines of specific types are obtained by moving constant terms (e.g. ρ_f, g, D) into the 'constant' on the right-hand sides of the above equations to give

$$\frac{Q}{N} = \text{constant}, \qquad \frac{gH}{N^2} = \text{constant}, \qquad \frac{T}{N^2} = \text{constant} \tag{3.59}$$

These are usually known as the affinity laws but are, of course, no longer non-dimensional. This can be overcome for present purposes by dividing through the various parameters by the appropriate values at rated conditions, i.e.

$$h = \frac{H}{H_R}, \qquad q = \frac{Q}{Q_R}, \qquad \alpha = \frac{N}{N_R}, \qquad \beta = \frac{T}{T_R} \tag{3.60}$$

The affinity laws in equation (3.59) may now be rewritten in the form

$$\frac{q}{\alpha} = \text{constant}, \qquad \frac{h}{\alpha^2} = \text{constant}, \qquad \frac{\beta}{\alpha^2} = \text{constant} \tag{3.61}$$

Some early computer codes made use of these parameters plotted directly as

$$\frac{h}{\alpha^2} \text{ versus } \frac{q}{\alpha} \qquad \text{and} \qquad \frac{\beta}{\alpha^2} \text{ versus } \frac{q}{\alpha} \tag{3.62}$$

The great disadvantage of this form is that at points of interest in transient flow analyses, when the flow and speed pass through zero as they change direction, these parameters contain discontinuities as the denominators pass through zero.

The key advance that Suter made was to utilize the fact that the flow and speed do not normally reverse at the same time. By incorporating both the non-dimensional flow and speed in the denominators of the head and torque equations, he was able to avoid divisions by zero and proposed modifying the dimensionless terms as follows (retaining Suter's nomenclature)

$$W(h) = \text{sign of } h\sqrt{\frac{|h|}{\alpha^2 + q^2}} \tag{3.63}$$

$$W(m) = \text{sign of } \beta\sqrt{\frac{|\beta|}{\alpha^2 + q^2}} \tag{3.64}$$

$W(h)$ and $W(m)$ are the revised dimensionless head and torque, both of which Suter plotted against θ, where θ is defined as

$$\theta = \tan^{-1}\left(\frac{\alpha}{q}\right) \tag{3.65}$$

The advantage of this formulation is that the head and torque characteristics for rotodynamic pumps can be represented by smooth continuous lines (see Figs 3.37–3.42) in the range $0 \leqslant \theta \leqslant 2\pi$ covering all zones of operation as defined in the following table:

$Q+\ N+$	$0 \leqslant \theta \leqslant \pi/2$
$Q-\ N+$	$\pi/2 \leqslant \theta \leqslant \pi$
$Q-\ N-$	$\pi \leqslant \theta \leqslant 3\pi/2$
$Q+\ N-$	$3\pi/2 \leqslant \theta \leqslant 2\pi$

Square roots for $W(h)$ and $W(m)$ were proposed in order to provide greater accuracy at low numbers. It should, however, be noted that some texts [e.g. Chaudhry (1987), Swaffield and Boldy (1993) and Wylie and Streeter (1993)] do not, in fact, follow this strategy and omit the square root as this makes it slightly

Fig. 3.37 *Head characteristics for pumps with Shape Numbers in the range 0.46–1.61*

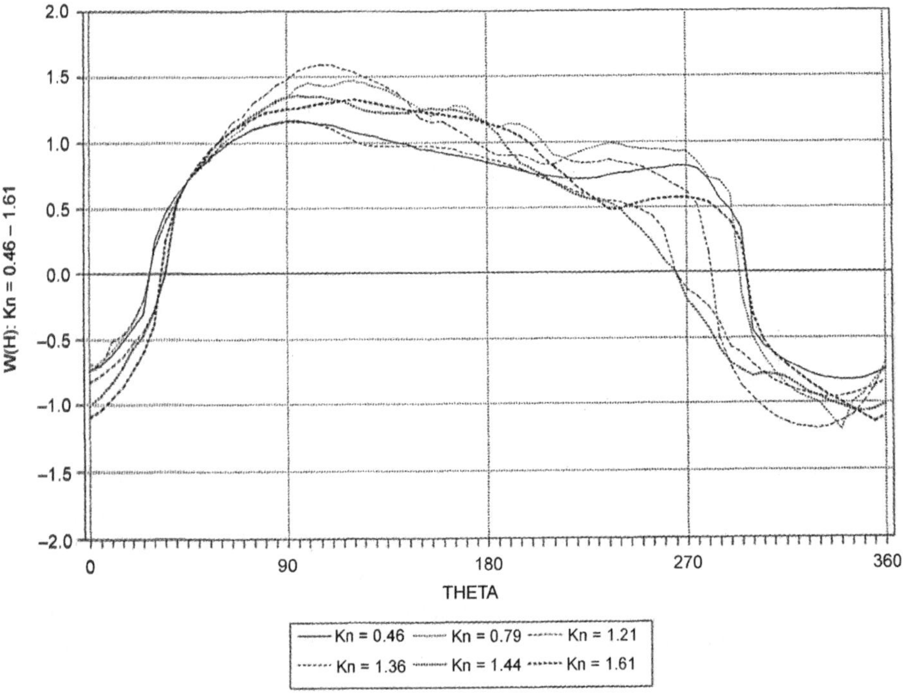

Fig. 3.38 *Torque characteristics for pumps with Shape Numbers in the range 0.46–1.61*

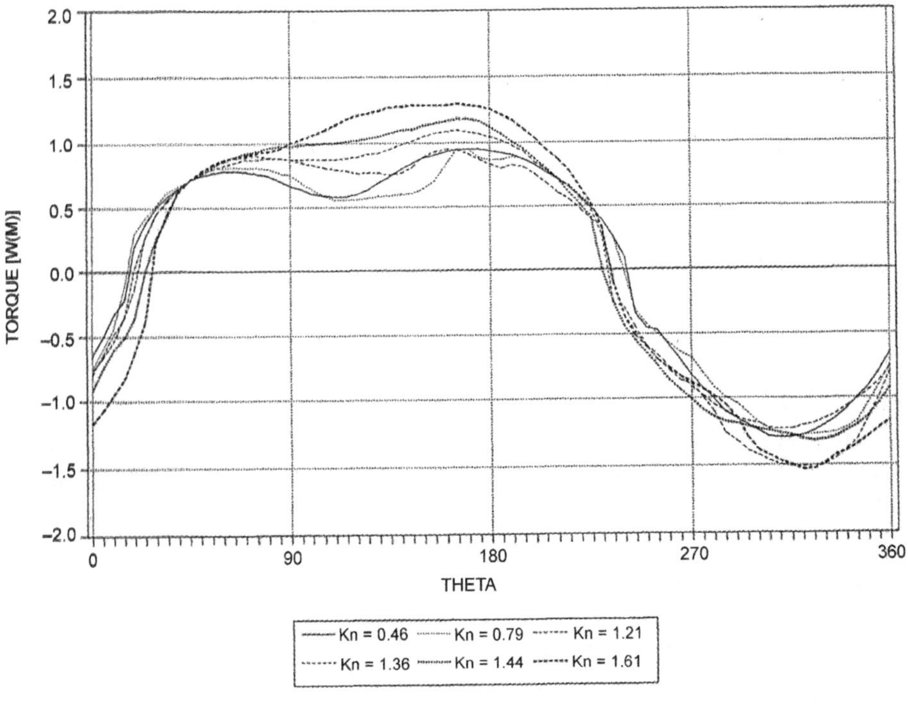

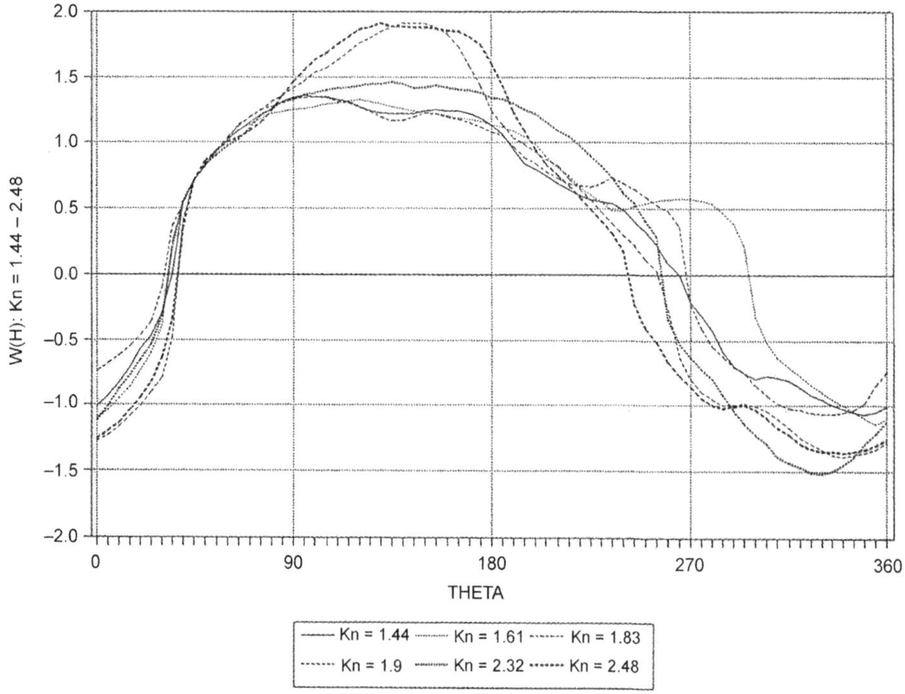

Fig. 3.39 *Head characteristics for pumps with Shape Numbers in the range 1.44–2.48*

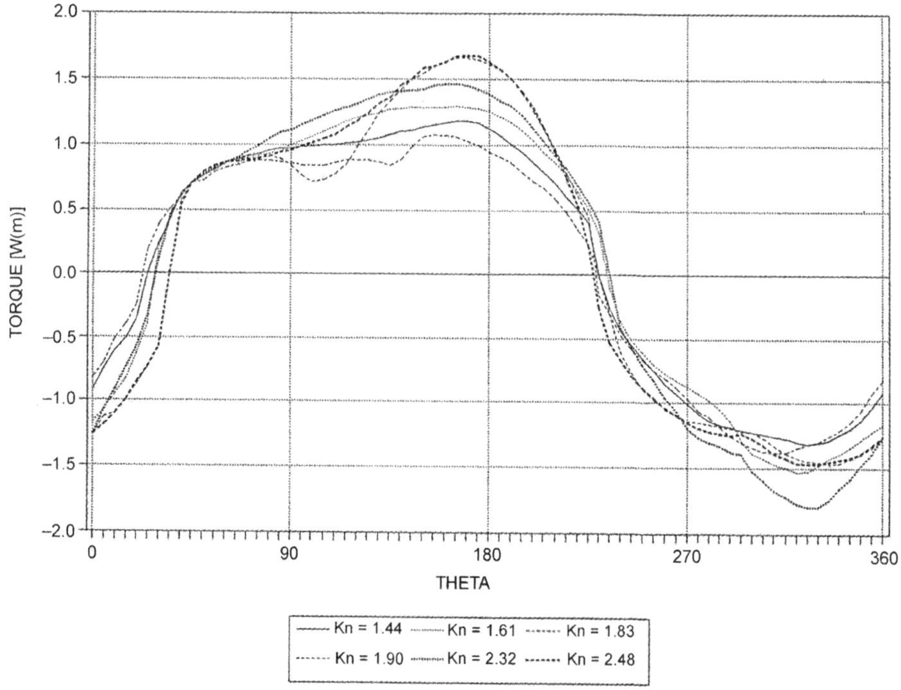

Fig. 3.40 *Torque characteristics for pumps with Shape Numbers in the range 1.44–2.48*

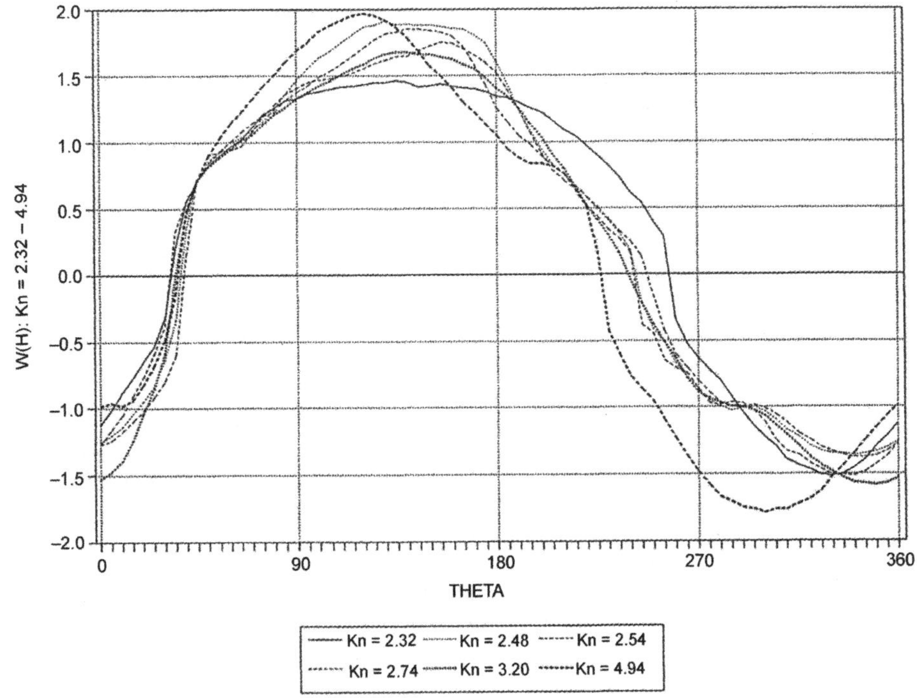

Fig. 3.41 *Head characteristics for pumps with Shape Numbers in the range 2.32–4.94*

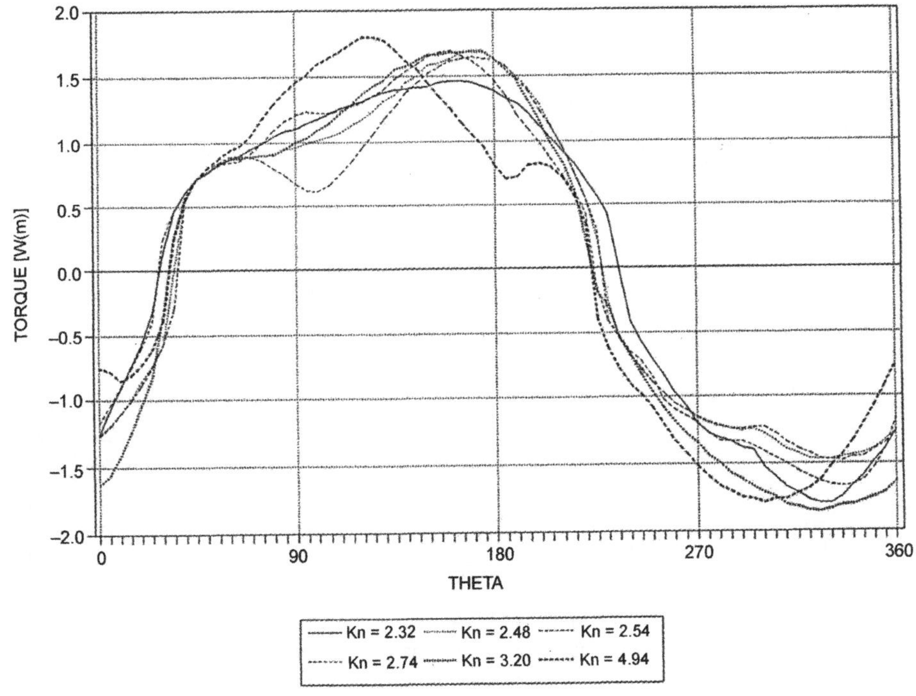

Fig. 3.42 *Torque characteristics for pumps with Shape Numbers in the range 2.32–4.94*

easier to incorporate this procedure into transient flow codes. Note also that some analysts define θ as the inverse of equation (3.65), i.e.

$$\theta = \tan^{-1}\left(\frac{q}{\alpha}\right) \quad (3.66)$$

In this text, however, the original definitions of Suter are used.

As previously mentioned, 14 sets of data are included here. How to choose between them if data for pumps of interest for a specific project are not available? In collating these data, another parameter known as the Shape Number was also determined, as this should, in principle, help define the pump type, i.e. radial, mixed or axial flow.

Hence, if no other data are available and these curves become the only resource available, the procedure is to determine the probable Shape Number of the pump(s) to be used and select the most appropriate set of curves. The Shape Number, K_n, in its preferred non-dimensional form is

$$K_n = \frac{\omega Q^{1/2}}{(gH)^{3/4}} \quad (3.67)$$

The only new term here is ω, the angular speed in terms of rad/s. Some writers and pump manufacturers use a simplified form of equation (3.67) by dropping the acceleration due to gravity term and using the speed of rotation (in r/min) to give the specific speed N_s

$$N_s = \frac{NQ^{1/2}}{H^{3/4}} \quad (3.68)$$

This is not, of course, non-dimensional and its numerical value depends very much on the system of units used. It is mentioned here so that readers may be reminded of it, but all datasets in this text are referred to the form given in equation (3.67) for which low values relate to purely radial flow (centrifugal) machines whereas high values relate to machines tending towards axial flow through the impeller.

Good data for real pumps are extremely difficult to acquire owing to the expense and time required to collate them. However, complete sets have been accumulated [Thorley and Chaudry (1996)] for all zones of operation for 14 pumps over a range of shape numbers extending between 0.46 and 4.94.

For ease of use, the data are presented in both tabular and graphical form. In tabular form the data are presented for intervals of 5° over the complete range of 0–360°.

In graphical format, the 14 sets of head and torque characteristics are distributed across three graphs each containing six sets. This is to reconcile the need for clear diagrams with the aim of continuity over the whole range.

It had been hoped that a better correlation between the general drift of the curves and shape numbers would have been apparent. Although the correlation is quite good in the normal zone of operation, it is less so outside this zone and further research is desirable. Meanwhile, users of this data should find them a useful resource, including the option either of using the dataset having the nearest

Fig. 3.43 *Tabulated data, Suter style, for head and torque characteristics for rotodynamic machines with Shape Numbers in the range 0.46–1.83*

	Kn = 0.46		Kn = 0.79		Kn = 1.21		Kn = 1.36		Kn = 1.44		Kn = 1.61		Kn = 1.83	
THETA	W(H)	W(M)	W(H)	W(M)	W(H)	W(M)	W(H)	W(M)	W(H)	W(M)	W(H)	W(M)	W(H)	W(M)
0	-0.74	-0.66	-0.74	-0.75	-0.69	-0.80	-0.83	-0.76	-1.01	-0.92	-1.10	-1.16	-0.74	-0.82
5	-0.69	-0.51	-0.66	-0.60	-0.71	-0.65	-0.77	-0.67	-0.93	-0.75	-1.03	-1.07	-0.68	-0.70
10	-0.62	-0.33	-0.58	-0.45	-0.52	-0.49	-0.71	-0.54	-0.83	-0.61	-0.95	-0.95	-0.62	-0.51
15	-0.52	-0.22	-0.49	-0.11	-0.46	-0.36	-0.63	-0.35	-0.71	-0.51	-0.86	-0.83	-0.55	-0.40
20	-0.41	0.20	-0.38	0.30	-0.37	-0.13	-0.54	0.04	-0.56	-0.36	-0.73	-0.63	-0.46	-0.25
25	-0.30	0.37	-0.20	0.42	-0.21	0.27	-0.44	0.27	-0.47	0.00	-0.60	-0.39	-0.36	0.20
30	0.24	0.50	0.18	0.52	0.19	0.45	-0.27	0.46	-0.29	0.25	-0.39	0.22	-0.10	0.40
35	0.47	0.58	0.40	0.62	0.41	0.56	0.00	0.55	0.00	0.42	0.24	0.46	0.37	0.51
40	0.61	0.65	0.54	0.66	0.57	0.65	0.56	0.65	0.54	0.63	0.54	0.62	0.54	0.61
45	0.71	0.71	0.71	0.71	0.71	0.71	0.71	0.71	0.71	0.71	0.71	0.71	0.71	0.71
50	0.80	0.74	0.79	0.75	0.81	0.78	0.81	0.76	0.82	0.76	0.84	0.77	0.81	0.73
55	0.88	0.77	0.87	0.80	0.92	0.82	0.89	0.79	0.93	0.81	0.94	0.83	0.93	0.79
60	0.95	0.78	0.98	0.81	1.03	0.85	0.97	0.82	1.01	0.86	1.02	0.86	1.01	0.82
65	1.01	0.78	1.05	0.81	1.15	0.88	1.02	0.84	1.09	0.88	1.09	0.89	1.14	0.85
70	1.06	0.77	1.11	0.81	1.20	0.89	1.07	0.87	1.16	0.92	1.14	0.90	1.19	0.87
75	1.10	0.76	1.19	0.81	1.30	0.90	1.10	0.87	1.21	0.94	1.18	0.92	1.25	0.88
80	1.13	0.74	1.25	0.80	1.36	0.89	1.13	0.88	1.26	0.96	1.22	0.93	1.28	0.88
85	1.15	0.71	1.30	0.77	1.43	0.88	1.15	0.88	1.31	0.97	1.24	0.95	1.30	0.87
90	1.16	0.66	1.35	0.75	1.48	0.87	1.16	0.87	1.34	0.98	1.25	0.99	1.33	0.86
95	1.17	0.64	1.42	0.69	1.54	0.85	1.16	0.87	1.36	0.98	1.26	1.03	1.34	0.85
100	1.16	0.61	1.45	0.64	1.57	0.83	1.15	0.87	1.35	0.99	1.28	1.06	1.35	0.85
105	1.14	0.59	1.43	0.58	1.59	0.81	1.14	0.87	1.35	0.99	1.29	1.10	1.34	0.85
110	1.13	0.58	1.43	0.56	1.59	0.79	1.10	0.87	1.33	1.00	1.30	1.14	1.31	0.86
115	1.10	0.58	1.46	0.56	1.56	0.78	1.07	0.88	1.31	1.01	1.32	1.18	1.30	0.89
120	1.08	0.60	1.47	0.56	1.53	0.76	1.02	0.88	1.28	1.02	1.32	1.20	1.27	0.89
125	1.06	0.63	1.45	0.57	1.49	0.77	0.98	0.91	1.24	1.03	1.31	1.22	1.24	0.89
130	1.04	0.69	1.42	0.58	1.46	0.77	0.98	0.92	1.23	1.06	1.30	1.25	1.20	0.87
135	1.02	0.73	1.40	0.59	1.42	0.75	0.97	0.96	1.22	1.08	1.28	1.27	1.17	0.85
140	0.99	0.79	1.35	0.60	1.37	0.76	0.98	0.98	1.22	1.10	1.26	1.28	1.17	0.89
145	0.98	0.84	1.31	0.61	1.29	0.81	0.98	1.02	1.22	1.11	1.25	1.28	1.20	0.98
150	0.96	0.88	1.25	0.64	1.18	0.89	0.97	1.06	1.24	1.13	1.23	1.29	1.23	1.06
155	0.94	0.91	1.20	0.70	1.15	0.93	0.97	1.07	1.25	1.15	1.22	1.29	1.22	1.08
160	0.92	0.93	1.24	0.83	1.16	0.95	0.96	1.09	1.24	1.16	1.21	1.29	1.20	1.07
165	0.91	0.94	1.28	0.95	1.10	0.94	0.95	1.10	1.24	1.18	1.19	1.29	1.18	1.06
170	0.89	0.95	1.27	0.95	1.06	0.91	0.92	1.08	1.22	1.18	1.18	1.29	1.16	1.03
175	0.87	0.95	1.20	0.88	1.00	0.86	0.90	1.07	1.19	1.16	1.16	1.28	1.12	1.00
180	0.84	0.94	1.15	0.86	0.94	0.83	0.87	1.05	1.14	1.13	1.14	1.25	1.08	0.95
185	0.82	0.92	1.11	0.86	0.90	0.80	0.86	1.02	1.06	1.07	1.12	1.22	1.05	0.91
190	0.81	0.91	1.14	0.89	0.90	0.82	0.82	0.97	0.95	1.01	1.09	1.17	0.97	0.86
195	0.78	0.86	1.13	0.87	0.90	0.81	0.80	0.93	0.84	0.95	1.05	1.11	0.89	0.79
200	0.76	0.82	1.09	0.82	0.87	0.75	0.76	0.84	0.80	0.87	0.99	1.04	0.84	0.73
205	0.74	0.77	1.02	0.78	0.83	0.69	0.73	0.79	0.75	0.80	0.89	0.95	0.78	0.68
210	0.73	0.71	0.90	0.71	0.83	0.61	0.69	0.72	0.69	0.71	0.81	0.87	0.73	0.59
215	0.73	0.65	0.88	0.66	0.87	0.55	0.66	0.65	0.64	0.61	0.74	0.77	0.69	0.49
220	0.72	0.57	0.91	0.58	0.85	0.48	0.63	0.56	0.60	0.52	0.66	0.65	0.67	0.36
225	0.72	0.49	0.95	0.50	0.84	0.40	0.60	0.47	0.57	0.41	0.61	0.52	0.66	0.24
230	0.73	0.40	0.97	0.39	0.85	0.35	0.57	0.28	0.55	0.00	0.55	0.33	0.69	-0.01
235	0.74	0.26	0.99	0.26	0.86	-0.10	0.55	-0.26	0.54	-0.25	0.49	-0.10	0.74	-0.30
240	0.75	0.10	0.98	-0.02	0.85	-0.35	0.54	-0.30	0.49	-0.44	0.49	-0.36	0.69	-0.47
245	0.77	-0.35	0.96	-0.31	0.83	-0.48	0.52	-0.45	0.40	-0.55	0.52	-0.51	0.65	-0.58
250	0.78	-0.46	0.95	-0.42	0.81	-0.60	0.48	-0.59	0.32	-0.65	0.54	-0.61	0.59	-0.67
255	0.79	-0.47	0.94	-0.50	0.77	-0.70	0.43	-0.67	0.23	-0.75	0.56	-0.70	0.53	-0.75
260	0.80	-0.59	0.94	-0.57	0.72	-0.78	0.33	-0.76	0.10	-0.86	0.57	-0.77	0.49	-0.81
265	0.81	-0.71	0.93	-0.65	0.66	-0.86	-0.01	-0.83	0.00	-0.94	0.57	-0.83	0.37	-0.89
270	0.81	-0.82	0.93	-0.69	0.62	-0.89	-0.13	-0.91	-0.21	-1.01	0.57	-0.88	-0.22	-0.97
275	0.79	-0.92	0.85	-0.81	0.52	-0.99	-0.19	-0.97	-0.31	-1.09	0.56	-0.93	-0.41	-1.06
280	0.71	-1.01	0.73	-0.91	0.15	-1.10	-0.26	-1.05	-0.42	-1.14	0.54	-0.98	-0.55	-1.11
285	0.57	-1.10	0.70	-0.99	-0.50	-1.25	-0.38	-1.09	-0.59	-1.17	0.47	-1.05	-0.62	-1.19
290	0.48	-1.15	0.60	-1.04	-0.72	-1.32	-0.57	-1.14	-0.70	-1.19	0.39	-1.14	-0.70	-1.25
295	0.33	-1.20	-0.16	-1.11	-0.87	-1.39	-0.63	-1.17	-0.75	-1.21	0.22	-1.29	-0.75	-1.31
300	-0.45	-1.25	-0.50	-1.19	-0.96	-1.43	-0.70	-1.19	-0.79	-1.23	-0.32	-1.39	-0.85	-1.34
305	-0.56	-1.28	-0.65	-1.23	-1.04	-1.47	-0.78	-1.22	-0.77	-1.24	-0.52	-1.43	-0.92	-1.37
310	-0.62	-1.29	-0.76	-1.27	-1.10	-1.49	-0.83	-1.23	-0.78	-1.26	-0.63	-1.47	-0.98	-1.38
315	-0.69	-1.29	-0.86	-1.28	-1.14	-1.51	-0.87	-1.23	-0.81	-1.28	-0.71	-1.50	-1.02	-1.38
320	-0.73	-1.28	-0.93	-1.28	-1.17	-1.51	-0.90	-1.20	-0.86	-1.30	-0.77	-1.53	-1.03	-1.35
325	-0.77	-1.25	-0.97	-1.27	-1.18	-1.50	-0.93	-1.19	-0.92	-1.31	-0.84	-1.53	-1.05	-1.32
330	-0.80	-1.20	-1.00	-1.27	-1.19	-1.49	-0.94	-1.15	-0.95	-1.31	-0.89	-1.48	-1.06	-1.30
335	-0.81	-1.15	-1.10	-1.25	-1.17	-1.45	-0.96	-1.12	-0.99	-1.29	-0.95	-1.43	-1.06	-1.26
340	-0.82	-1.09	-1.20	-1.22	-1.13	-1.38	-0.95	-1.07	-1.02	-1.24	-1.00	-1.40	-1.05	-1.21
345	-0.82	-1.00	-1.00	-1.17	-1.09	-1.31	-0.94	-1.01	-1.05	-1.20	-1.05	-1.34	-1.03	-1.15
350	-0.81	-0.91	-0.93	-1.04	-1.00	-1.15	-0.90	-0.94	-1.06	-1.13	-1.10	-1.28	-0.98	-1.06
355	-0.78	-0.80	-0.82	-0.90	-0.88	-0.99	-0.87	-0.87	-1.05	-1.04	-1.14	-1.22	-0.85	-0.93
360	-0.74	-0.66	-0.74	-0.75	-0.69	-0.80	-0.83	-0.76	-1.01	-0.92	-1.10	-1.16	-0.74	-0.82

	Kn = 1.9		Kn = 2.32		Kn = 2.48		Kn = 2.54		Kn = 2.74		Kn = 3.2		Kn = 4.94	
THETA	W(H)	W(M)	W(H)	W(M)	W(H)	W(M)	W(H)	W(M)	W(H)	W(M)	W(H)	W(M)	W(H)	W(M)
0	-1.28	-1.27	-1.12	-1.26	-1.25	-1.26	-1.27	-1.27	-1.27	-1.17	-1.53	-1.64	-0.98	-0.75
5	-1.23	-1.13	-1.01	-1.07	-1.19	-1.19	-1.24	-1.19	-1.16	-1.04	-1.47	-1.56	-0.96	-0.78
10	-1.16	-1.10	-0.88	-0.91	-1.13	-1.11	-1.18	-1.10	-1.05	-0.91	-1.39	-1.41	-0.98	-0.85
15	-1.09	-0.99	-0.77	-0.74	-1.04	-0.99	-1.09	-1.01	-0.91	-0.75	-1.24	-1.24	-0.94	-0.81
20	-1.01	-0.88	-0.66	-0.57	-0.94	-0.85	-1.01	-0.89	-0.77	-0.61	-1.04	-1.04	-0.82	-0.73
25	-0.90	-0.73	-0.53	-0.33	-0.82	-0.73	-0.91	-0.75	-0.59	-0.40	-0.84	-0.83	-0.68	-0.62
30	-0.79	-0.57	-0.32	0.13	-0.64	-0.57	-0.79	-0.58	-0.37	0.24	-0.56	-0.35	-0.49	-0.39
35	-0.48	0.00	0.21	0.46	-0.34	0.00	-0.60	-0.28	0.33	0.47	0.00	0.27	-0.14	0.24
40	0.41	0.58	0.55	0.62	0.35	0.54	0.16	0.56	0.56	0.61	0.45	0.58	0.49	0.55
45	0.71	0.71	0.71	0.71	0.71	0.71	0.71	0.71	0.71	0.71	0.71	0.71	0.71	0.71
50	0.83	0.79	0.82	0.76	0.85	0.79	0.84	0.79	0.90	0.77	0.81	0.77	0.89	0.80
55	0.94	0.84	0.90	0.82	0.93	0.84	0.92	0.84	0.93	0.82	0.89	0.82	1.03	0.87
60	1.04	0.87	0.96	0.86	1.00	0.86	1.01	0.88	0.94	0.84	0.95	0.86	1.14	0.94
65	1.13	0.89	1.02	0.90	1.04	0.87	1.07	0.90	0.96	0.85	1.01	0.88	1.22	0.97
70	1.20	0.91	1.11	0.95	1.09	0.88	1.14	0.89	1.07	0.91	1.06	0.89	1.32	1.05
75	1.25	0.93	1.20	0.99	1.16	0.89	1.19	0.86	1.19	0.99	1.14	0.89	1.41	1.18
80	1.31	0.92	1.25	1.05	1.25	0.91	1.22	0.80	1.28	1.10	1.20	0.90	1.50	1.29
85	1.36	0.89	1.31	1.09	1.36	0.94	1.31	0.74	1.36	1.17	1.26	0.94	1.59	1.37
90	1.42	0.85	1.32	1.11	1.46	0.96	1.38	0.67	1.41	1.21	1.33	1.00	1.68	1.45
95	1.47	0.77	1.36	1.15	1.55	0.99	1.42	0.62	1.45	1.24	1.37	1.03	1.75	1.51
100	1.53	0.72	1.38	1.18	1.63	1.02	1.48	0.62	1.47	1.23	1.43	1.10	1.82	1.59
105	1.57	0.73	1.39	1.22	1.69	1.04	1.53	0.66	1.48	1.23	1.46	1.15	1.87	1.64
110	1.63	0.77	1.42	1.26	1.75	1.08	1.59	0.75	1.49	1.24	1.51	1.22	1.92	1.68
115	1.69	0.86	1.42	1.28	1.80	1.13	1.65	0.85	1.52	1.28	1.55	1.28	1.95	1.74
120	1.75	0.99	1.43	1.31	1.86	1.18	1.71	0.96	1.55	1.30	1.58	1.36	1.97	1.80
125	1.80	1.11	1.45	1.35	1.88	1.24	1.77	1.06	1.59	1.35	1.63	1.42	1.95	1.80
130	1.85	1.24	1.45	1.38	1.90	1.32	1.81	1.18	1.61	1.40	1.66	1.47	1.92	1.77
135	1.87	1.33	1.46	1.39	1.89	1.36	1.83	1.27	1.64	1.47	1.68	1.55	1.86	1.70
140	1.90	1.42	1.45	1.41	1.89	1.45	1.85	1.38	1.65	1.53	1.68	1.57	1.78	1.61
145	1.90	1.49	1.42	1.42	1.88	1.50	1.85	1.45	1.69	1.59	1.66	1.61	1.69	1.55
150	1.90	1.54	1.42	1.42	1.88	1.58	1.84	1.52	1.72	1.65	1.66	1.66	1.57	1.45
155	1.87	1.60	1.44	1.46	1.87	1.60	1.83	1.56	1.75	1.68	1.64	1.66	1.50	1.35
160	1.83	1.63	1.42	1.46	1.85	1.63	1.80	1.61	1.74	1.69	1.62	1.68	1.40	1.25
165	1.73	1.66	1.41	1.46	1.84	1.67	1.72	1.63	1.70	1.65	1.58	1.68	1.30	1.15
170	1.58	1.66	1.40	1.46	1.80	1.67	1.58	1.65	1.66	1.62	1.55	1.69	1.22	1.05
175	1.43	1.64	1.38	1.42	1.74	1.67	1.43	1.63	1.59	1.55	1.48	1.69	1.13	0.96
180	1.24	1.61	1.34	1.39	1.61	1.63	1.25	1.62	1.52	1.47	1.40	1.63	1.04	0.81
185	1.14	1.56	1.34	1.34	1.44	1.57	1.15	1.58	1.39	1.36	1.34	1.55	0.95	0.70
190	1.04	1.49	1.30	1.29	1.23	1.50	1.05	1.49	1.24	1.20	1.26	1.47	0.88	0.72
195	0.98	1.39	1.25	1.22	1.10	1.40	0.98	1.39	1.07	1.10	1.14	1.32	0.84	0.81
200	0.93	1.26	1.22	1.12	0.95	1.28	0.89	1.25	0.92	0.98	1.04	1.17	0.84	0.82
205	0.87	1.12	1.17	1.03	0.84	1.14	0.83	1.13	0.79	0.86	0.91	0.98	0.82	0.80
210	0.82	0.96	1.08	0.92	0.75	0.94	0.75	0.95	0.71	0.73	0.80	0.83	0.76	0.71
215	0.72	0.79	1.05	0.82	0.67	0.75	0.67	0.75	0.64	0.60	0.67	0.62	0.64	0.57
220	0.65	0.61	0.98	0.70	0.59	0.54	0.59	0.55	0.53	0.42	0.53	0.21	0.51	0.35
225	0.57	0.46	0.90	0.57	0.50	0.22	0.51	0.29	0.44	-0.17	0.42	-0.19	0.17	-0.39
230	0.48	-0.16	0.82	0.42	0.41	-0.25	0.42	-0.41	0.35	-0.41	0.30	-0.30	-0.42	-0.62
235	0.39	-0.31	0.74	0.00	0.32	-0.53	0.33	-0.55	0.28	-0.53	0.17	-0.54	-0.61	-0.78
240	0.30	-0.60	0.62	-0.42	0.19	-0.67	0.25	-0.68	0.17	-0.66	-0.04	-0.69	-0.77	-0.90
245	0.23	-0.77	0.54	-0.57	-0.23	-0.80	0.14	-0.79	-0.37	-0.73	-0.23	-0.83	-0.86	-0.98
250	0.12	-0.90	0.41	-0.71	-0.42	-0.89	-0.15	-0.88	-0.45	-0.85	-0.38	-0.96	-0.95	-1.08
255	0.03	-0.98	0.28	-0.83	-0.52	-0.97	-0.43	-0.97	-0.65	-1.01	-0.51	-1.08	-1.09	-1.21
260	-0.26	-1.04	-0.34	-0.97	-0.67	-1.05	-0.60	-1.05	-0.70	-1.10	-0.63	-1.18	-1.23	-1.32
265	-0.63	-1.10	-0.53	-1.07	-0.78	-1.11	-0.70	-1.10	-0.74	-1.14	-0.78	-1.28	-1.36	-1.42
270	-0.80	-1.14	-0.63	-1.20	-0.85	-1.16	-0.80	-1.15	-0.87	-1.20	-0.86	-1.34	-1.48	-1.52
275	-0.90	-1.15	-0.72	-1.27	-0.94	-1.19	-0.88	-1.19	-0.97	-1.27	-0.93	-1.44	-1.58	-1.59
280	-0.95	-1.16	-0.80	-1.31	-1.00	-1.22	-0.96	-1.21	-0.98	-1.32	-0.96	-1.49	-1.67	-1.67
285	-0.99	-1.18	-0.92	-1.34	-1.02	-1.23	-1.00	-1.23	-0.96	-1.33	-0.99	-1.57	-1.71	-1.71
290	-0.99	-1.20	-1.04	-1.38	-1.01	-1.25	-1.00	-1.24	-0.97	-1.33	-1.00	-1.62	-1.75	-1.75
295	-0.98	-1.22	-1.14	-1.40	-1.00	-1.24	-0.98	-1.23	-1.02	-1.36	-1.03	-1.68	-1.76	-1.76
300	-1.01	-1.23	-1.22	-1.52	-1.03	-1.28	-1.00	-1.23	-1.11	-1.41	-1.06	-1.72	-1.79	-1.79
305	-1.04	-1.27	-1.30	-1.60	-1.09	-1.32	-1.05	-1.27	-1.24	-1.45	-1.13	-1.75	-1.76	-1.76
310	-1.10	-1.33	-1.39	-1.68	-1.16	-1.39	-1.11	-1.34	-1.32	-1.49	-1.21	-1.80	-1.76	-1.76
315	-1.17	-1.39	-1.44	-1.72	-1.22	-1.43	-1.19	-1.39	-1.36	-1.56	-1.30	-1.83	-1.72	-1.72
320	-1.24	-1.42	-1.47	-1.77	-1.29	-1.46	-1.25	-1.42	-1.42	-1.59	-1.37	-1.86	-1.69	-1.69
325	-1.30	-1.44	-1.51	-1.79	-1.33	-1.47	-1.30	-1.46	-1.47	-1.63	-1.44	-1.86	-1.62	-1.62
330	-1.34	-1.46	-1.51	-1.79	-1.35	-1.47	-1.34	-1.47	-1.51	-1.66	-1.48	-1.84	-1.52	-1.52
335	-1.37	-1.47	-1.50	-1.75	-1.35	-1.45	-1.36	-1.48	-1.51	-1.67	-1.53	-1.82	-1.44	-1.44
340	-1.39	-1.46	-1.46	-1.66	-1.36	-1.44	-1.38	-1.45	-1.52	-1.66	-1.57	-1.80	-1.34	-1.33
345	-1.38	-1.43	-1.39	-1.58	-1.35	-1.42	-1.37	-1.44	-1.49	-1.62	-1.57	-1.78	-1.25	-1.21
350	-1.36	-1.39	-1.30	-1.48	-1.33	-1.39	-1.36	-1.40	-1.43	-1.53	-1.58	-1.75	-1.15	-1.07
355	-1.33	-1.35	-1.21	-1.39	-1.30	-1.33	-1.33	-1.35	-1.36	-1.39	-1.57	-1.72	-1.06	-0.92
360	-1.28	-1.27	-1.12	-1.26	-1.25	-1.26	-1.27	-1.27	-1.27	-1.17	-1.63	-1.64	-0.98	-0.75

Fig. 3.44 *Tabulated data, Suter style, for head and torque characteristics for rotodynamic machines with Shape Numbers in the range 1.9–4.94*

shape number to the machine in which they are interested or interpolating between the curves on either side.

3.4.2 Moment of Inertia of Pumps and Motors

The combined inertia of pumps and the motors driving them, including the connecting shafts and couplings, is often required for computer-based analyses of transient flow situations associated with the starting and stopping of pumps. Unfortunately, reliable information on these parameters is frequently not available at the time that many transient studies are undertaken.

The charts and equations presented in this section are intended to provide an initial guide to the inertia values that may be used as a reasonable first approximation, when more accurate data are not available. The discussion that follows is developed from work that was initiated by Faithfull (1989 and 1990) coupled with more recent experience of modern pumping installations.

3.4.2.1 Pump Inertias

Data from several pump manufacturers are plotted in Fig. 3.45. These cover a wide range of rotodynamic pumps used in the water supply, sewage, process and petrochemical fields. They include horizontal spindle, single- and double-entry, split-case machines as well as vertical spindle borehole and wet-well pumps.

The general style of the various impellers is radial and mixed flow. The only group of data points that clearly separates out from the others applies to a small, quite slim type of machine that is very strongly radial flow in design. Current indications are that this may be an evolving trend among more recent designs of machine that utilize lighter impellers, and is noted if appropriate.

Over 300 data points are plotted in Fig. 3.45 in the form of inertia, including entrained water, against a Power Coefficient. The 'Power Coefficient' in the present context is defined as (P/N^3) where P denotes the shaft power (in kW) supplied to the pump at its rated condition and maximum efficiency, and N is the rotational speed (in thousands of r/min).

The rationale for this correlation follows from the homologous laws described earlier in this section for rotodynamic pumps, plus an assumption that the inertia is related to their size and, hence, power requirements. From equation (3.57), the full non-dimensional power coefficient is

$$\frac{P}{\rho_f N^3 D^5} = \text{constant} \tag{3.69}$$

The inertia I may be defined in terms of the mass, m, of a rotating component and its radius of gyration, k, i.e.

$$I = mk^2$$

From the laws of geometric similarity, this relationship may be re-expressed as follows

$$I \propto (\rho_m D^3) \times D^2, \qquad \text{i.e. } I \propto \rho_m D^5 \tag{3.70}$$

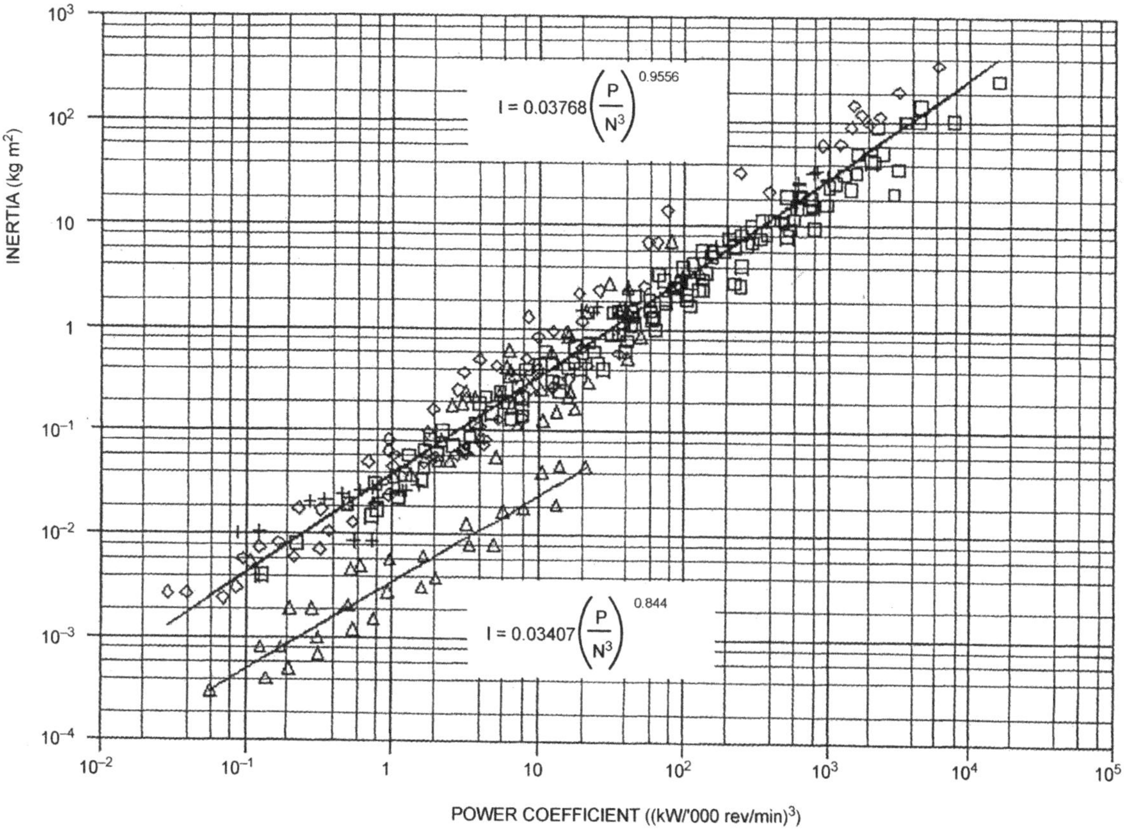

Fig. 3.45 *Moment of inertia of impellers, including entrained water, for pumps of radial and mixed flow design. The upper set of data covers single- and double-entry impellers, single and multistage, and horizontal and vertical, spindle machines. The lower set is for relatively small, single-entry, radial flow impellers of a lightweight design*

where

ρ_m = density of the impeller material

Using equation (3.70) to replace D in equation (3.69), and merging the two densities into the constant, leads to

$$I \propto \frac{P}{N^3} \tag{3.71}$$

The only reason for using N in terms of thousands of r/min is to have convenient numbers for the graphs and equations.

From linear regression analyses of the data points shown in Fig. 3.45, the following equations were developed for predicting the inertia I of pump impellers, including the entrained water and the shaft on which the impeller is mounted:

(a) The main bulk of the data, comprising some 284 data points provided by five pump manufacturers, and covering the variety of pump types described above, yielded

$$I = 0.037\,68 \times \left(\frac{P}{N^3}\right)^{0.9556} \tag{3.72}$$

(b) A set of 28 data points for one particular type of pump from one manufacturer did not fall in line with the general mass of data, and these have been characterized by a separate equation

$$I = 0.034\,07 \times \left(\frac{P}{N^3}\right)^{0.844} \tag{3.73}$$

This last equation applies to relatively small pumps of a lightweight design and for which the maximum inertia is only 0.0465 kg m^2. It does suggest, however, that there is a secondary influence on the inertia due to Shape Number.

Equations (3.72) and (3.73) were developed from linear regression analyses of the logarithmic plots and gave correlation coefficients of 0.960 and 0.903 respectively. These are quite high, encouragingly so, considering the variety of manufacturers and pump types.

Despite these apparently good correlations, it will be noted from the graphs that the actual range of inertias above and below the mean line is of the order of +100 and –50 per cent. This will only be important in those systems where the rate at which pumps change speed is significant, such as in networks or short pipelines of, say, 5 km or less. If this is a matter of concern, it can easily be checked by an analysis with the predicted inertia and then, say, doubling and halving it.

To estimate the pump inertia for a particular application, it is necessary to estimate the shaft power and rotational speed. Shaft power P is related to the operating head H and flowrate Q through the efficiency equation

$$\eta = \frac{\rho_f g Q H}{P} \tag{3.74}$$

and so shaft P power is

$$P = \frac{\rho_f g Q H}{\eta} \tag{3.75}$$

Ideally, the operating conditions H_o and Q_o will be the rated conditions H_R and Q_R. For a given application, the design conditions H_o and Q_o, will be known, but an assumption of the likely efficiency of the pump will be required. For a modern pump this will normally be in the range 70–85 per cent, with the higher values being associated with the larger machines. Sewage and other pumps designed to handle solids will have lower efficiencies.

More judgement will be required concerning the pump speed N. The speeds of the pumps represented by the data points in Fig. 3.45 were, typically, 590, 740, 980, 1450 and 2900 r/min, which is one cause of the scatter in the data. The most common speeds were 1450 and 2900 r/min, with 740 r/min being the next most common. If there is uncertainty over the likely running speed, it may be prudent to consider inertias for two, or perhaps three, possible speeds for a particular case.

An example calculation follows:

Design flowrate Q_o 0.25 m^3/s
Operating head H_o 70 m
Assumed efficiency η 0.72
Assumed speed $N1$ 450 r/min

$$P = \frac{\rho_f g Q H}{\eta} = \frac{1000 \times 9.81 \times 0.25 \times 70}{0.72 \times 1000} = 238.44\,\text{kW}$$

From equation (3.72)

$$I = 0.037\,68 \times \left(\frac{238.44}{1.45^3}\right)^{0.9556} = 2.428\,\text{kg}\,\text{m}^2$$

which can be taken as 2.45 kg m^2 as the basic prediction for the pump. The motor inertia must be added to this, with allowance made for any gearing connecting the two together.

3.4.2.2 Motor Inertias

Motor inertia data are plotted in a similar fashion to pump data, except that the speed is not cubed, i.e. the inertia I (kg m^2) is correlated with (P/N), as shown in Fig. 3.46.

The linear regression of 272 data points yields the following equation, having a correlation coefficient of 0.97

$$I = 0.0043 \times \left(\frac{P}{N}\right)^{1.48} \tag{3.76}$$

where, as above, P denotes the shaft power (in kW) and N is the speed (in thousands of r/min).

The data follow the linear regression equation quite well. However, it will be

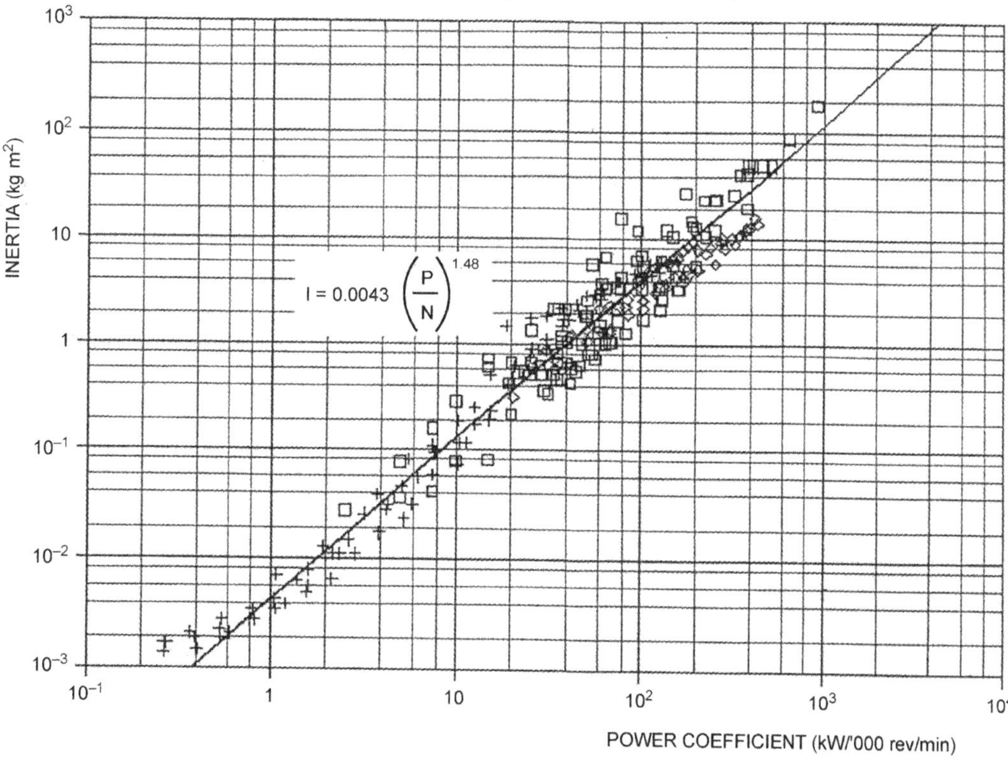

Fig. 3.46 *Moment of inertia for electric motors of various types and from different manufacturers as used for driving rotodynamic pumps*

seen from the graph that there is a tendency to drift away from the regression line for very small motors, i.e. at inertias of 0.002–0.005 kg m^2.

An interesting point about this set of data is that, in addition to the data collected originally by Linton (1972), about 60 per cent of the data points are for modern motors. This not only extends the range of data but also indicates that, contrary to popular belief, there is not a significant difference between the inertias of motors of old and new designs.

From the point of view of collating data for a transient analysis, it may be noted that, while pump manufacturers often have difficulty in providing a value for the inertia of pumps, motor manufacturers can usually do so provided they are given the power and speed. Nevertheless, equation (3.76) is quite adequate initially.

Continuing the example calculation from above, the estimated inertia of the motor will be

$$I = 0.0043 \times \left(\frac{238.44}{1.45}\right)^{1.48} = 8.2\,\mathrm{kg\,m^2}$$

The total estimated inertia for this pump/motor unit would therefore be in the region of 10.65 kg m^2. This could be rounded up to, say, 11 kg m^2 to make some allowance for the coupling.

As previously stated, it should be recognized that this is only a prediction until such time as the true data become available. If it is felt that the magnitude of the inertia is important, the predicted value should be doubled and halved to give a reasonable indication of the range within which the actual value is likely to fall. Furthermore, any design and operating conditions proposed in order to minimize fluid transient effects should be based on the worst-case scenario – normally the lower values of predicted inertia

$$P = \frac{\rho g Q H}{\eta} = \frac{1000 \times 9.81 \times 0.25 \times 70}{0.72 \times 1000} = 238.44\,\mathrm{kW}$$

SECTION 3.5

Pressure rises following valve closure

When a valve at the downstream end of a pipeline is closed in a time less than a pipeline period T $(= 2L/c)$, the increase in pressure in the vicinity of the valve is

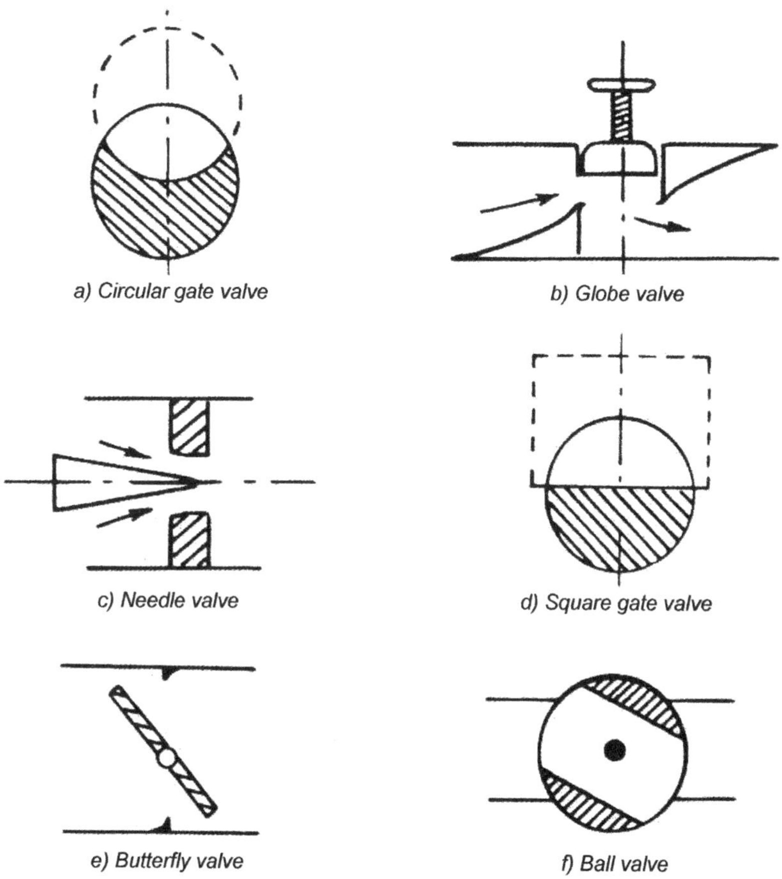

Fig. 3.47 *Schematics of the common valve types referred to in Figs 3.48 to 3.55*

easily found from Joukowsky's equation [equations (3.2) and (3.3)]. How far this increase propagates upstream from the valve depends on the duration of the actual closure as a proportion of the pipeline period. When valve closures extend over more than a pipeline period, giving time for reflected pressure waves to return to the valve, the maximum pressures depend very much on the type of valve and the manner in which it is closed.

A number of approximate formulae are available in the literature for estimating the pressure changes, but they are generally inaccurate and should be treated with caution. An alternative approach is a selection of charts for various types of valve in common use [Wood and Jones (1972)]. The principal valve types are shown schematically in Fig. 3.47.

Most valves exhibit non-linear relationships between the area of the opening and the position of the valve spindle, as shown in Fig. 3.48. Consequently, the change in flow through the valves and the associated pressure drop are also highly non-linear. For example, many circular gate valves will be as much as 85 per cent closed before there is a 15–20 per cent reduction in flow.

The charts reproduced, from Wood and Jones (1972), as Figs 3.49 to 3.55 enable estimates to be made of the maximum pressures on the upstream side of a closing valve at the downstream end of a pipeline.

A constraint on the application of the charts is that they were devised for a

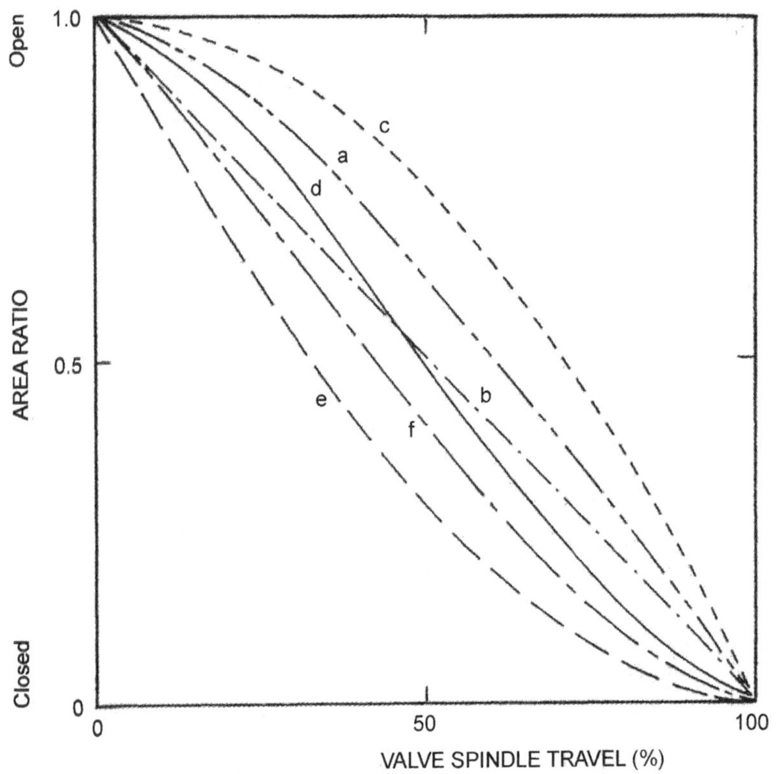

Fig. 3.48 *Variation in area ratios with percentage of opening for the six valve types shown in Fig. 3.47*

pipeline with a negligible frictional head loss. To make some allowance for the line friction that occurs in real pipe systems, some proportion of the head loss should be added to the predicted maximum pressure change due to the transients to improve the estimate of the overall peak pressure. It should also be borne in mind that, ultimately, the final pressure head on the upstream side of the valve will be the static head – either from a reservoir or the shut-off head of a pump that continues running.

The charts were, in fact, developed assuming that the upstream source was a constant head reservoir. However, they are also suitable for lines supplied by pumps, provided that the *H–Q* characteristic for the pump is fairly flat between the operating point and the shut-off head.

The curves shown in Figs 3.49 to 3.55 are of dimensionless maximum transient pressures and pressure heads, ΔH_m and Δp_m, plotted against a dimensionless valve closure time t_c.

The pressure changes, rendered non-dimensional by dividing through by the Joukowsky pressure change, are defined as

$$\Delta H_m = \frac{gH_{max}}{cv_o} \quad \text{and} \quad \Delta p_m = \frac{p_{max}}{\rho c v_o} \tag{3.77}$$

where H_{max} and p_{max} are the maximum pressure head, or pressure, rise upstream of the valve.

The actual valve closure time, T_c, is converted to a non-dimensional form with

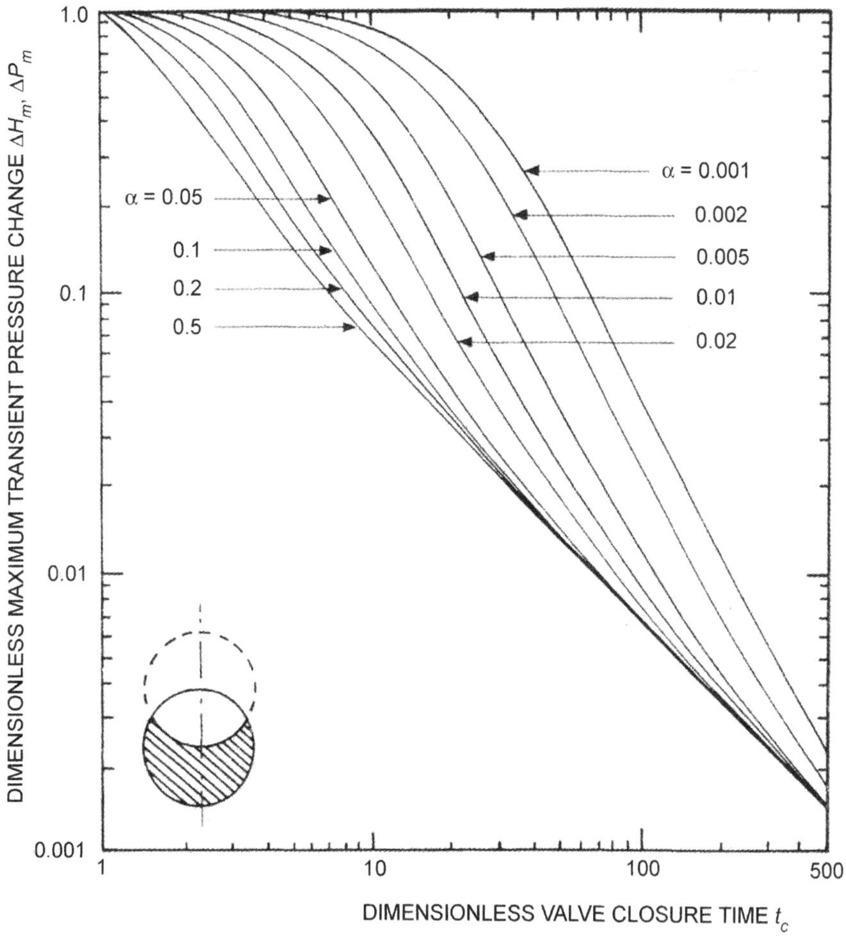

Fig. 3.49 *Maximum change in pressure in a pipeline owing to closure of a circular gate valve at the downstream end with linear movement of the valve spindle*

the pipeline period $(2L/c)$, i.e.

$$t_c = \frac{T_c}{2L/c} \tag{3.78}$$

The pipeline upstream of the valve is represented by a dimensionless initial condition parameter, α, defined as

$$\alpha = \frac{gH_o}{cv_o} \tag{3.79}$$

where H_o is the head drop across the valve under the initial steady flow conditions.

With the exception of Fig. 3.55, all the charts are based on a linear rate of valve spindle movement, and are not applicable therefore to two-stage closures.

The procedure for using the charts is as follows, and is demonstrated in Part 1:

1. Select the chart that is most appropriate for the valve in question.
2. Evaluate equations (3.78) and (3.79) for the parameters t_c and α.

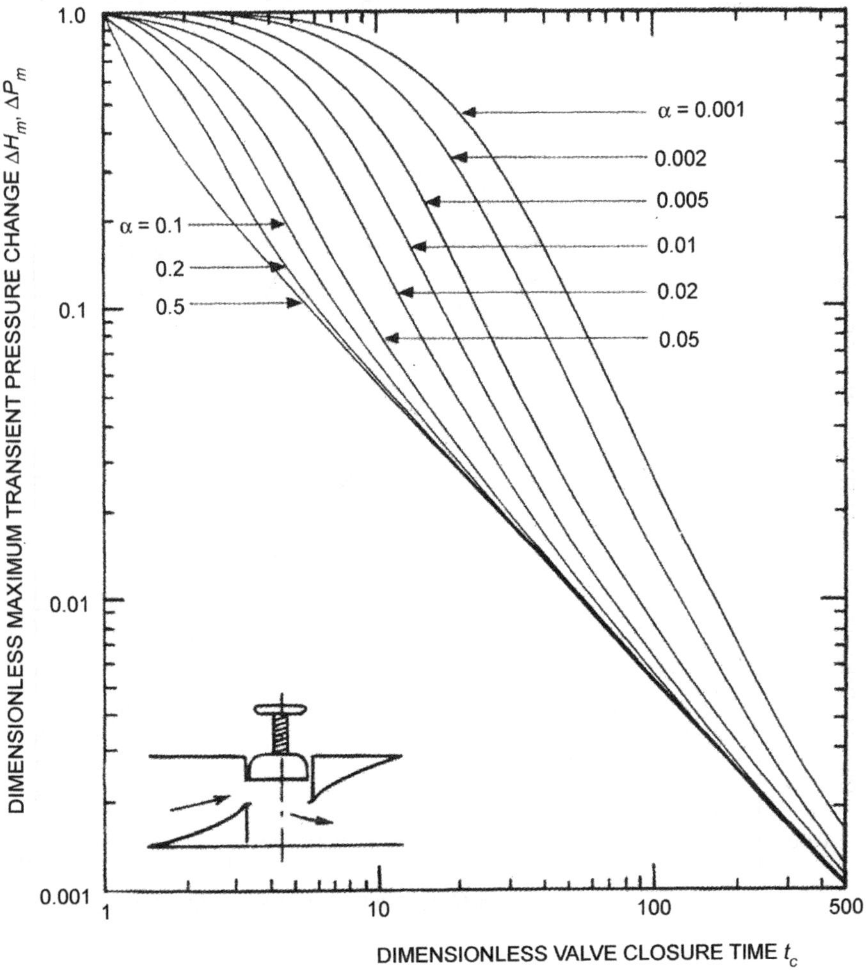

Fig. 3.50 *Maximum change in pressure in a pipeline owing to closure of a globe valve at the downstream end with linear movement of the valve spindle*

3. Read up from the baseline, i.e. from t_c, to the relevant curve for α, interpolating if necessary.
4. Then read across to take off a value for ΔH_m (or Δp_m).
5. Re-arranging equation (3.77), calculate H_{max} or p_{max}.

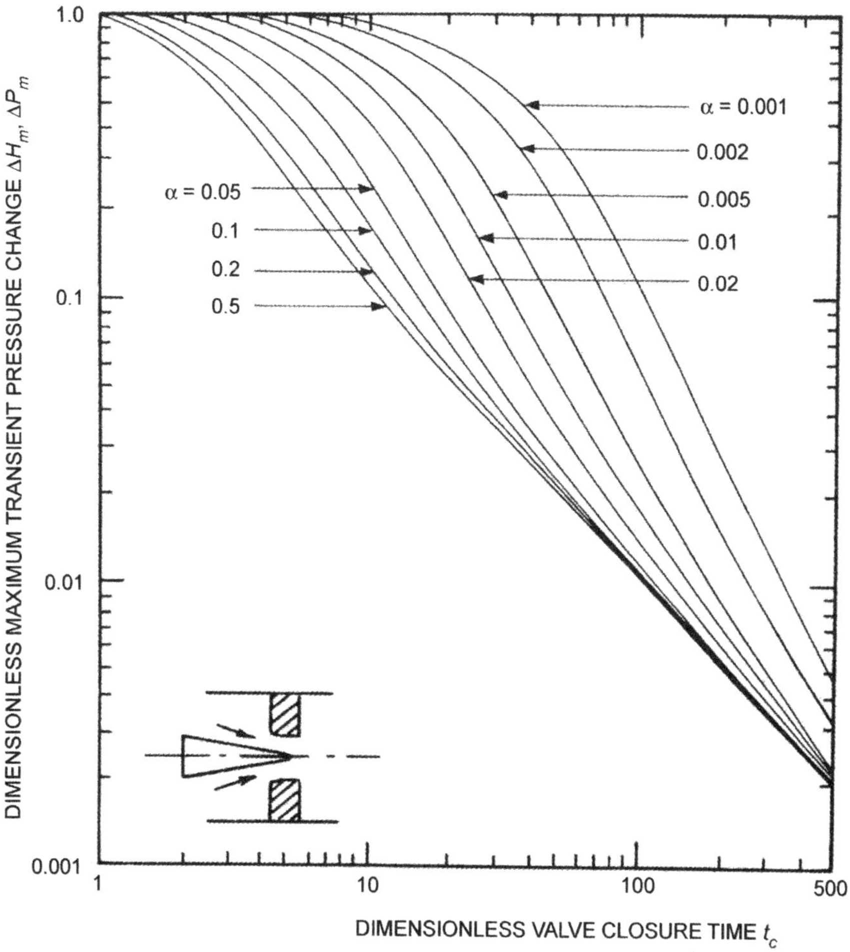

Fig. 3.51 *Maximum change in pressure in a pipeline owing to closure of a needle valve at the downstream end with linear movement of the valve spindle*

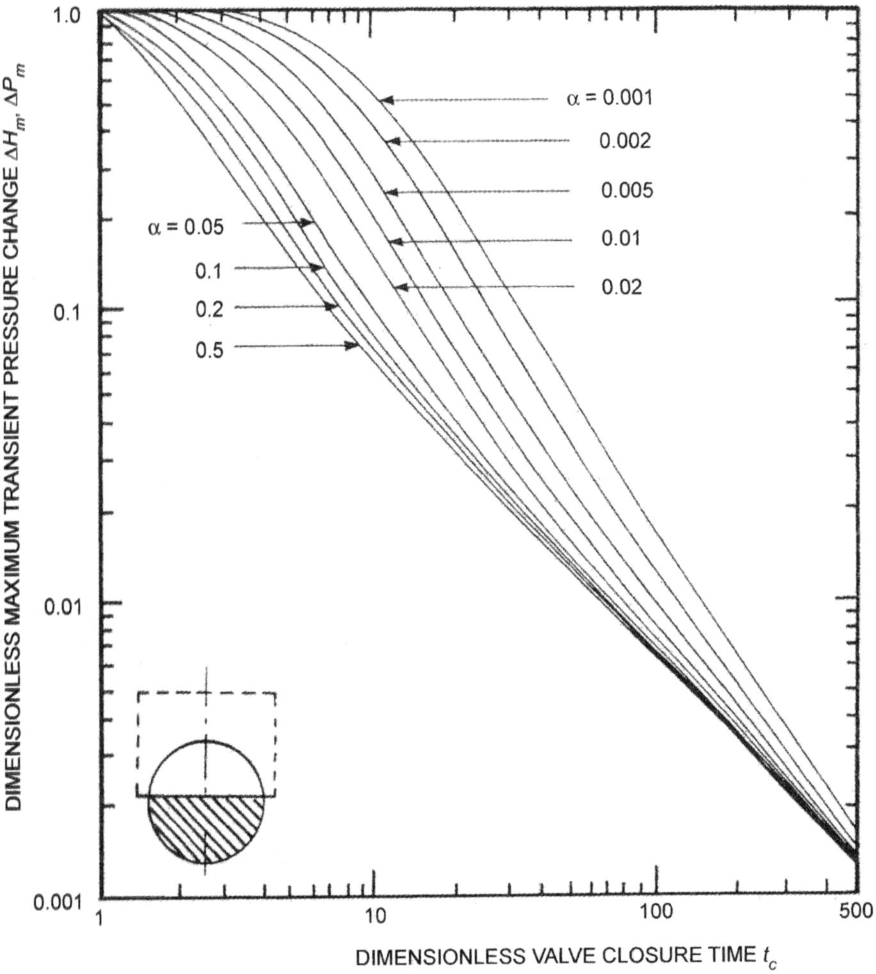

Fig. 3.52 *Maximum change in pressure in a pipeline owing to closure of a square gate valve at the downstream end with linear movement of the valve spindle*

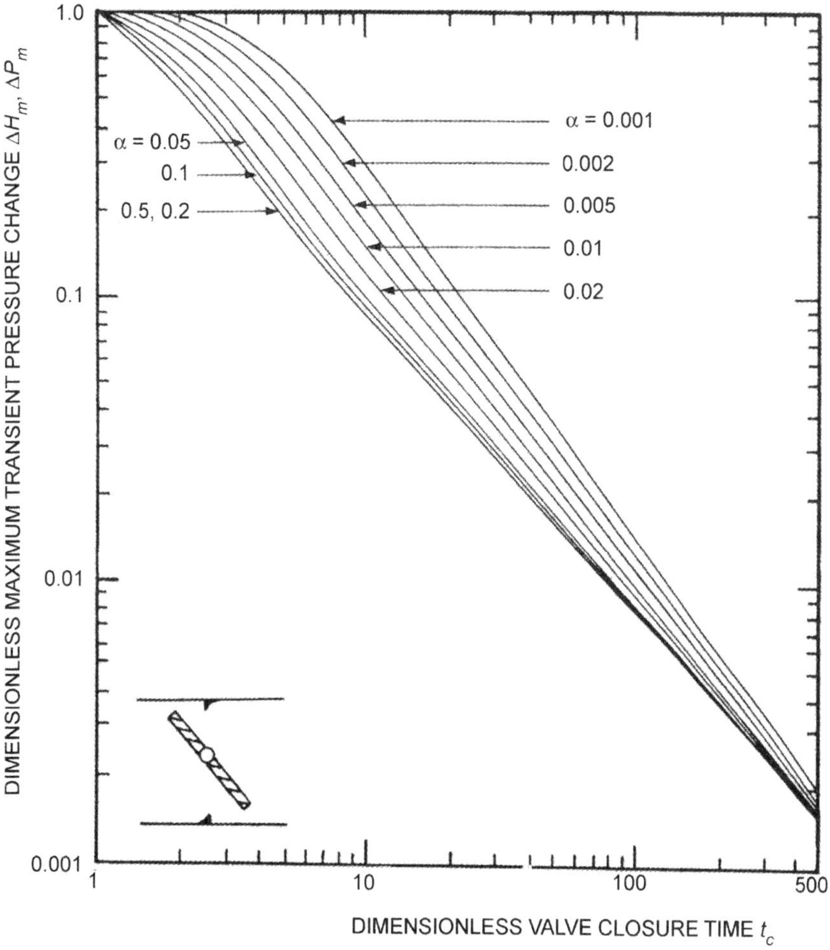

Fig. 3.53 *Maximum change in pressure in a pipeline owing to closure of a butterfly valve at the downstream end with linear movement of the valve spindle*

Fig. 3.54 *Maximum change in pressure in a pipeline owing to closure of a ball valve at the downstream end with linear movement of the valve spindle*

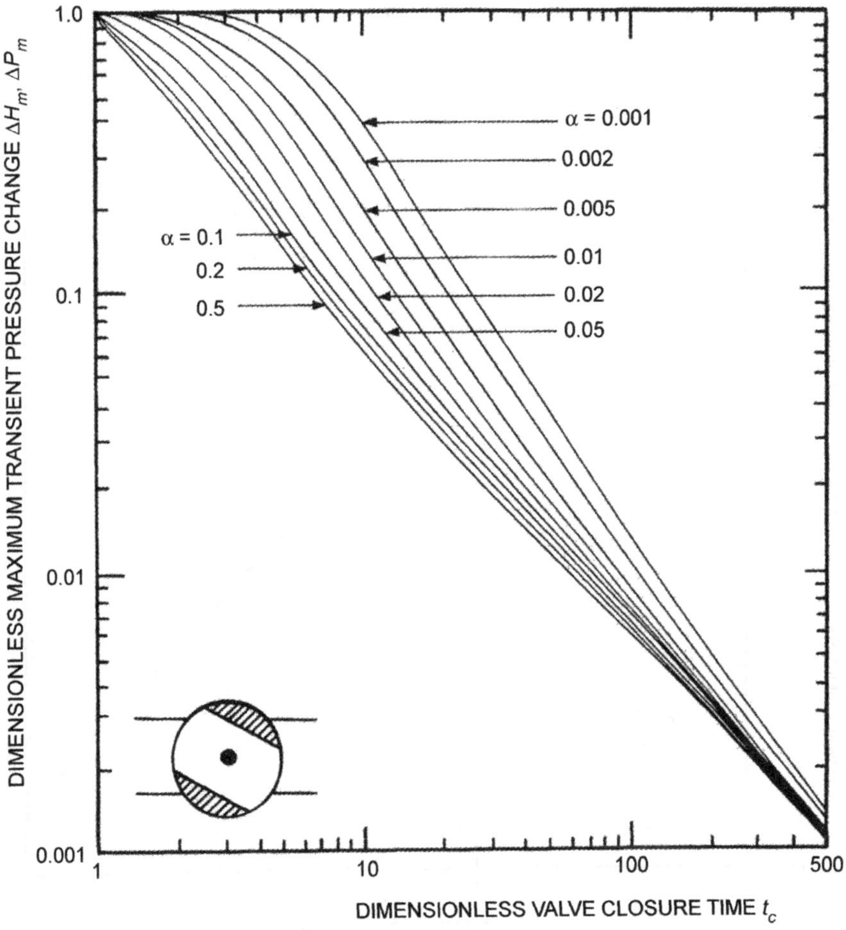

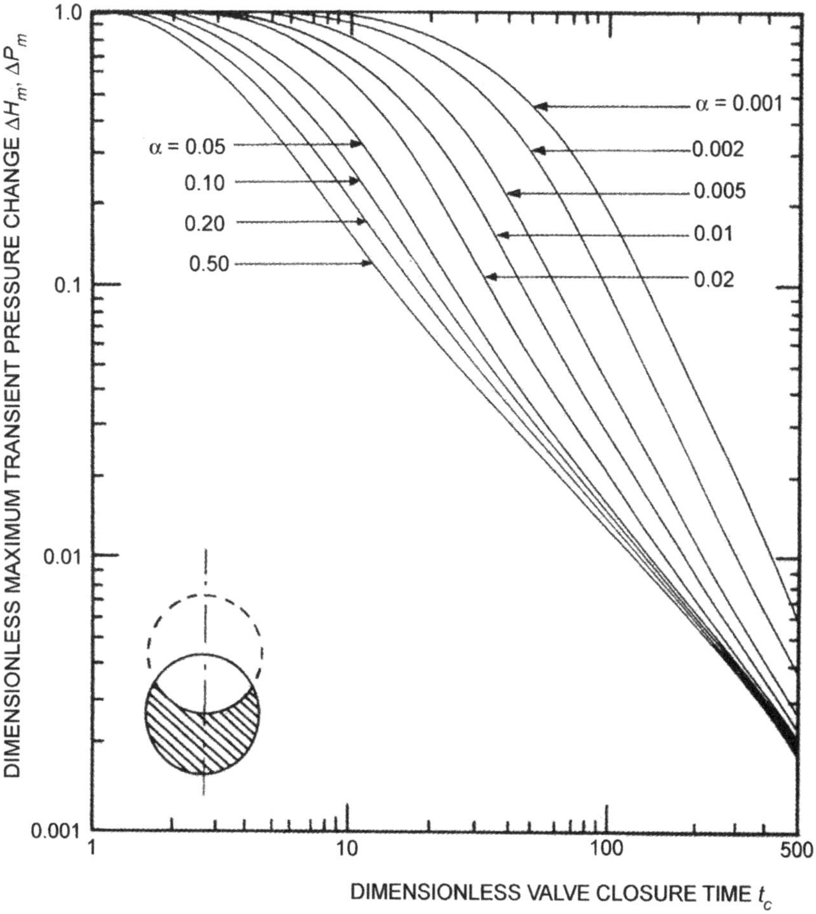

Fig. 3.55 *Maximum change in pressure in a pipeline owing to closure of a circular gate valve at the downstream end with accelerated movement of the valve spindle*

SECTION 3.6

Air relief and vacuum-breaking valves

Valves for air admission, air release and vacuum breaking are available in a proliferation of different types and styles. After check valves, they are perhaps the system component over which there is most confusion and uncertainty when it comes to selecting the valve or valves most appropriate for a given system. Part of this confusion arises because the valves have two rather different functions to fulfil, and it is important to be able to distinguish between them since incorrect choices can create more problems than they solve.

The two principal reasons why engineers consider incorporating air and vacuum valves in a pipeline system are for ventilation and the control and suppression of fluid transients.

The ventilation requirements are threefold:

- rapid expulsion of air when filling the system;
- rapid admission of air when draining the system;
- slow release during normal operation, under pressure, of air and other gases that are given up by the fluid, or that enter the system at low-pressure points.

Virtually all pipelines are fitted with air valves for ventilation to help ensure their efficient operation since the presence of air, even under normal conditions, can inhibit the flow and, should it move around the system, give rise to unacceptable shock loads at bends and junctions, etc.

The control and suppression of fluid transients with the aid of vacuum-breaking and air release valves usually requires that:

- air will be admitted rapidly at appropriate points in the system to limit subatmospheric pressures;
- the subsequent behaviour of the air cavity will not cause any unacceptable conditions;

- the air and any released gases are vented very slowly so that the cavity created by air admission acts as a cushion when it is repressurized owing to the moving liquid column(s).

3.6.1 Ventilation of Pipelines

Ventilation is not strictly a feature of fluid transient control. However, since valves to achieve this are often fitted, their presence in a system must be recognized and their behaviour understood. It may be that, in some systems, there is a contribution that they can make in alleviating some of the risks due to fluid transients, so the following brief discussion is included as background to the later subsections.

During the commissioning (and recommissioning) of pipelines, air and vapour must be removed from the system, often as quickly as is reasonably possible. Likewise, in preparation for maintenance work, the system may need draining and, in the interests of a speedy return to service, will also be required to be completed quickly. Furthermore, during normal operation, the presence of pockets of air or vapour is generally unwelcome since they have an adverse effect on the throughput and on the economical operation of the system.

Valves used for the rapid ventilation of pipes, i.e. for filling and draining down, usually have a single large orifice and are float operated. The large orifice may be 80–100 mm in diameter, and occasionally even larger, depending on the line size and flow required. The same orifice is used for both inflow and outflow, and generally it is the float that meets the seat to close off the valve, although in some designs the float closes a disc or plug against a seat via a lever mechanism.

Note that, if the closure of the valve is accompanied with a rapid reduction in liquid flow velocity, as the float is driven up onto its seat, there is the potential for a significant, and unacceptable, pressure rise, as per the Joukowsky equation. This can occur in lines where only one air release valve is fitted, or it is the last valve to close in a line, especially on a high peak [Blum (1994) and Marcos (1998)]. Some manufacturers provide variations in their basic design to try to overcome this, such as providing air valves fitted with a vented non-return, or surge check, valve that, in effect, reduces the speed of the liquid as it enters the lower part of the valve body.

Once a float has closed onto its seat, the valve will remain shut while there is a pressure difference across it, even if the liquid level in the valve drops. To avoid the float becoming seated prematurely, on account of the outgoing air tending to roll or blow the float shut, the so-called Kinetic and Aerokinetic designs of air valve have evolved.

A consequence of the valves remaining closed when pressurized, even after the water level drops, creates a potential problem during the normal operation of a pipeline since air and/or vapour will almost certainly come out of solution or may be drawn into the system. To maintain the efficient and safe operation of the line, it is necessary for this air, gas and vapour to be removed. The type of air valve required for this is one that will provide for the slow or intermittent release of air, and valves are available for this purpose. Some designs have the float attached to an arm, with the valve plug being attached near to the fulcrum, others use a float that incorporates the orifice. The orifice sizes vary with application, but are usually within the range 1.5–12.5 mm.

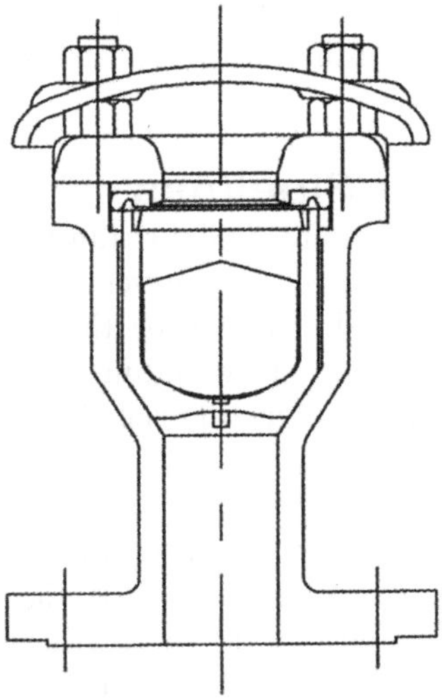

Fig. 3.56 *Typical large-orifice air valve for the rapid ventilation of pipelines*

It is often desirable that valves for the slow release of air under pressure are located at the same point as those for rapid ventilation. This may be achieved by combining in one unit an air valve having a large orifice for inflow with a release valve having a small orifice. Such valves are known as Dual-Orifice, or Combination, Air Valves. They can be selected such that the large orifice will be either open or closed during air release according to their intended function. If it is to remain open, an aerokinetic valve should be used, otherwise a 'kinetic' design is required.

The key point in this section is that air valves intended for the ventilation of pipelines have a legitimate role to play, but they may not provide adequate protection against fluid transient phenomena. In particular, some large-orifice

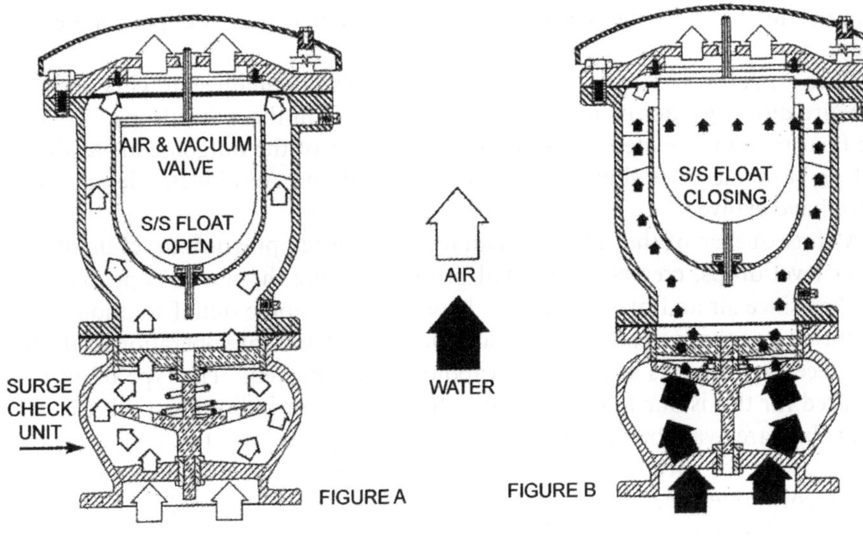

Fig. 3.57 *An example of a large-orifice valve used for rapid ventilation, fitted with a check valve to limit excessive transient pressures as filling nears completion*

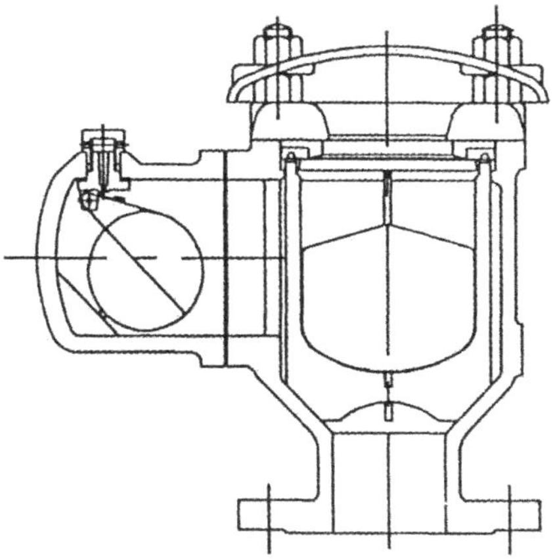

Fig. 3.58 *A typical example of a Dual-Orifice, or Combination, Air Valve. Rapid ventilation occurs through the large orifice, while air given up by the water under normal operation is released periodically through the small unit on the left*

exhaust valves may be unsuitable and can actually present a distinct hazard if located incorrectly.

Among the problems that have been known to occur are floats becoming deformed as a result of wear, leading to leakage out of the system when under pressure and contamination of the fluid in the pipe when shut down. Other examples include becoming totally inoperative as a result of the floats being jammed against, and sometimes forced through, their seats at high pressure.

3.6.2 Air Valves for Surge Control

Many design engineers are very wary of using air admission for surge control as there have been several cases reported in the literature [e.g. Arregui *et al.* (2003), AWWA (2001), Balutto (1998), Balutto and Muller (1995 and 1996), Bournaski *et al.* (2003), Burrows and Qiu (1996), Kruisbrink *et al.* (2004) and Thorley and Main (1986)] of serious failures. These have been due to pipelines bursting or the failure of components attributed to either air admission itself, malfunctioning of air valves and/or the occurrence of pressures that significantly exceed the Joukowsky pressure changes as the air or vapour cavities reduce to a small, but finite, size when the kinetic energy of the decelerating liquid columns is converted into strain energy.

It can be argued that these cases are the result of using the wrong type of valve or that, in the systems being cited, the use of vacuum-breaking and air release valves is fundamentally inappropriate. Examples would include pipelines for which the profile of the line is such that air cavities, once formed, move in an uncontrolled fashion, and systems conveying liquids that could be contaminated, or in which the air–vapour mixture may be explosive, flammable or toxic. In such cases the relevant design criterion for the fluid transient risk assessment should be that system pressures should remain at all times, say, 2 or 3 m head above atmospheric, or the vapour pressure of the fluid, whichever is the higher, at all points along the line, including under transient flow conditions.

Nevertheless, there are situations where air valves can contribute to the control and suppression of excessive fluid transient pressures. These will primarily be those systems where the normal static head on the system, or sections of it, is low, say less than 5 m head, and possibly even negative. Examples of these include:

- nearly horizontal or falling pumped mains where the pump is mainly overcoming friction;
- pipelines that will tend to drain down following a pump trip or loss of power;
- undulating lines but ones where air, if admitted, will remain close to the point of admission;
- risers on (some) deep-well and borehole pumping sets;
- effluent and slurry lines to settling ponds or treatment reservoirs.

A common feature of these examples is that the driving forces on the liquid columns adjacent to any cavities are low, except possibly the start-up operations of deep-well systems and borehole installations.

Since generally the objective is to admit air freely and then retain it for a while at a specific location within the system, it is necessary to control the rate at which the air is expelled as the separated columns of liquid move towards each other. This requires a valve, or combination of valves, that provides free flow in, but the aperture through which air enters then closes and the air is exhausted more slowly through a constricted passage.

Several designs are currently available for achieving this. One example is a valve described as a dual-orifice valve fitted with a vented non-return valve that is mounted above the large orifice and is illustrated in Fig. 3.59. Since the float is in the lower part of the valve body, it is also suitable for sewage systems.

Another example is the dual-orifice valve, i.e. the vacuum-relief/air release

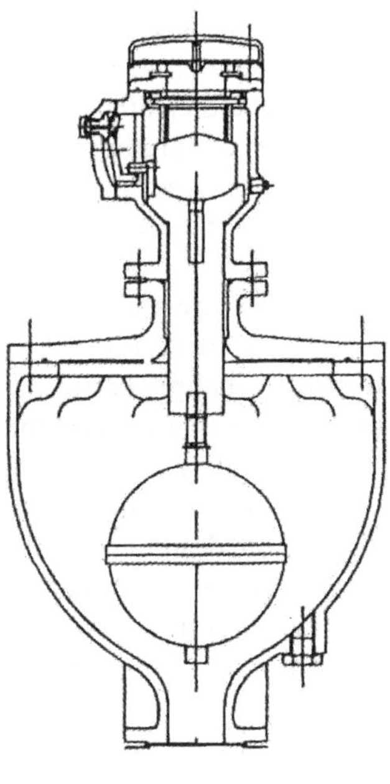

Fig. 3.59 *A dual-orifice air valve with a vented non-return valve mounted at the top. The small orifice is located below, and to the left of, the spherical float*

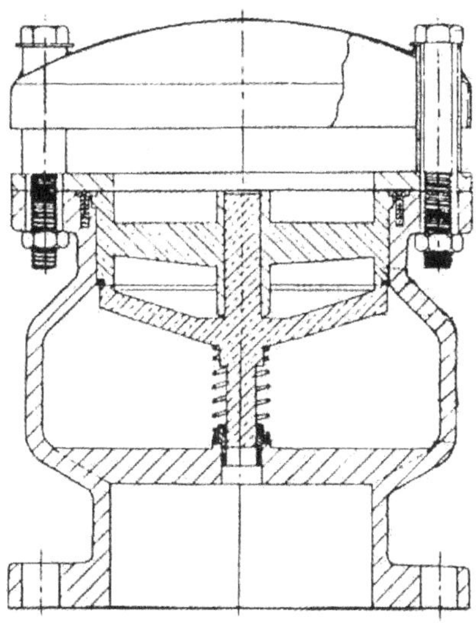

Fig. 3.60 *A vacuum-breaking valve based on a spring-loaded disc. To achieve slow air release, a small-orifice air valve is added to this unit, usually being bolted onto the side*

valve. This too is a combination valve, but in this case the moving element that permits air inflow may be a disc instead of a float. When the pressure is stabilized either side of the disc, it is closed by the action of a light spring or balance weight. The natural position of this valve with atmospheric pressure on both sides is closed, whereas a normal air admission valve would be in the open position if the local pressure and fluid were the same on both sides. The associated air release valve is float controlled and is intended to release air at a very slow rate.

A variation on the float design that makes use of three cylindrical floats is shown in Fig. 3.61. This triple-orifice design is a relatively new addition to the field [Balutto and Muller (1995 and 1996)]. The principle of operation is that all three floats drop down when rapid air admission is sought, and air flows through the annular space between the floats and the valve body.

For venting of air during refilling, air will travel in the opposite direction along the same path until, at a predetermined flowrate and as the internal pressure rises, the top float closes. From this point on, the air must pass through the smaller-diameter orifice in the centre of this float. As the last of the air is exhausted, the liquid entering the valve body lifts the lower two floats until they meet with, and seal against, the uppermost float.

The middle float of the three also has a small hole through it, smaller than that in the top one. This forms the small orifice through which air and gas that come out of solution can be vented under pressure during normal operation of the line. As this air and gas accumulate within the valve body, the bottom cylinder no longer floats and so drops away from the seal against the small orifice in the float above it. Since the pressure within the valve body will be above atmospheric, the pressure difference across the top two floats keeps them in the closed position. When this small volume of air has been released, the lower float resumes its position up against the middle float and the valve is again sealed.

Yet another variation on the theme of rapid air admission under low-pressure conditions is the hydraulically controlled air valve. Basically, this is a large-orifice

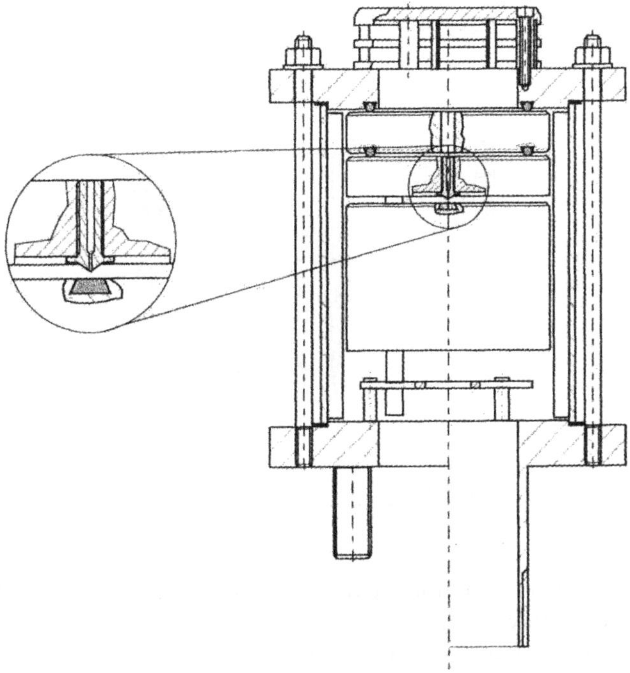

Fig. 3.61 *By using three floats in this design, rapid air admission is permitted, but air can be released more slowly. This Triple-Orifice design also provides for slow release of air that builds up during normal pipeline operation*

air valve that opens rapidly to admit air as the local pressure falls below atmospheric. However, as the liquid columns move back together, the air is permitted to leave the system relatively freely, followed by water. The flow of water, running to drain or to a convenient sump, is then shut off automatically and slowly, as a weighted plunger falls at a rate which is designed to protect the system from excessive pressures.

3.6.3 Selection and Siting of Air Valves

The basic criteria for siting air valves are:

- Valves for air admission should be installed at locations where the pressures first tend to become subatmospheric.
- Valves to release air should be installed at those points where air will tend to collect naturally. Great care must be taken in deciding whether air and gas release should be fast or slow, and then only appropriate valves (or combinations of valves) should be installed.

The locations chosen will generally be at peaks in the line and at significant changes in slope. These criteria are sometimes described as 'peaks with respect to the hydraulic grade line', rather than with respect to the usual horizontal datum.

Also, lines having long sections with a uniform gradient, which can be either falling or rising slowly, and especially when the pressure is falling along the line, can benefit from the installation of valves at regular intervals.

The initial choice of air valve for a particular location is governed by its primary function. However, the way it interacts with the rest of the system, of which it is an integral part, must be considered – particularly with regard to transient flows.

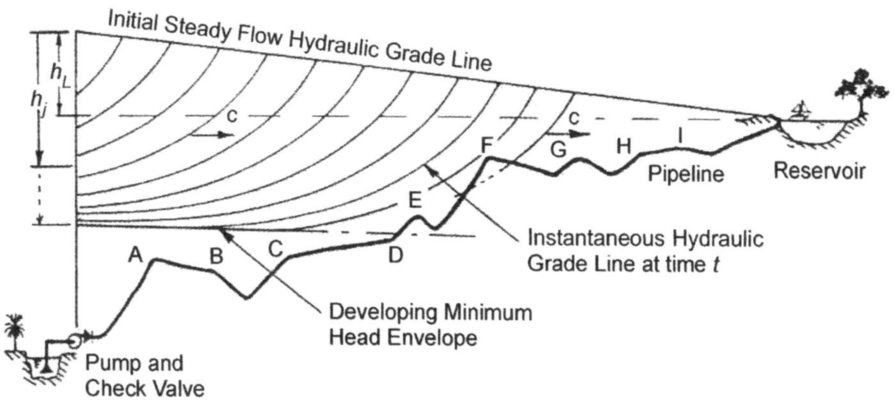

Fig. 3.62 *A typical rising main pipeline showing the principal locations for air valves. Snapshots of the falling hydraulic grade line during the first few seconds after a pump trip are also shown, together with the evolving minimum head envelope*

There are likely to be situations where it will be necessary to reconcile the conflicting needs of rapid ventilation with those of transient suppression and control.

More specific comments follow with respect to some typical locations for air valves, such as those illustrated in Fig. 3.62. This longitudinal section of a rising main shows the initial steady state hydraulic grade line together with an indication of the way it changes during the first few seconds after a full pump trip. It is assumed in the discussion that follows that air admission for surge control is under consideration.

With respect to the initial filling of the line, its steady state operation and draining down for inspection and maintenance, air valves for the ventilation of the line will be recommended for many of the locations A to I – but, should they all be the same and can they contribute to surge control?

Consider the individual locations:

A Ventilation only is required here since, following a pump trip, the pressure remains above atmospheric. The ventilation requirements are: rapid admission for draining and rapid release on filling – but with the constraint that the water flow should be slow. Consider a Dual- or Triple-Orifice valve.

B Ventilation only is required as the normal operating pressures will always be above atmospheric. If A and B are reasonably close, e.g. less than 100 m, rapid admission at A when draining could be acceptable, with only a small-orifice release valve at B. Otherwise the ventilation requirements will be similar to A.

C Ventilation is again the only requirement, principally in the form of rapid admission during draining and the slow release under pressure of dissolved gases given up by the liquid as the pressure falls. As this point is only a reduction in the positive slope of the pipeline, a large-orifice air valve may not be essential here, provided the distances B–C–D–E are not great and provision is made at D or E.

D Between C and D, where there is a risk of more dissolved gases emerging from solution, air valve manufacturers generally recommend the fitting of small-orifice air release valves at intervals of 500–800 m or so. The number needed really depends upon the amount of air and other gases that will come out of solution and that may be entrained owing to unsatisfactory pump sump conditions. This latter reason is why the frequency of small-orifice air release valves often tends to be greater near to the upstream end of a line.

E This is an obvious peak and in the context of ventilation the requirements are: rapid admission when draining, slow release during normal operation and fairly fast release, here or in the vicinity, when filling. However, as indicated in Fig. 3.62, the pressure will fall below atmospheric under transient flow conditions without some remedial action being taken. No data are given in Fig. 3.62 to indicate the normal steady flow pressure heads. However, because they appear greater than, say, 5 m or so, this is just the kind of location where it may be considered that the relatively high static head driving a returning liquid column towards an air valve is too great. The relatively high kinetic energy that the liquid column acquires is likely to be converted into an unacceptably high pressure head as the air cavity is finally driven out. A strategy that would normally be considered with this line profile is to incorporate an air vessel at the pump discharge and/or possibly a feed tank (one-way surge tank) at F such that the local pressure at E always remains at, or above, atmospheric, except when draining down when it may be allowed to become very slightly subatmospheric. If, despite these reservations, vacuum breaking with air release is still deemed acceptable, the air must be released in a controlled manner.

F The ventilation requirements here are similar to those at locations A and E. However, there is a strong tendency for vacuum conditions to develop following an uncontrolled pump trip and, as indicated in the previous paragraph, an air vessel at the pump discharge would often be the preferred approach to surge control in a system with the profile of Fig. 3.62. A feed tank may also be located at F to help reduce the size of air vessel required at the pump to avoid subatmospheric conditions at F, as well as the surge control strategy downstream of this point. Nevertheless, if reliance is to be placed on the use of vacuum relief and subsequent controlled air release at F, rapid admission, followed by very slow release, is the requirement. This can be met with a triple-orifice valve or a dual-orifice valve fitted with a vented non-return valve. Air release through the large orifice should be restricted, so that the air remains trapped long enough to suppress the fluctuations in pressure.

G, H, I Assuming adequate control is provided at F following a pump trip, transient effects in the elevated portion of this line may not be very significant. Triple- or Dual-Orifice air valves, as used for ventilation, would be a common choice here. With the latter choice, they would have fast inflow and slow release, either through a vented non-return valve or fast air outflow but with a surge check valve fitted on the entry to prevent the float being hammered onto its seat as the last of the air is expelled.

Although the discussion above has focused on individual locations with respect to the choice of air admission and release valves, and their potential role in surge control and suppression, the following essential points must be reiterated:

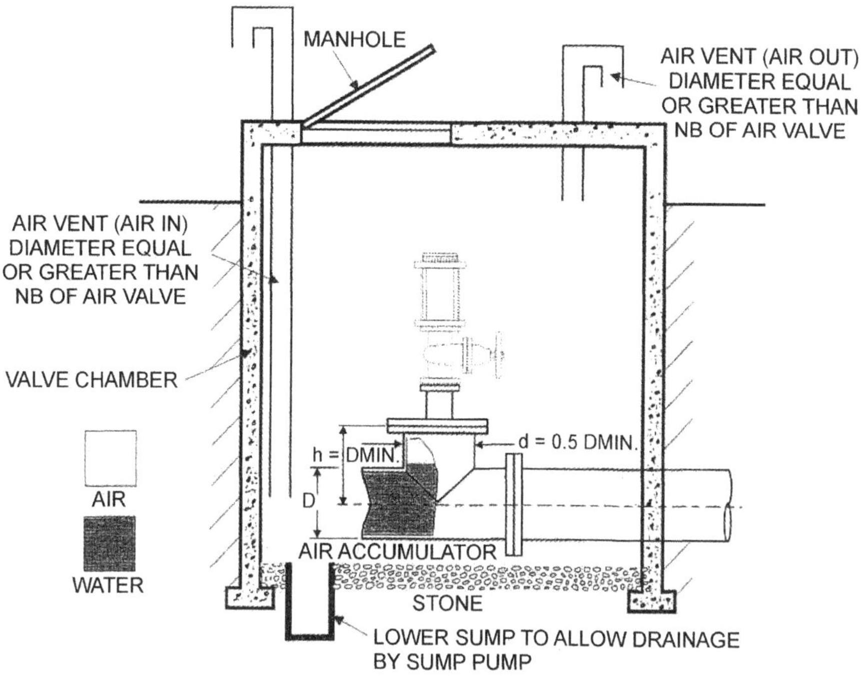

Fig. 3.63 *It is often desirable that air valves be mounted within chambers to avoid unauthorized tampering and to avoid contamination*

- Each system must be considered on its own merits – no two systems are alike.
- The control and suppression strategy must be devised *for the whole system.*
- Even if air valves are not specified for transient flow control, their presence in the system for ventilation purposes must be recognized in the risk assessment of planned and unplanned changes to flows in the system to ensure they do not aggravate any problems.

The physical location of air valves for whatever purpose should be on the upper surface of the pipeline. Consideration must be given to their security and to contamination of both the environment and to the fluid in the pipe. In some cases they can be left in the open if the pipe is laid above ground, but otherwise they should be mounted within chambers through which the main pipeline passes. Figure 3.63 [courtesy of Dynamic Fluid Control (Pty) Ltd] is one example, also illustrating the need for adequate ventilation of the chamber itself.

Many engineers prefer to have air valves mounted on top of a tee-section which will act as a trap for air that comes out of solution during normal operation, rather than directly on top of the pipe itself. When air valves are used for transient flow control, they should ideally be mounted at locations where air, once admitted to the system, will remain as the air cavity so formed reduces in size and requires venting.

3.6.4 Air Valves in Fuel and Petrochemical Lines

Special considerations will often apply with respect to pipelines that convey liquids other than water, especially if the liquid is flammable, explosive and/or toxic. As with all pipelines, provision must be made for venting during both initial commissioning and drain-down operations and appropriate valve types selected. Standard valves designed for water systems should not normally be used. Points to consider include:

- Materials for the various internal components must be compatible with the liquids in the system.
- Moving parts, such as floats, may need reviewing, e.g. with hydrocarbon fuels the floats should be more buoyant to ensure proper sealing against their seats.
- Air, gas and vapour release must be suitably contained and controlled to avoid fire, explosion, contamination of the environment, danger to personnel, etc.
- Air admission would normally be inadmissible for fear of contamination of the liquid being transferred or due to the risk of explosion [Main (1985) and Thorley and Main (1986)].

In summary, air valves for both inflow and outflow will normally be incorporated in many pipe systems and may have a useful role to play in the control of unsteady flows. On some systems, air admission will not be acceptable and other forms of surge control must be used. The choice of air valves on these systems will therefore be dictated by the ventilation requirements only, but they must still be selected carefully in relation to the profile of the line they are protecting to ensure that they do not themselves initiate unacceptable transients through rapid closure.

3.6.5 Air Valves for Sewage and Industrial Effluents

Pipelines to convey sewage and various industrial effluents, including fine slurries such as ash from coal-burning power stations and some chemical plants, are often pumping against low or even negative pressure heads. The air valve requirements for ventilation are basically the same as for conventional water lines and, because the static head on the system is low, even if the operating pressures are quite high near the pumps owing to the need to overcome friction, vacuum relief following pump trip becomes an attractive option for surge control.

However, owing to the presence of solids in the liquid the valve design and mounting should be such that the moving parts, valve seats and smaller flow passages remain clear of the liquid. They should be in no way affected by accumulations of solid matter or corrosion products. The materials of construction should be compatible with the chemistry of the material being transported, and, as with valves for fuel line applications, it may be necessary to reconsider the weight of the float if the density of the liquid in the system is significantly different to that of water.

Vacuum-relief and air release valves for these fluids are often characterized by having the floats that activate the mechanism mounted very low down in the valve body. Figure 3.59 is one example of a valve that is suitable for use in sewage lines; for the valve type shown in Fig. 3.61 the lower float is designed to be significantly longer so that, even under transient flow conditions, the liquid level never rises more than about 75 per cent from the bottom of this particular float.

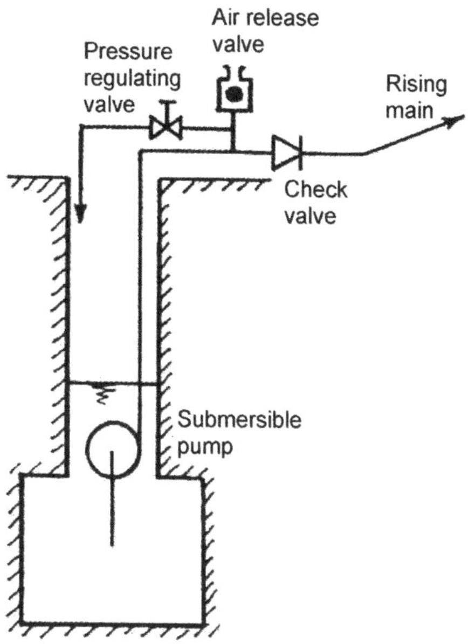

Fig. 3.64 *An air valve fitted to the top of the riser on a deep-well pump shaft, just upstream of the check valve. A PRV is closed off after air in the riser has been vented to pressurize the system*

3.6.6 Air Valves for Deep-Well Installations

As discussed previously, in Part 1, the start-up of deep-well and borehole pumps can present transient flow problems as the air in the riser is compressed and expelled. The ideal scenario is for the air to be expelled rapidly and completely, and for the water flow in the main pipeline to be initiated gradually. At the same time, air should not be forced into the main line, nor should the air valve slam shut onto its seat and generate a high transient pressure. This is a prime example of where the requirements of valves for ventilation must be reconciled with those for transient flow control.

One practical solution is to use a valve that permits rapid evacuation of the air on start-up so that the riser is primed, with the water initially being diverted to drain or back into the well as illustrated in Fig. 3.64. This pipe is fitted with a valve that can then be closed slowly to pressurize the system and initiate flow in a controlled manner.

3.6.7 The Sizing of Air Valves

The sizing of air valves is somewhat empirical, and particularly so with respect to transient suppression and control. The *theoretical* relationships between the pressure difference across the valves and the steady state flowrate through them is quite straightforward. The problem comes in having sufficient information to solve the relevant equations, and iteration is usually required. For unsteady flow the task is much more complicated, though attempts to resolve this have been initiated [e.g. see Arregui *et al.* (2003), Bournaski *et al.* (2003) and Kruisbrink *et al.* (2004)].

By way of background, small-orifice valves for air release operate almost entirely in the choked flow mode, for which the outflow capacity is given by an equation of the form

$$Q = 0.686 C_s d^2 p_1 \tag{3.80}$$

where

Q = flowrate of dry air (m^3/min, ft^3/min or m^3/s)
d = orifice diameter (mm, m, ft or in)
p_1 = absolute pressure (N m/s, lbf/in^2 or bar) at the valve inlet
C_s = a flow coefficient for the particular valve

The numerical value of the flow coefficient C_s is highly dependent on both the shape and size of the valve and the various unit systems in use around the world.

Large-orifice air admission and vacuum-relief valves should provide minimum resistance to flow into the system, and the corresponding relationship between flowrate and pressure drop is

$$Q = C_s\sqrt{(\Delta p \times p_1)} \tag{3.81}$$

in which Δp is the pressure drop across the valve (bar). The other symbols are as defined above, and the flow coefficient C_s (which is not the same as for outflow) must again be found by experiment for each size and type of valve. For example, some typical values for large-orifice valves are given in Table 3.5 and were used for a specimen example earlier in Part 2. Compare these with a typical value of $C_s = 0.0115$ (in metric units and, like Table 3.5, provided by Glenfield Valves Ltd) for a small-orifice valve used for surge control, and also is for use in equation (3.80).

The equations and flow coefficients given above illustrate the theoretical principles behind air valve selection for their various roles. An alternative way of presenting the pressure drop versus flowrate data is through charts of the form shown in Figs 3.65a and 3.65b.

The development of the World Wide Web has encouraged valve manufacturers to provide data on their products through this medium. In some cases (e.g. www.apcovalves.com, www.ventomat.com, etc.) this has extended to making available to potential clients computer software that enables them to identify for themselves the valves that the manufacturers would recommend should be used. The software can either be downloaded directly from the web or is provided by the manufacturers on a CD.

Irrespective of whichever of the above procedures is to be adopted to determine the size, and number, of air admission valves needed to limit subatmospheric pressures in a pipeline, e.g. at point F in Fig. 3.62, it will be necessary to know the minimum pressure that is permitted in the line and the airflow required to achieve this.

Table 3.5 Some Typical Flow Coefficients for Large-orifice Air Admission Valves (Courtesy of Glenfield Valves Ltd). The Nominal Sizes Refer to the Pipeline

Nominal size (mm)	400	600	900
Flow coefficient C_s	27–37	53–74	97–35

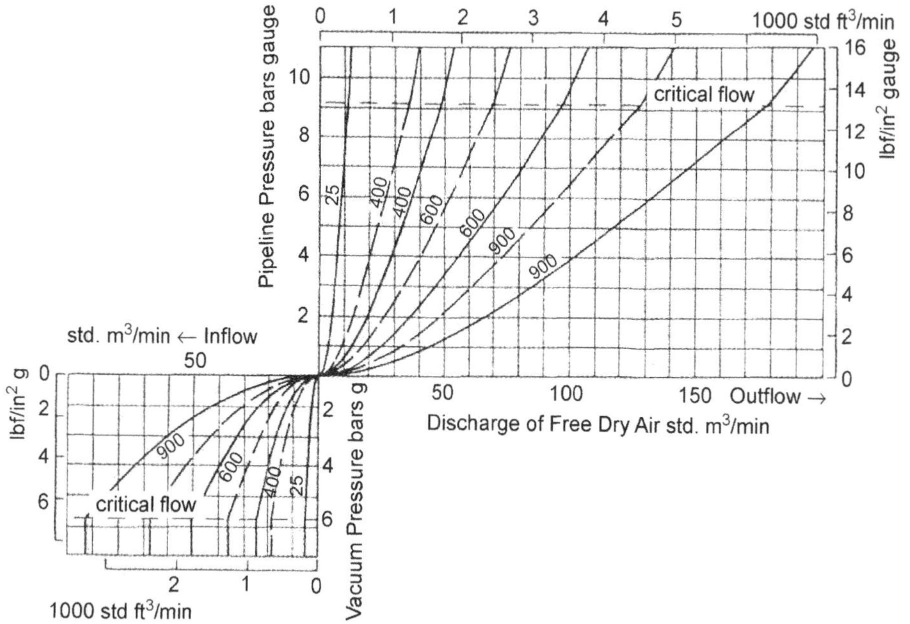

Fig. 3.65a *An air flow chart, typical of those provided by manufacturers, for large-orifice air admission valves*

The former will usually be known from the design specification; the latter will be governed by the maximum rate at which the liquid columns move apart after separation – and estimating this can present a problem.

The most extreme case would generally be to assume that the liquid column upstream of the peak (e.g. at F) stops rapidly, while the downstream column continues, initially, at the steady flow velocity. This yields an air flow requirement into the cavity behind it which, together with the permitted pressure difference between the inside and outside of the line, enables a valve size to be selected by reference to a manufacturer's data such as tables of discharge coefficients, charts

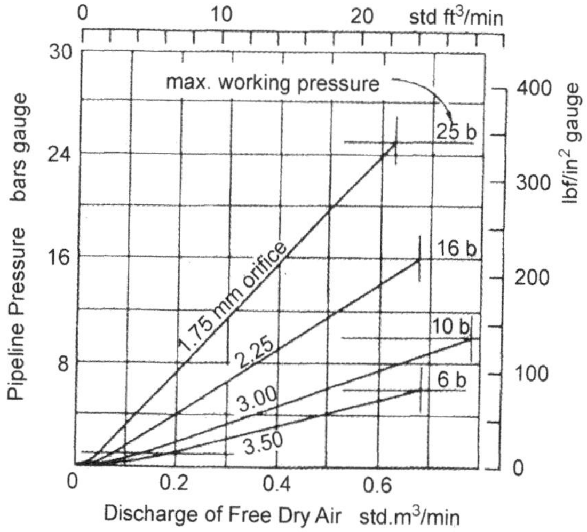

Fig. 3.65b *Similar to Fig. 3.65a, but for choked flow through small-orifice air release valves*

such as those in Fig. 3.65a or the various computer programs now being made available by valve manufacturers.

The outflow orifice size is equally difficult to estimate. It is necessary to retain the air in the system long enough to limit, and to damp out, the fluctuations in pressure, but also to release as much air as possible before the system is restarted.

The system operating pressure is also a factor that needs to be considered, as implied by Fig. 3.65b. In the first instance, a suitable initial estimate might be the size chosen for normal venting of air and undissolved gases at the working pressure. Then, finally, all the large- and small-orifice air valves should be incorporated in a computer model of the system for a transient analysis, and their suitability checked.

3.6.8 Care and Maintenance

Air and vacuum valves are automatic devices upon which the safe and economic operation of the system depends. Failure to operate when required could be catastrophic for the system, and leakage into or out through badly fitting or corroded seals could lead to contamination of the environment and/or the product.

Regular inspection and maintenance, including periodic testing, is required. Isolating valves should be fitted on the inlet to enable these tasks to be undertaken without interrupting the normal operation of the system.

SECTION 3.7

Pressure relief and safety valves

In contrast to air admission and vacuum-breaking valves which limit pressure reductions following a pump trip, pressure relief and safety valves can help restrict the increase in pressure on the upstream side of closing valves. Typical applications include petrochemical applications, delivery lines to oil tankers and high-pressure (and high-temperature) feed water lines to power station boilers.

In many cases the fluid discharged must be stored and reprocessed to avoid environmental contamination and for economic reasons. This can lead to a lower total flow through the exit line from the relief valve (owing to the additional friction losses) when compared with predictions based purely on manufacturers' data for the flow through the valves alone.

Several designs are available, ranging from simple spring-loaded, or weighted, safety valves that are either fully open or shut to much more complex designs with the valves being controlled by remote sensors activating pilot valves, and which can have automatically varied set points dependent upon prevailing operating conditions in the main pipeline.

The simple spring-loaded safety valve is a relatively cheap and reliable device. It is frequently set to open when the upstream pressure is 10–20 per cent above the normal line pressure. Potential drawbacks with the use of valves of this type for surge suppression are valve chatter and the generation of a new transient when the valve closes.

Valve chatter may be inhibited by specifying valves that close at a pressure lower than that at which they open. The initiation of new transients from rapid closure of the safety valve can be avoided through the use of controlled closures. This is sometimes achieved by restricting the flow of a hydraulic oil, or the working fluid, through an orifice, control valve or similar constriction. Figures 3.66 and 3.67 are examples of two such pressure relief valves used in the water and process industries. They can be tuned to the requirements of the system they are protecting, including opening just enough to pass the flow needed to limit the increase in pressure.

Quite often, in order to permit adequate discharge from the system to limit the pressure rise, pressure relief valves are mounted in groups. It is important that their individual operating pressures are set so that they do not 'hunt' or otherwise interact with each other.

The additional features incorporated in the more sophisticated valves increase their cost and often the inertia of the moving elements. In the latter case, especially, they may be slower in responding to the rising pressure in the system.

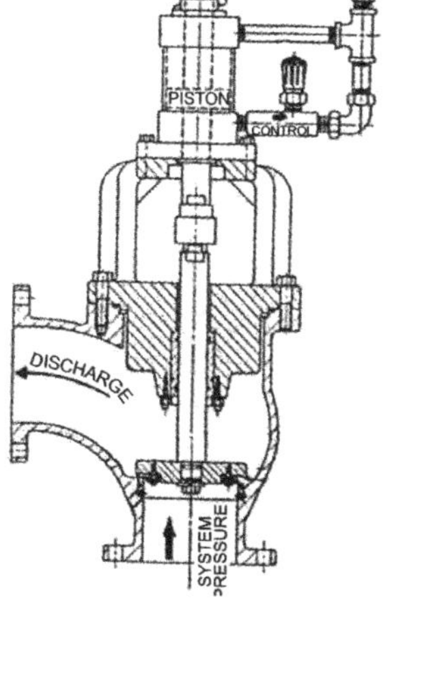

Fig. 3.66 *A spring-loaded pressure relief valve fitted with a hydraulic dashpot to restrict the rate at which it closes*

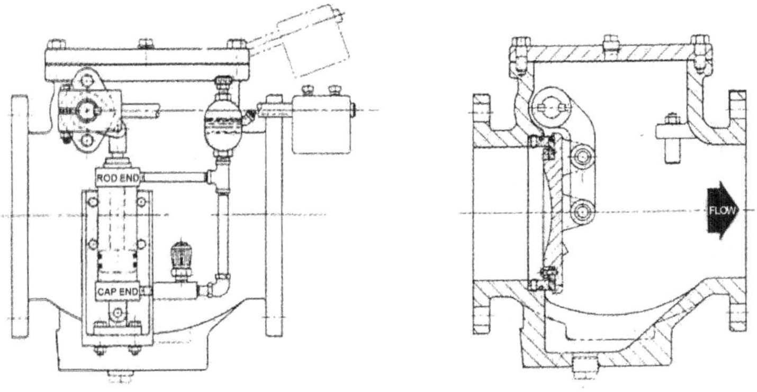

Fig. 3.67 *An example of a weighted pressure relief valve based on the principle of a swing check valve. Closure is restricted by the hydraulic dashpot shown on the side view*

Whether or not this matters depends upon the characteristics of the device causing the transient and the safety margins incorporated in the design of the plant.

Figure 3.68 illustrates a pressure transient relief system in which the set pressure is modified by changing conditions in the main pipeline. The principal moving element is the flexible seal around the pipe (see Fig. 3.69). This is quite light and can respond quickly once the pressure reaches the set point. If the latter is also changing, the response time of the overall relief system may, however, be quite slow.

Alternatively, if the pilot valve is activated by a sensor close to the point where a transient is initiated by, say, a closure of the downstream valves, the relief system can operate before the arrival of the transient pressure waves at the relief valve itself, i.e. it becomes a surge anticipatory valve. In this event, a pressure reduction wave is propagated into the pipeline towards the advancing pressure rise and will help alleviate it that much sooner.

This type of relief system is quite well known in the oil and petrochemical industries, where it may form a part of the overall operation and control system. The fluid in these pipelines cannot simply be discharged to waste on either

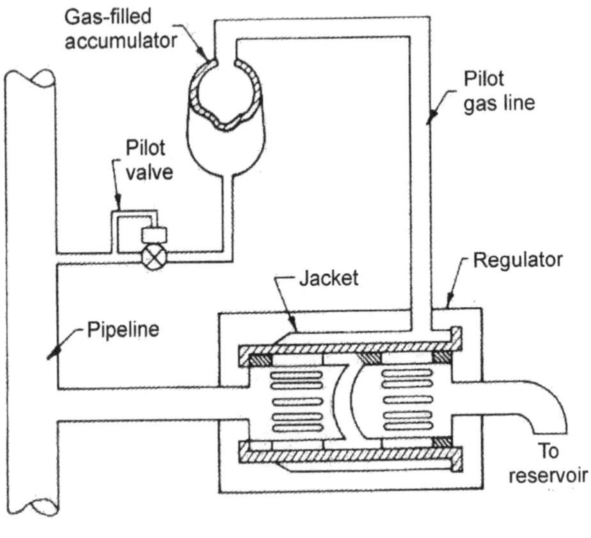

Fig. 3.68 *A pressure transient relief system for which conditions in the main pipeline influence its response*

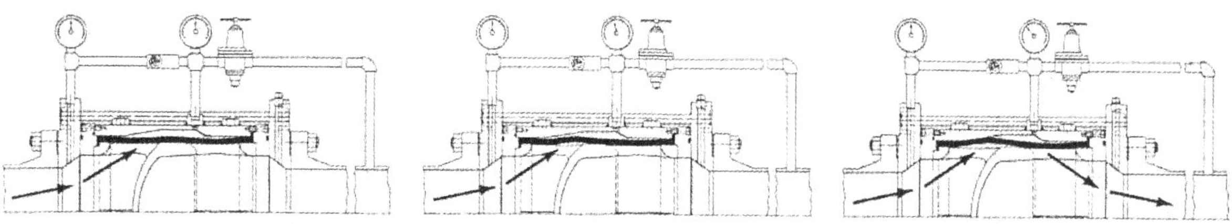

Fig. 3.69 *In this relief valve, increasing fluid pressure forces the flexible membrane away from the pipe against the set pressure holding the valve closed*

economic or environmental grounds. Full provision must be made to store it and to recover it at a later time when the transient event has passed.

3.7.1 Sizing Considerations

It is not always necessary, with respect to the control of excessive transient pressures, to provide sufficient relief capacity to discharge the normal full line flow. The discharge must be adequate, but need only be sufficient, to maintain the line pressure within its pressure rating.

When a discharge valve at the end of a line closes rapidly, the pressure increases by the Joukowsky head $h_{\rm j}$ $(= cv_{\rm o}/g)$. However, additional pressure increases occur owing to the build-up of the frictional head drop in the line as the pump, or upstream reservoir, continues to drive more fluid into the system – a process known as 'line packing'. It can be very significant in long pipelines such as sub-sea oil pipelines and can be many times the Joukowsky pressure head. The net effect on the downstream end of an unprotected line is shown in Fig. 3.70.

As shown here, the pressure will exceed the permitted pressure head by $h_{\rm ex}$. To remain below the permitted level, some flow should continue in the main line, but it must be discharged through the relief system – the question is, how much?

For a simple relief system, a first approximation may be obtained as follows. First, assume that the frictional addition will remain almost the same – this will be a conservative assumption. The allowable increase, ΔH, in pressure head at the valve owing directly to the closure is $\Delta H = h_p - h_L$, where h_p is the total permitted

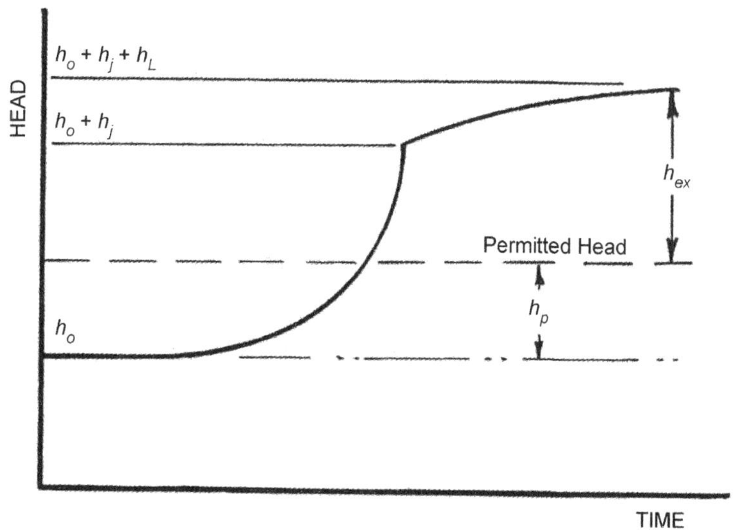

Fig. 3.70 *The pressure at the downstream end of a pipeline increases by the Joukowsky head h_j following valve closure, but the frictional head may also be 'packed' on top*

increase in pressure. Hence, from Joukowsky's equation (3.3)

$$\Delta H = \frac{c}{g}(v_r - v_o) \tag{3.82}$$

where v_r is the velocity in the main line when the relief system is operative, and the volumetric flowrate Q_r is therefore

$$Q_r = A \times v_r \tag{3.83}$$

in which A is the cross-sectional area of the main pipeline.

Designing the relief system around this flow and using valves that open only sufficiently to limit the pressure will leave some margin in hand. It is important to consider the whole relief system and not just the valves themselves when arriving at the size and number required. As Fig. 3.71 indicates, a significant back pressure can arise from the pipes downstream of the relief valves themselves owing to the high line friction losses that inhibit the flow.

Assume the relief line is initially empty and that it is less than, say, 30–40 m in length, so that dynamic effects are not significant. Applying the steady flow energy equation across the relief system shown in Fig. 3.71 gives

$$-h_{in} + (z_2 - z_1) + h_{LP} + h_{LM} + h_{RV} = 0 \tag{3.84}$$

where

h_{in}	=	pressure head in the main line
$(z_2 - z_1)$	=	exit level of the relief pipe minus the inlet level
h_{LP}	=	distributed head loss in the relief pipe
h_{LM}	=	minor losses from bends, isolating valve, etc.
h_{RV}	=	head drop across the relief valve

The first term in equation (3.84) will be known from the permitted pressure head in the system, while the second one relates to the geometry of the plant. The distributed pressure drop due to friction along the pipe is calculated from the normal head drop equation [e.g. equation (3.24)], but does require a decision on the diameter of the pipe to be used for conveying the relief discharge Q_r.

The minor loss calculations follow, since the pipe bends, fittings and isolating valve dimensions are governed by the pipe diameter.

Therefore, the unknown, that is left is the permitted head drop across the relief

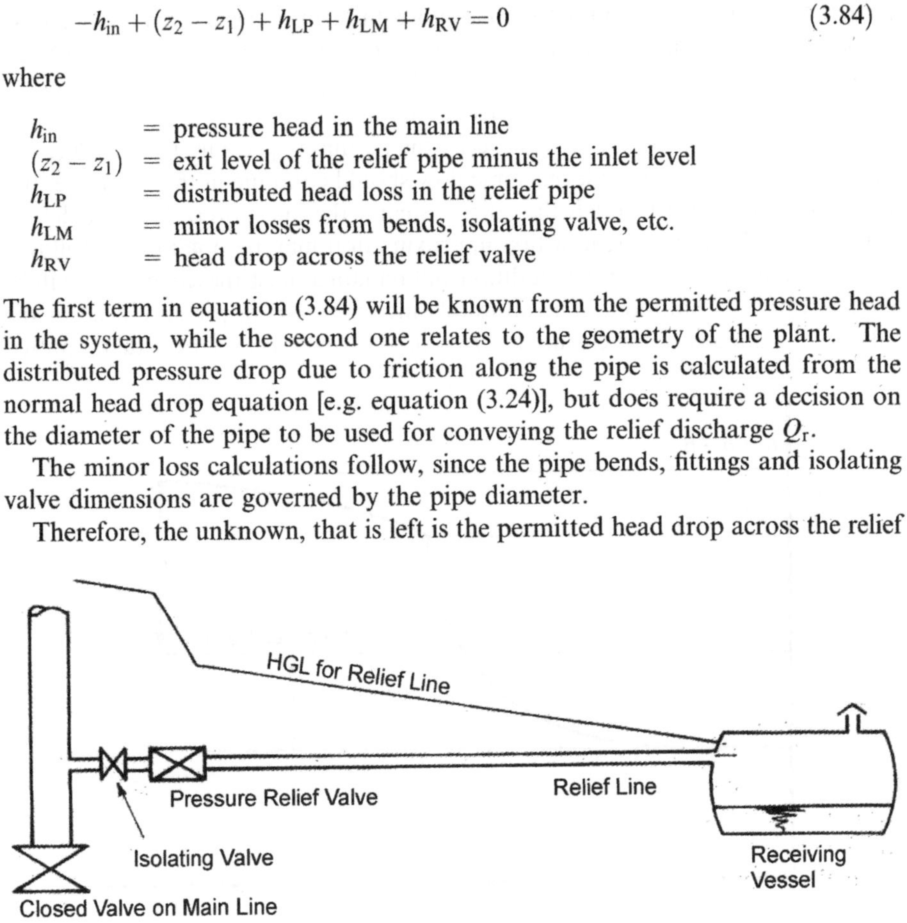

Fig. 3.71 *A significant back pressure can arise on the downstream side of relief valves*

valve(s). One may be adequate, though often it will be necessary to use two or more in parallel to give the required throughput, commensurate with not exceeding the allowed pressure drop across them.

Recourse to manufacturer's performance data is necessary to resolve this with the aid of either charts or tabulated discharge coefficients for use in an equation of the form

$$Q_r = k \times A_v \times \sqrt{2gh_{RV}} \tag{3.85}$$

in which

k = non-dimensional discharge coefficient
A_v = nominal area of flow through the valve(s)

The non-dimensional discharge coefficient may be expected to lie within the range 0.6–0.8. A test programme [Kruisbrink (1990)] on a typical simple spring-loaded type of valve yielded an experimental value of 0.729 when in the fully open position. Similar data for other valve types of specific interest may be found from manufacturers' data listed on the web.

The volume of the storage capacity needed for the relieved flow will depend upon how long it takes to drop the supply pressure to the system by shutting down the delivery pumps. The sooner this can be done, the smaller the storage capacity required.

If two or more valves are installed, they should be set to open at slightly different pressures to avoid the risk of hunting. Ideally, one valve should be included as a standby so that, individually, they can be isolated for regular inspection, maintenance and testing, without the need to shut down the plant.

The foregoing discussion is intended to provide reasonable and conservative estimates for the design of pressure relief systems. They should, however, be confirmed through suitable computer-based modelling.

For the more sophisticated sensor-controlled relief systems there is scope for considerable refinement in the design of surge protection schemes. For example, on long oil pipelines the required volume of the receiving vessel will be quite large, since it must absorb the relief flow Q_r until such time as the relief system can be shut down. This storage capacity can be quite costly, especially on offshore terminals, and a modest investment in refining the design of the relief system may be recovered from the saving in cost on the storage system.

Since this involves computer modelling, in conjunction with manufacturers of these systems, it is outside the scope of this text. However, the provisional capacities estimated by the methods described previously will help provide the initial data.

3.7.2 Bursting discs

Another form of pressure relief that has the advantage of permitting a high flow capacity very quickly is the bursting disc, also known as the rupture disc. These devices have the advantage over pressure relief valves that they can respond more quickly to permit flow. However, unless a duplicate set can be mounted in parallel, and provision is made for their isolation, there will inevitably be delays in restarting the system until they can be replaced after rupturing. The use of bursting discs is therefore likely to be attractive only in those systems that are not expected to experience excess pressures very often.

The discs may be made from metal or graphite. Metal discs have been known to suffer from fatigue, although this appears to be less common with non-metallic discs. Whatever material is used, it is essential that it is compatible with both the product being conveyed in the system and the environment in which the discs are used. For example, in marine and offshore environments it is not simply adequate to be compatible with, say, hydrocarbon products, they must also be able to withstand the effects of sea water spray and accumulations of salt.

Discharge to a receiving tank is often necessary for economic and environmental reasons, and friction losses in the line to this must be considered when determining the capacity of the bursting disc and its supports. The receiving, or blowdown, tank may, in some systems, be filled with an inert gas under pressure to offer an increasing resistance to flow as the vessel fills.

The use of an inert gas may also be required as a fire precaution since, if the liquid being discharged is volatile and flammable, there could be a significant risk of autoignition as the pressure, and hence temperature, in the receiving vessel rises as it fills. Note that, in gas pipelines being protected with bursting discs, as the pressure drops the temperature in the line can drop substantially and may cause significant changes to the physical properties of the pipe and component materials in the vicinity.

If space for a receiving vessel is limited, consideration should be given to mechanisms for controlling the flow. Unlike pressure relief valves that close as the pressure in the system being protected falls to, and can be maintained at, a safe level, a bursting disc once ruptured cannot itself be simply shut off. Either a full-bore valve in the relief line may be installed or the back pressure from the gas in the receiving vessel, combined with a selection of orifice restrictors in the relief line, can be used to control the flow out of the system.

Once the system has been stabilized, the receiving vessel must be emptied and the bursting disc replaced. This may introduce other hazards, depending on the nature of the fluid in the pipelines. Appropriate risk assessments should be undertaken and the relevant guidelines followed, bearing in mind that specialized codes of practice, regulations and standards apply to many of the systems where pressure relief valves and/or bursting discs are used [see, for example, AICE (1998), Inst. Pet. (2001) and Parry (1991)].

SECTION 3.8

Valve characteristics

Every effort should be made to obtain data on the operating and head loss characteristics of all valves and similar components used in piping systems to help ensure that realistic models are developed [Miller (1990)]. Should this not prove possible, which is usually the case during the early stages of design, recourse to 'typical' values has to be made. The charts and diagrams on the following pages are presented as a guide, for use when nothing more definite is available.

Sketches of various valve types have been included with many of the charts, but it is emphasized that they are only to indicate valve types and *not* manufacturers. Any resemblances to particular makes is unintentional and purely coincidental.

3.8.1 Head Losses through Valves

Figures 3.72 to 3.81 provide data on the head loss coefficient, K, for various types of control and isolating valves commonly encountered in pipeline systems. The loss coefficient is defined by

$$k = \frac{h_L}{v^2/2g} \tag{3.86}$$

in which

h_L = the head loss across the valve
v = the mean flow velocity

This formulation has the advantage that the loss coefficient K is non-dimensional and independent of the unit systems used. Valve manufacturers often use other

Fig. 3.72 *Values of loss coefficients for typical circular and square (sluice) gate valves as a function of valve open position*

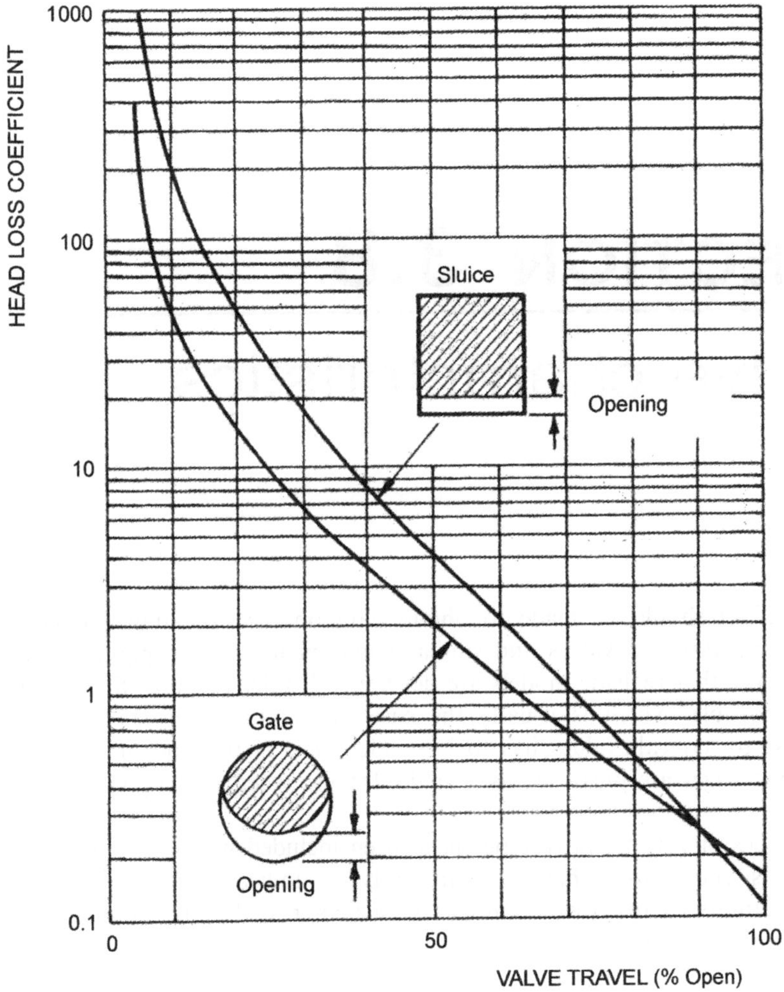

relationships, such as

$$C_v = Q \times \sqrt{\frac{s}{\Delta p}} \tag{3.87}$$

in which

C_v = flow coefficient
Q = volumetric flowrate
s = relative density (specific gravity) of the fluid
Δp = pressure drop across the valve

In this case, C_v is not non-dimensional. Care must be exercised to ensure that the various parameters are expressed in the appropriate units, bearing in mind that

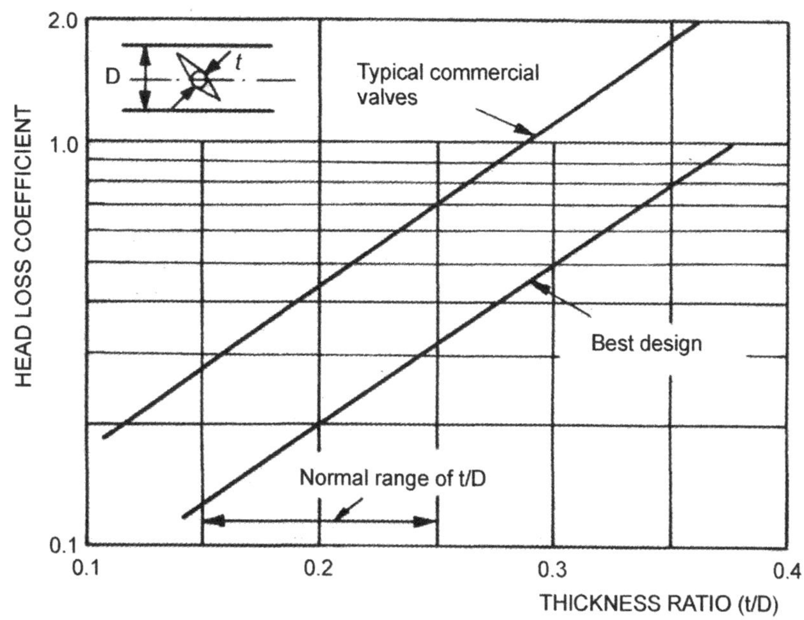

Fig. 3.73 *Values of loss coefficients for typical commercial butterfly valves*

different manufacturers do tend to use their own versions of equation (3.87), not always with the same systems of units.

Both the head loss coefficient K and the flow coefficient C_v vary with valve position, i.e. the extent to which the valve is open. They can be related to either spindle travel or the open area of the valve, and corresponding values for K_o or C_{vo} when in the fully open position, e.g.

$$C_v = C_{vo} \times \left(\frac{A}{A_o}\right) \quad \text{or} \quad C_v = C_{vo} \times \left(\frac{L}{L_o}\right)$$

$$K_v = K_{vo} \times \left(\frac{A}{A_o}\right) \quad \text{or} \quad K_v = K_{vo} \times \left(\frac{L}{L_o}\right) \tag{3.88}$$

Further, when the valve position, and hence open area, is changing, and giving rise to unsteady and transient flow conditions, the loss coefficients also become functions of time, since

$$A = A_o(t) \quad \text{or} \quad L = L_o(t)$$

Figures 3.82 to 3.90 provide similar data for a variety of check valves, with the loss coefficient K still defined according to equation (3.86). The data for these charts have been accumulated over a number of years from a variety of sources in Europe and North America, mainly from manufacturers' published literature and test reports. For comparison purposes, the loss coefficients for check valves are all plotted against Reynolds Number, *Re*. In most cases no information was available

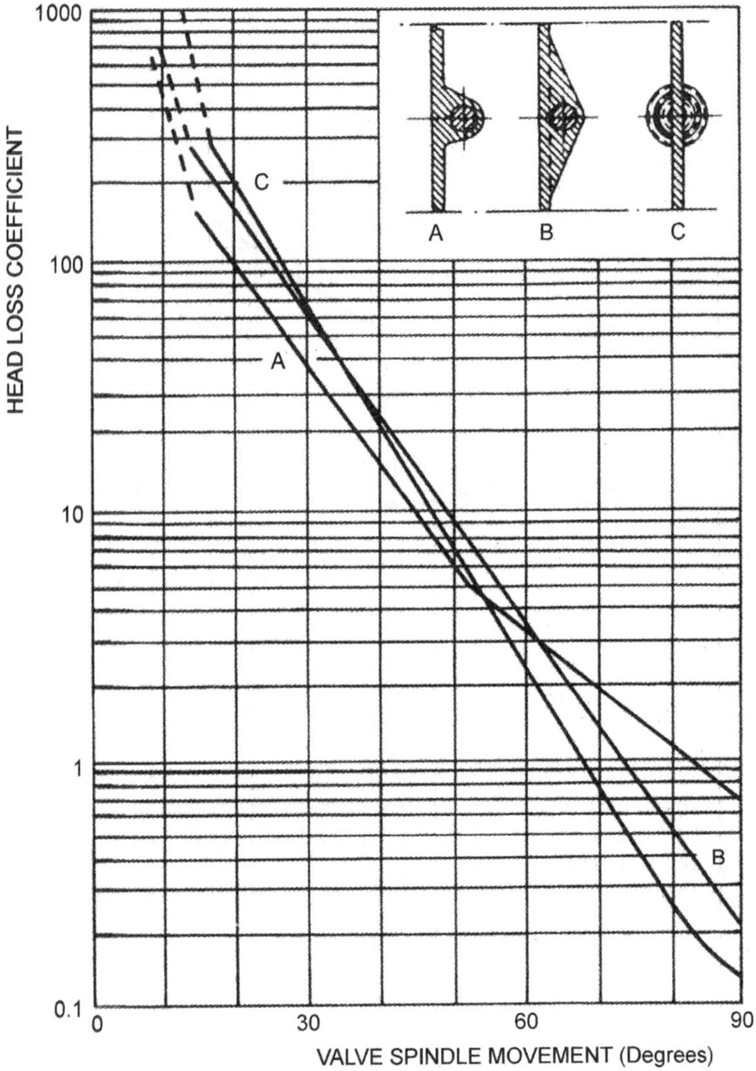

Fig. 3.74 *Values of loss coefficients for typical butterfly valves as a function of valve open position*

directly to link the loss coefficient with the percentage opening of these valves

$$\text{Reynolds Number } Re = \frac{vD}{\nu} \tag{3.89}$$

where

v = flow velocity
D = nominal diameter
ν = kinematic viscosity of the fluid

The general form of these curves shows the loss coefficient falling rapidly as the valve opens, tending towards a fairly constant value for most designs as they reach

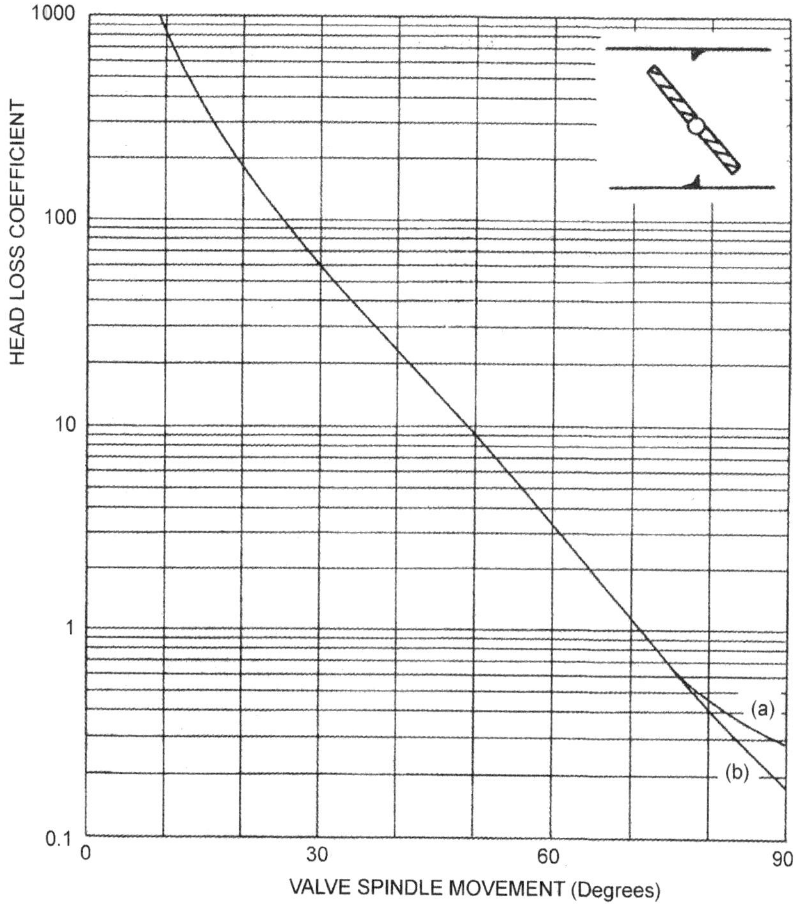

Fig. 3.75 *Loss coefficients for butterfly valves in the size range (a) 350–1140 mm and (b) 1200–1800 mm, courtesy of Glenfield Valves Ltd*

their fully open position. Several other points may be noted from a perusal of the various curves.

There is a clear influence of valve size on the loss coefficient. Valve type, or style, is also an important factor, as indicated by Figs 3.85 and 3.90. The loss coefficients are adversely affected during opening by fitting balance weights and strong springs to aid closing. Likewise, the 'recoil' designs tend to have higher losses. The slim wafer style of valve can give rise to relatively high losses, since the short designs tend to generate higher turbulence which inhibits good pressure recovery downstream and may also initiate cavitation.

To determine an estimate of the loss coefficient for a particular type of valve, calculate the Reynolds number based on the nominal diameter and the design flow velocity. The kinematic viscosity may be obtained from standard references on fluid properties. Values for water at normal temperatures are given in Table 3.6.

Table 3.6 Kinematic Viscosity of Water at Ambient Temperatures

Temperature (°C)	5	10	15
Kinematic viscosity (m^2/s) $\times 10^6$	1.519	1.308	1.141

Fig. 3.76 *Loss coefficients for typical globe valves when fully open*

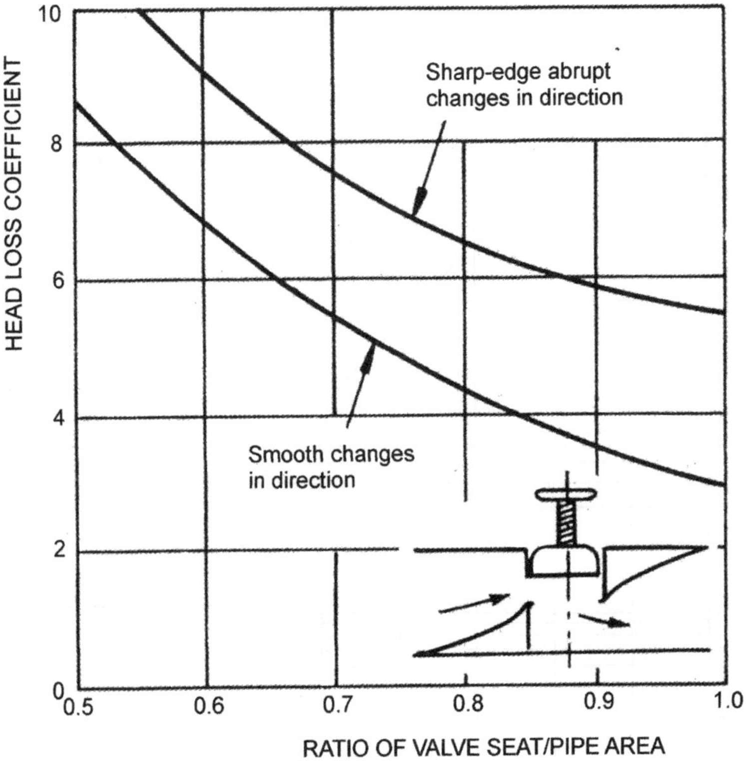

A specimen calculation for a check valve follows.

For a 600 mm nominal bore valve to be fitted into a line conveying water at 10 °C at a flow velocity of 1.9 m/s, the Reynolds Number from equation (3.89) would be

$$Re = \frac{1.9 \times 0.6}{1.308 \times 10^{-6}} = 8.716 \times 10^5$$

If the valve proposed is a simple swing check valve of the form shown in Fig. 3.82, the estimated loss coefficient would be approximately 1.5.

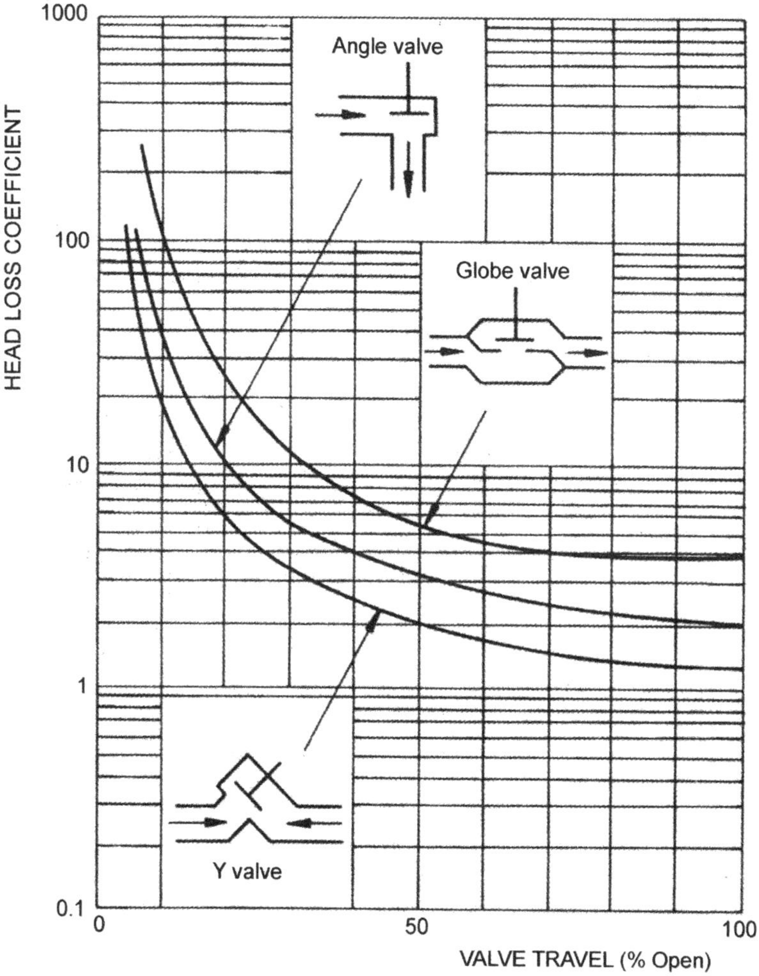

Fig. 3.77 *Loss coefficients for typical globe, Y and angle valves as a function of valve position*

Fig. 3.78 *Loss coefficients for typical ball valves as a function of the valve position*

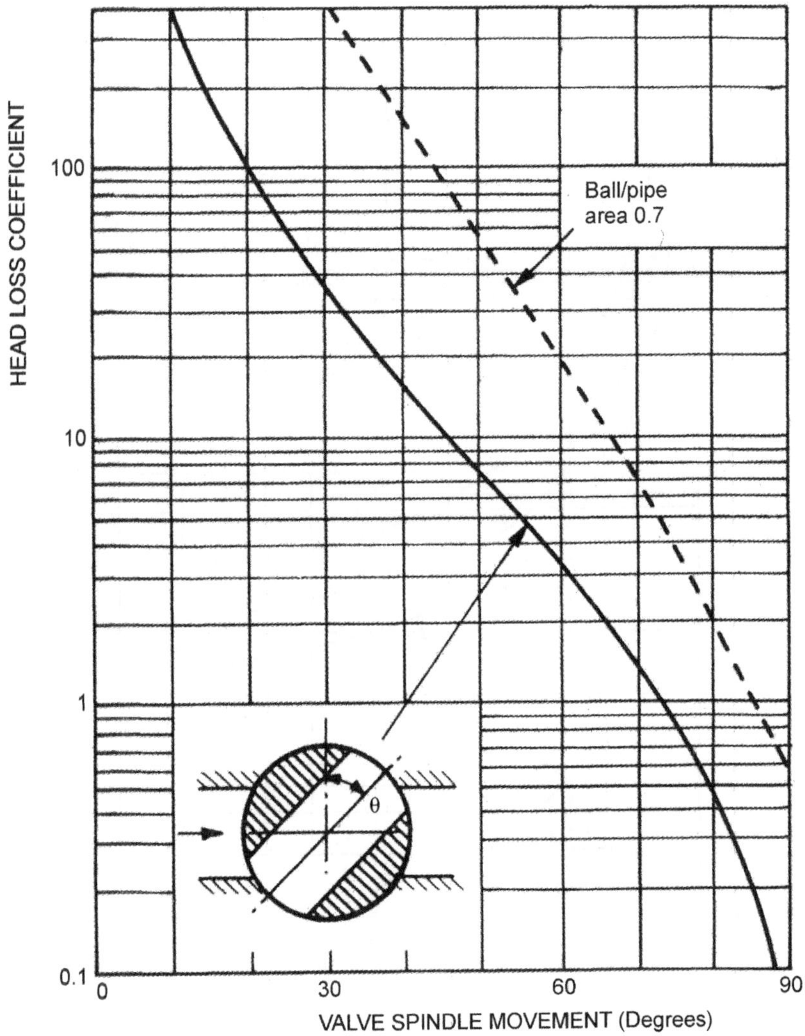

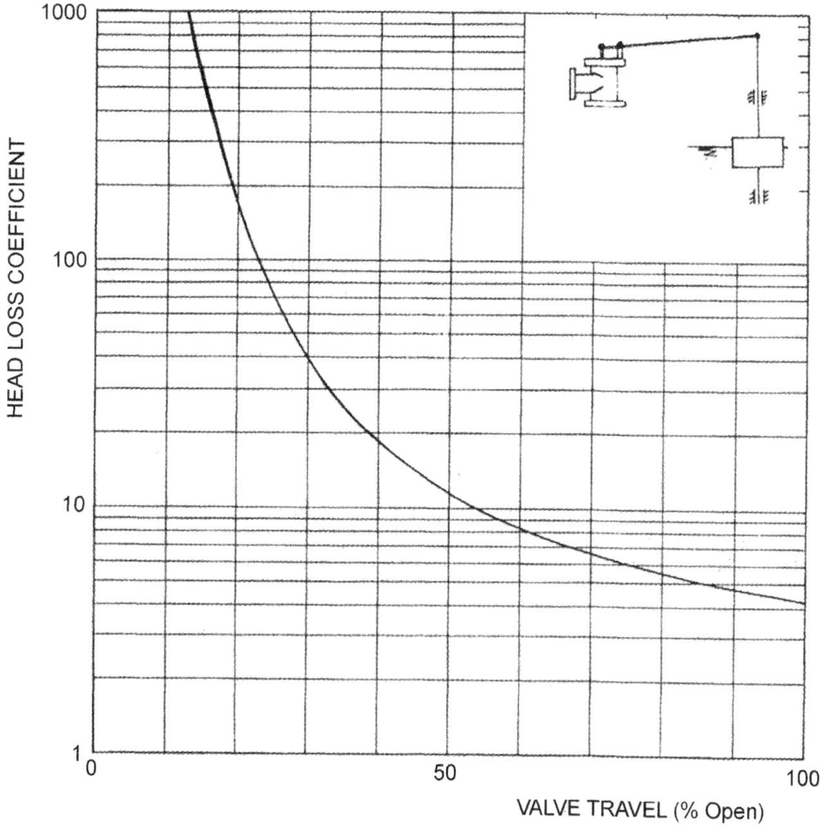

Fig. 3.79 *Loss coefficients for a typical right-angled float valve, in the size range 50–300 mm, as a function of the open position*

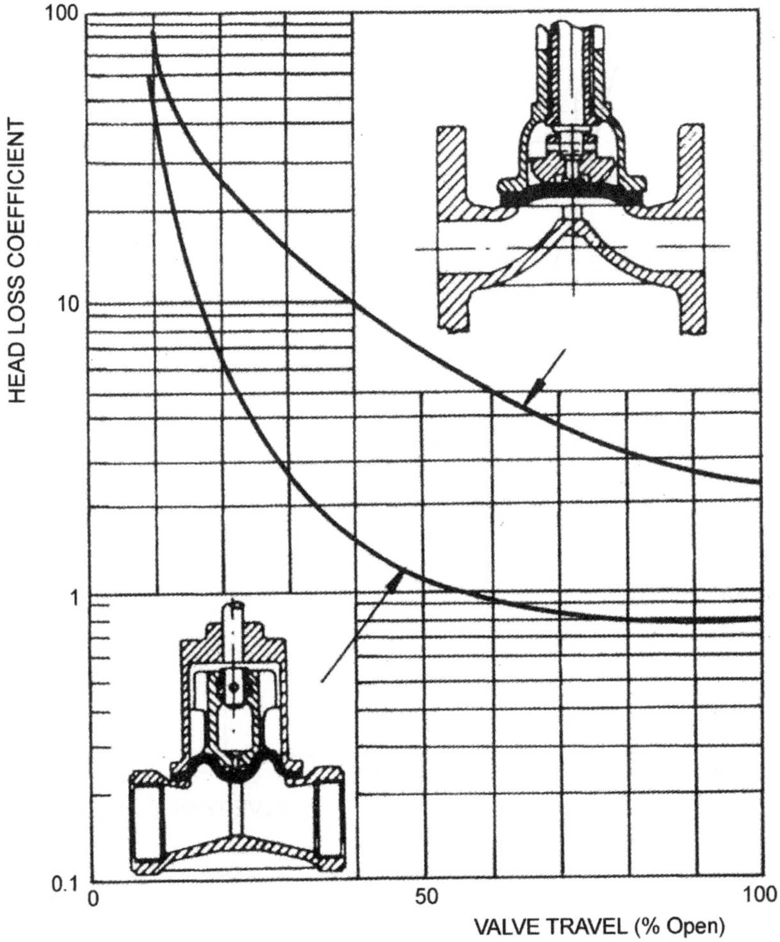

Fig. 3.80 *Loss coefficients for two typical diaphragm valves as a function of valve position*

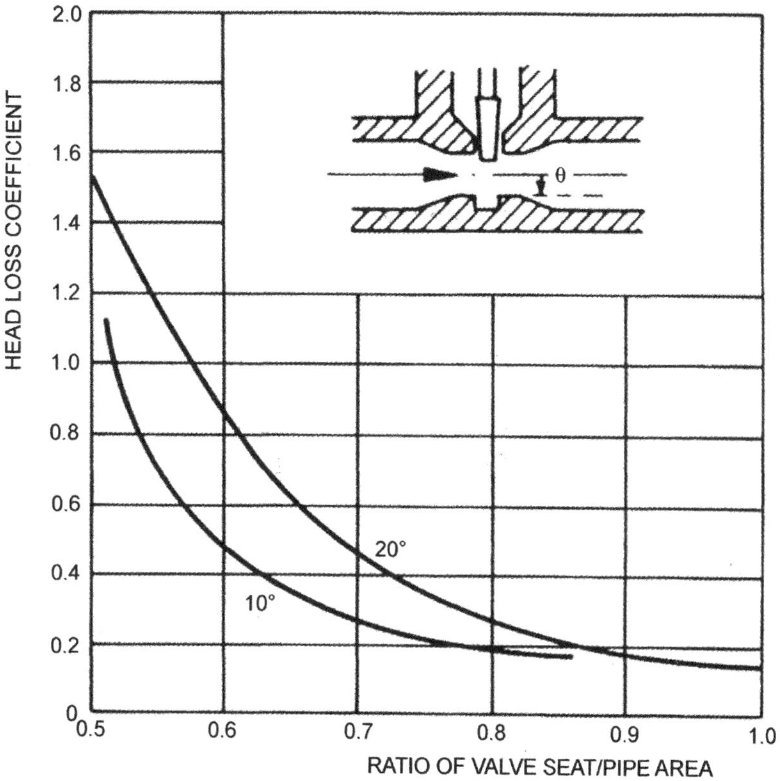

Fig. 3.81 *Loss coefficients for typical circular gate valves when fully open and having tapered approaches*

Fig. 3.82 *Loss coefficients for typical simple swing check valves of various nominal sizes in the range 200–600 mm*

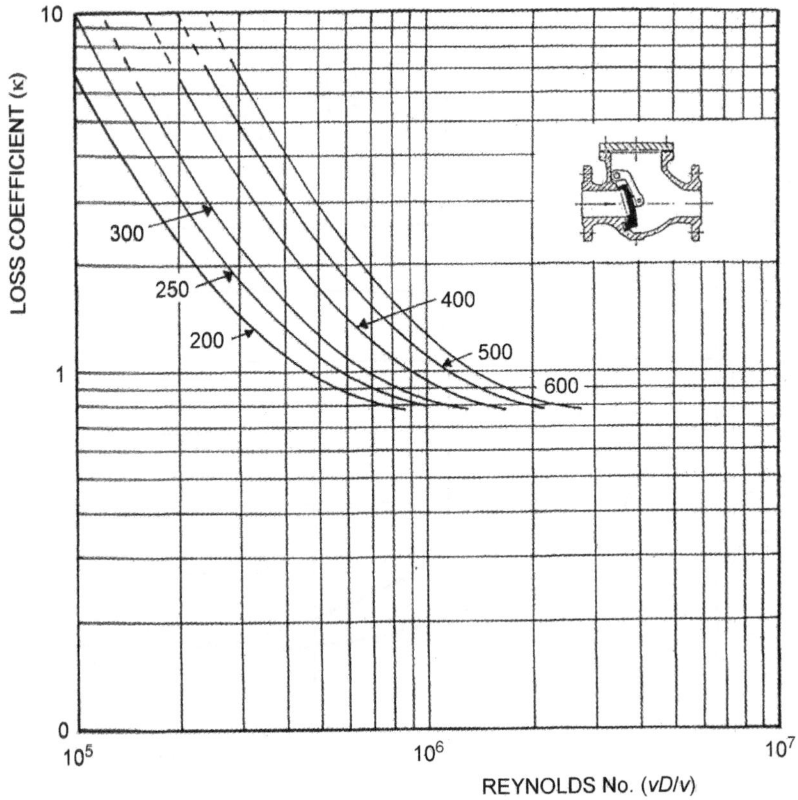

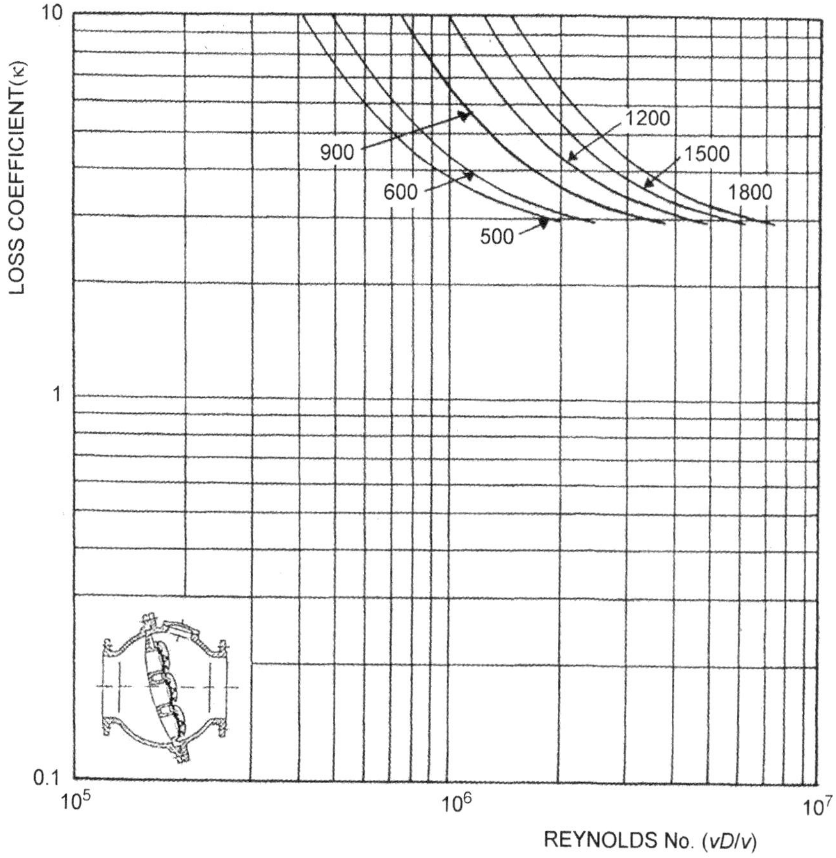

Fig. 3.83 *Loss coefficients for typical twin and multidoor pattern swing check valves in the nominal size range 500–1800 mm*

Fig. 3.84 *Loss coefficients for a 200 mm nominal bore swing check valve fitted with a counterweight*

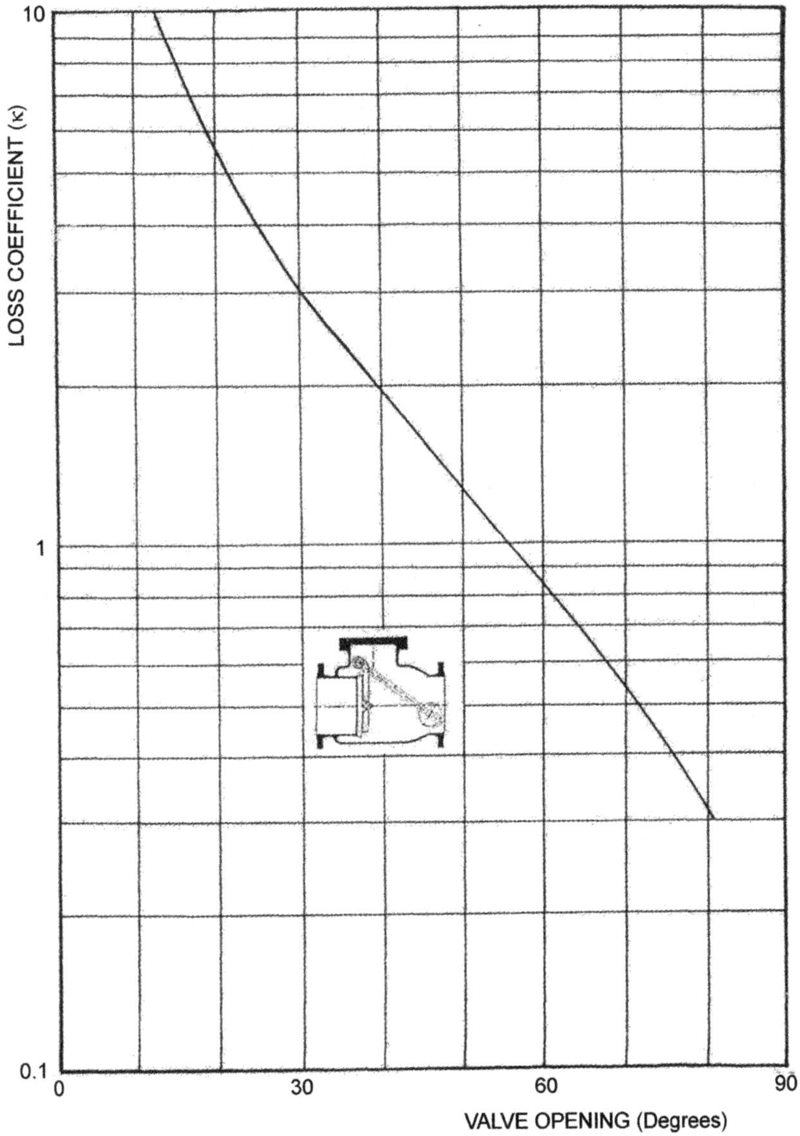

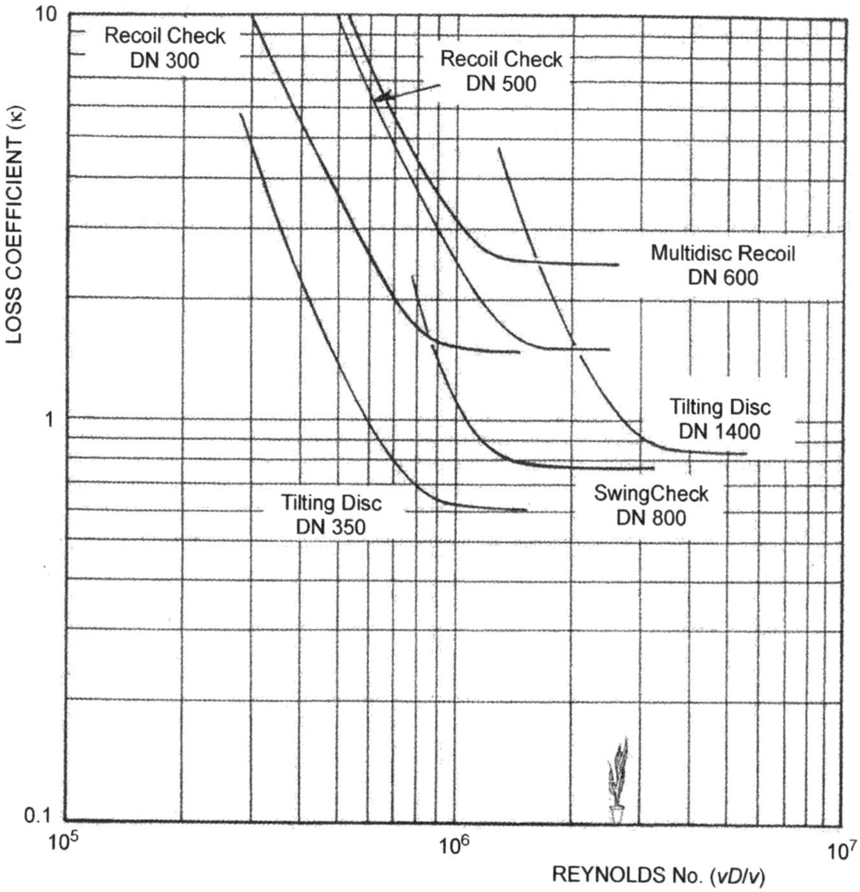

Fig. 3.85 *Comparison of the loss coefficients for a selection of simple swing check, recoil swing check and tilting disc check valves*

Fig. 3.86 *Loss coefficients for typical nozzle-type check valves. The dotted lines on the 300 and 500 mm curves show the effect of fitting relatively strong return springs*

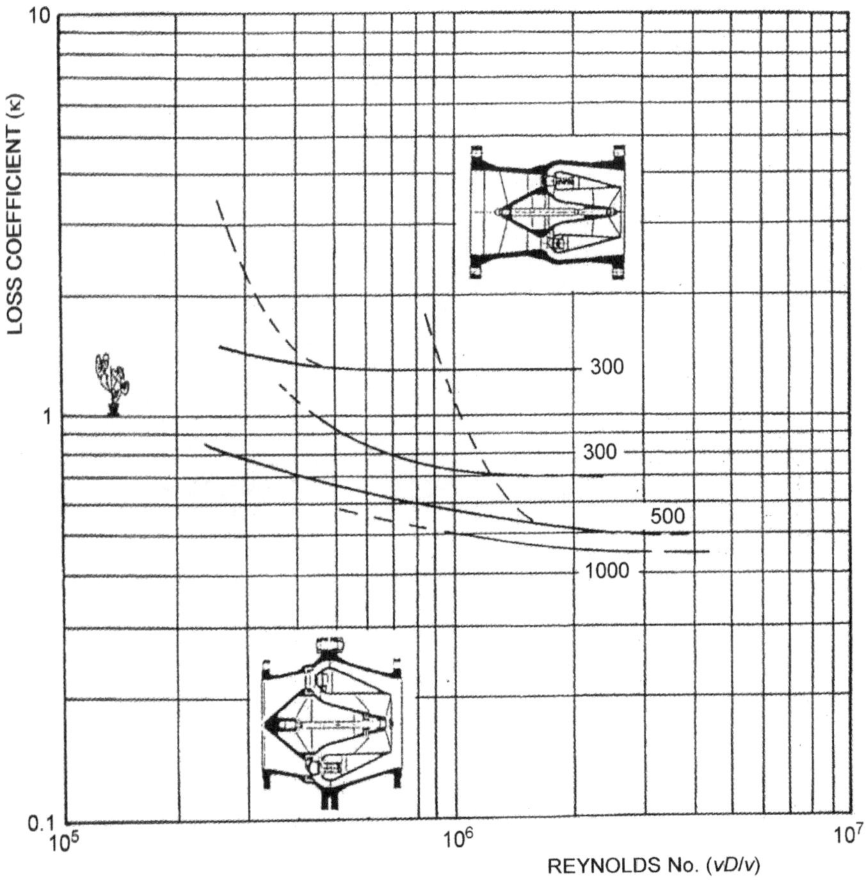

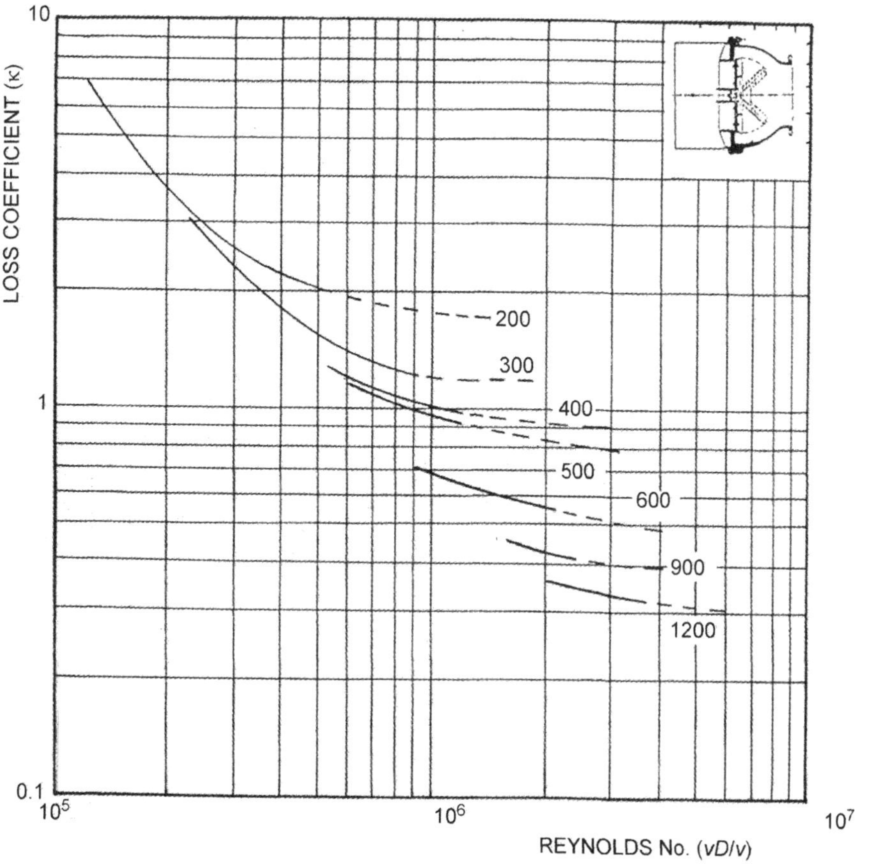

Fig. 3.87 *Loss coefficients for typical split disc check valves in the size range 200–1200 mm*

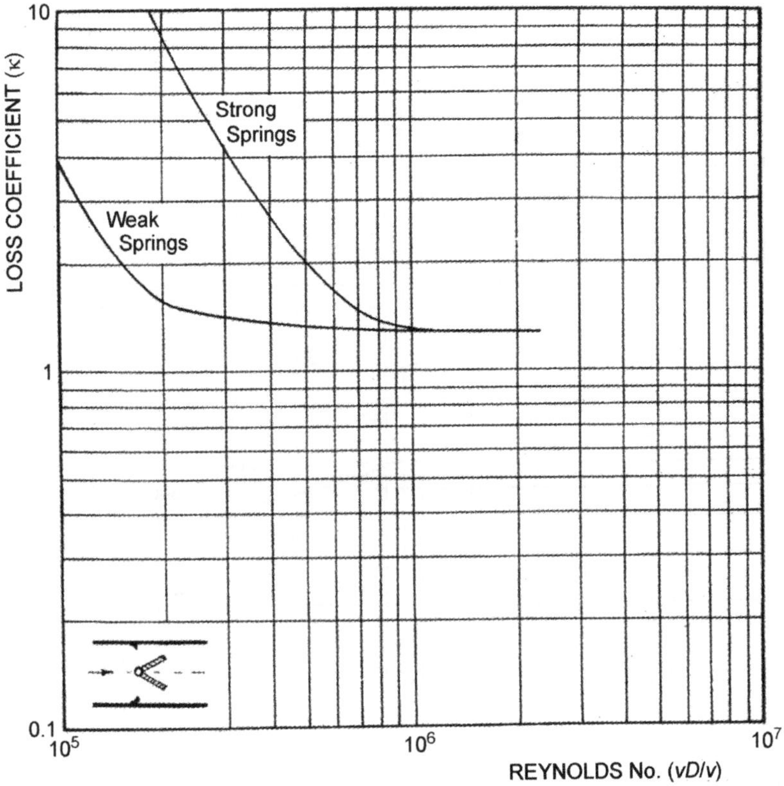

Fig. 3.88 *Comparison of loss coefficients for a 200 mm split disc valve fitted with weak and strong springs*

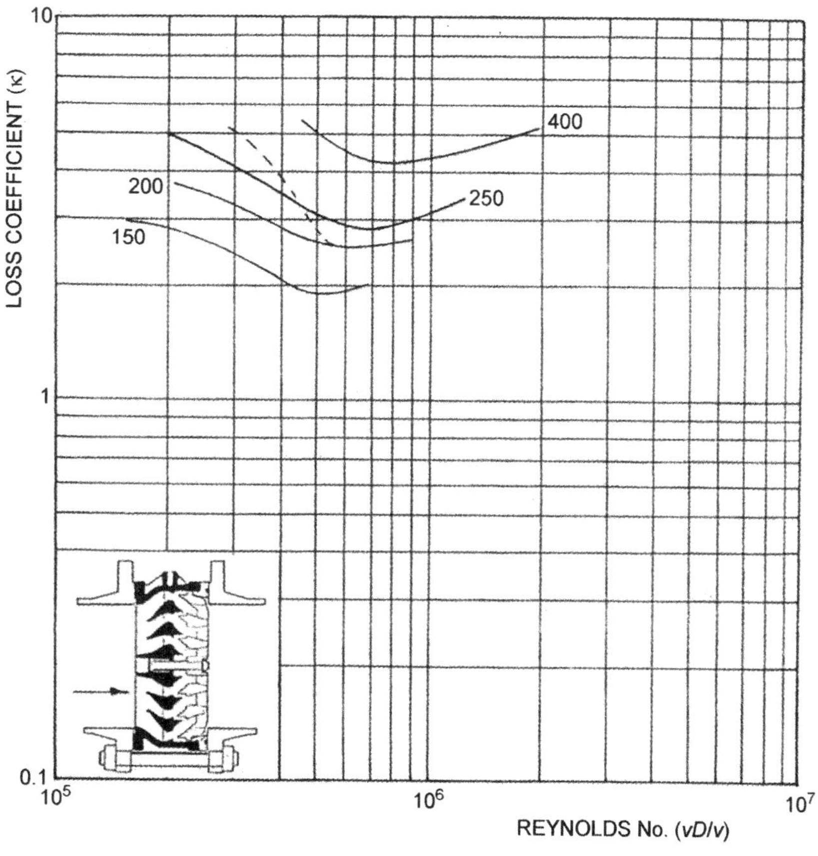

Fig. 3.89 *Loss coefficients for typical wafer-type plate check valves. Stronger springs delay the full opening, as shown on the 200 mm valve, until higher flow velocities are reached*

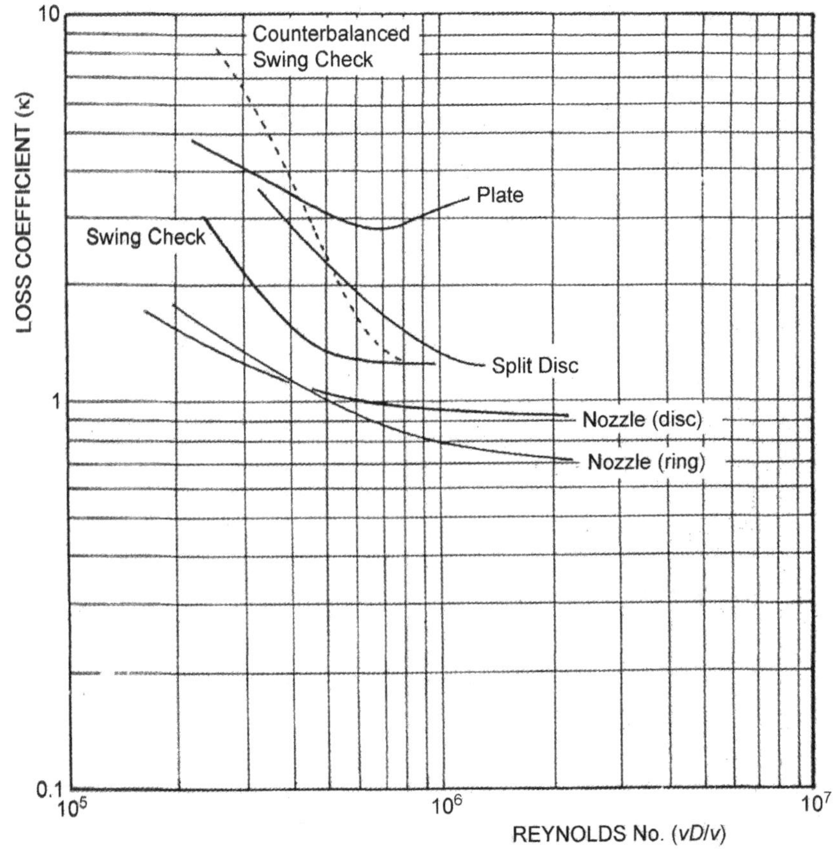

Fig. 3.90 *Comparison of the loss coefficients of 250 mm nominal bore check valves of various types. The effect of a counterweight on swing check valves is similar to that of fitting stronger springs on other designs*

3.8.2 Dynamic Performance of Check Valves

Check valves should be capable of responding quickly to changing flow conditions in a system and, in particular, should close rapidly if the flow falls to zero. The rate at which the flow velocity does change will vary from one system to another, and it is important when selecting check valves that they should be appropriate for the system of which they are an integral part.

The speed with which check valves respond may be represented by the Dynamic Performance Characteristic. Typical examples for four types of valve are shown in Fig. 3.91.

This diagram is a non-dimensional form of Fig. 1.13 but includes some additional data. It illustrates the relatively slow response of ball and swing type check valves. The two curves for split disc valves also show that, for this type of valve, stronger springs (in the lower of the two curves) can improve the dynamic response. Nozzle valves display the best performance. The two curves are for two different styles of nozzle valve, and in each case more than one spring strength is represented. For these particular valves, it appears that spring strength is less significant.

Pipe systems most at risk from the problem of check valve slam (see Subsection 1.3.2.4) are those in which a high energy source is maintained downstream of a pump, or pumps, being tripped. This includes other pumps continuing to run in parallel, and stored energy in an air vessel.

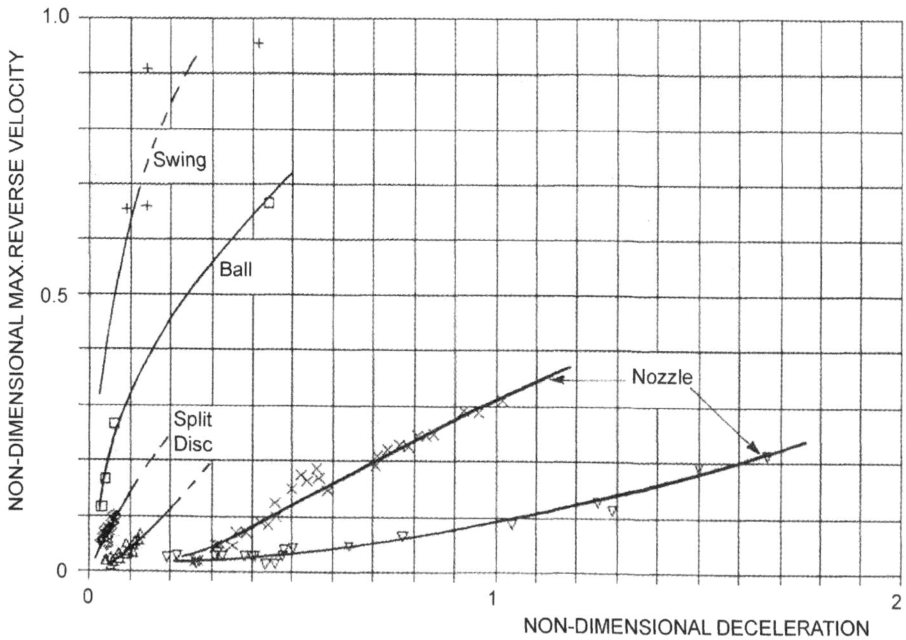

Fig. 3.91 *Comparison of the dynamic performance of four undamped check valves of different sizes and types. Similar work for damped check valves has been undertaken more recently [Kruisbrink (1994) and Kruisbrink (1996) and Thorley]*

Check valves that give rise to the least risk are those that respond most rapidly to changing flow conditions. The best response is characterized by a curve that is closest to the x axis on the Dynamic Performance Characteristic.

Non-dimensional versions of these characteristics can be used as it has been demonstrated [Thorley (1989)] that non-dimensional laws, analogous to the homologous laws for pumps, can be developed for undamped check valves. These provide scope to extend the application of the dynamic performance characteristics determined by experiment for one size of valve to another in the same series, subject to maintaining geometric similarity.

Eight non-dimensional groups can be identified, of which the two principal ones are based on the maximum reverse flow velocity through the valve and the mean deceleration of the liquid column immediately downstream of the valve. The non-dimensional parameters are

$$\text{Maximum Reverse Velocity} = \frac{v_R}{v_o} \tag{3.90}$$

$$\text{Deceleration} = \frac{dv}{dt}\frac{D}{v_o^2} \tag{3.91}$$

in which

dv/dt = deceleration of the liquid column immediately downstream of the valve
D = nominal diameter of the valve
v_R = maximum reverse velocity of the liquid through the valve at closure
v_o = steady flow velocity at which the valve is just fully open

The physical characteristics of check valves that have a 'good' response are:

- the mass of the moving element(s) should be low;
- the moving element(s) should not have far to travel from being fully open to shut;
- springs to initiate and assist closure can be beneficial.

For those pipeline systems where there is a real risk of check valve slam, suppliers of non-return valves that are being considered for use should be persuaded to provide dynamic performance characteristics such as those in Fig. 3.92 or 3.93. Figure 3.94 shows the recommended layout for the test certificate for spring-assisted check valves.

A procedure for using this information is as follows. Note first that the ideal check valve for a pump discharge is one that, without influencing the flow itself, will be closed the instant the flow velocity reaches zero. In practice, there would generally be a small reverse velocity since the moving element in the valve would be moving in that direction to effect the closure. When this reverse velocity is suddenly stopped, as the valve finally closes, there will be a step change in pressure, as previously shown in Fig. 1.12, but the objective is to ensure that this is within acceptable limits.

Estimate the permissible level to which the pressure can be allowed to rise without unacceptable conditions developing in the system. From Joukowsky's equation (1.1), calculate the corresponding maximum reverse velocity, v_R, and convert it to a non-dimensional form with the aid of equation (3.90).

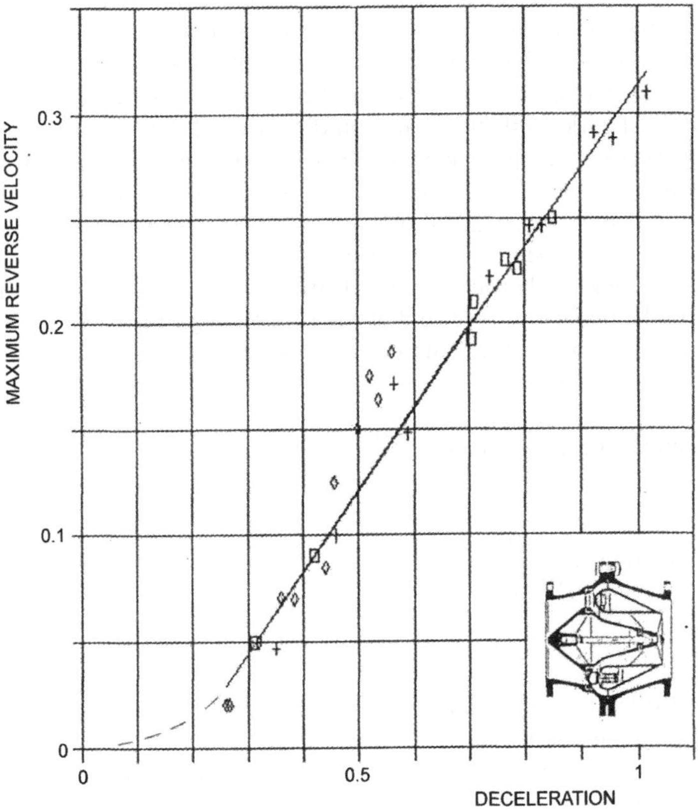

Fig. 3.92 *Non-dimensional dynamic performance characteristic for a 800 mm nozzle type of check valve with three different spring strengths (courtesy of NoReVal)*

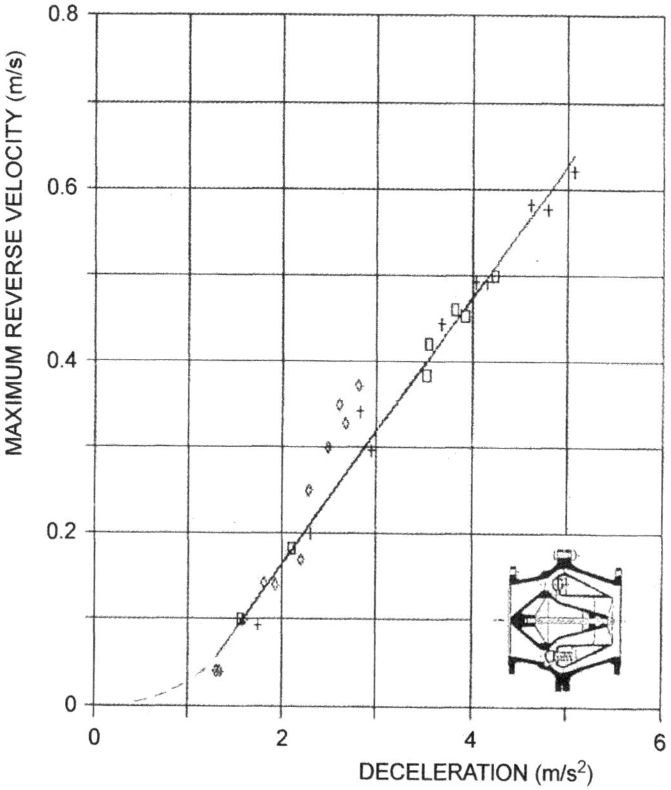

Fig. 3.93 *Similar to Fig. 3.92, but in a dimensional form*

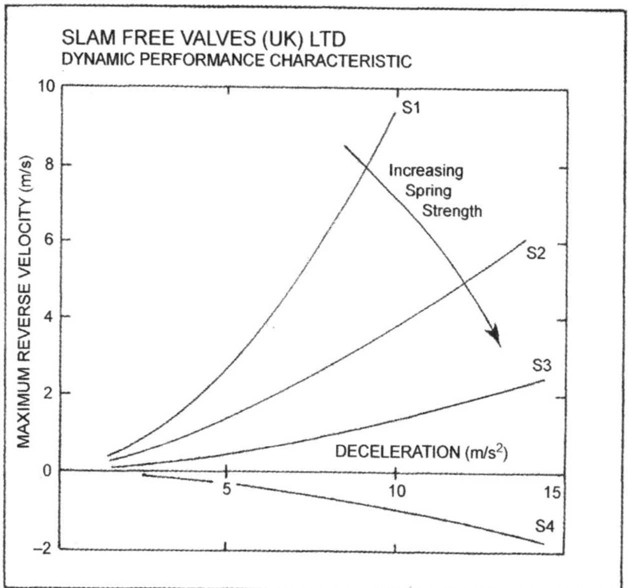

Valve Type	Model No.	Serial No.
Nominal Dia.	Rated Press.	
Test Press.	Initial Flow	Spring Strength
Axis Horiz/Vert.		Ctr. Wt.
Date of Test	Tested by	

Fig. 3.94 *Recommended style for the presentation of Dynamic Performance Characteristics for undamped check valves*

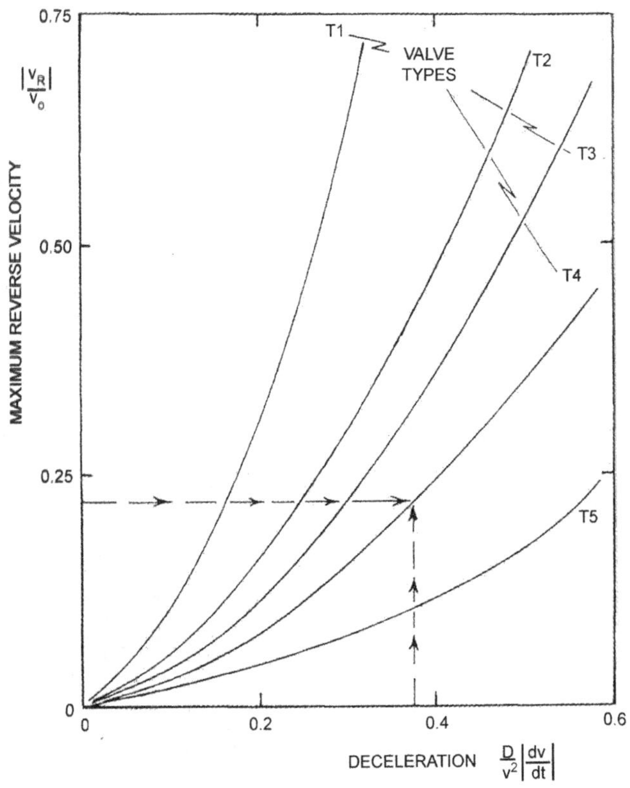

Fig. 3.95 *To identify suitable check valve types, determine the permissible maximum reverse velocity, v_R, following a pump trip and the steady forward flow velocity, v_o, at which the valve is just fully open. Then, determine the deceleration of the liquid column following loss of power to the pump driving motors. Valves whose characteristics cut the vertical line will generally be suitable*

Then, from a computer model of the system, take the average deceleration of the liquid column downstream of the pump from the results of a transient analysis and convert it to a non-dimensional form according to equation (3.91).

On the Dynamic Performance Characteristic graph, plot lines for these two values as indicated in Fig. 3.95. Valves having characteristics that cut the vertical line, T4 and T5 in this example, would be acceptable in the context of minimizing the risk of check valve slam problems.

Since the cost of these valves generally appears to increase as their dynamic response improves, a valve of type T4 would probably be selected as it is likely to be the cheaper option and still satisfy the criterion of being 'fit for purpose'.

The foregoing discussion has concentrated on undamped check valves. However, in the context of minimizing check valve slam, there are at least two other types of check valve that, in suitable circumstances, may avoid the development of unacceptable pressure rises on closure. These are the damped check valve and the 'antipressure surge', or APS, check valve.

The principle by which damped check valves avoid valve slam is that a mechanism such as a hydraulic or pneumatic damper (see Fig. 3.96) is incorporated into the valve to slow down its rate of closure, usually in the final stages as the moving element approaches it seat. These valves can normally open freely and the initial stages of closure are unrestrained. However, and bearing in mind that virtually all valves have a highly non-linear closure characteristic, the last stages of closure are very slow – with the aim of extending the closure time

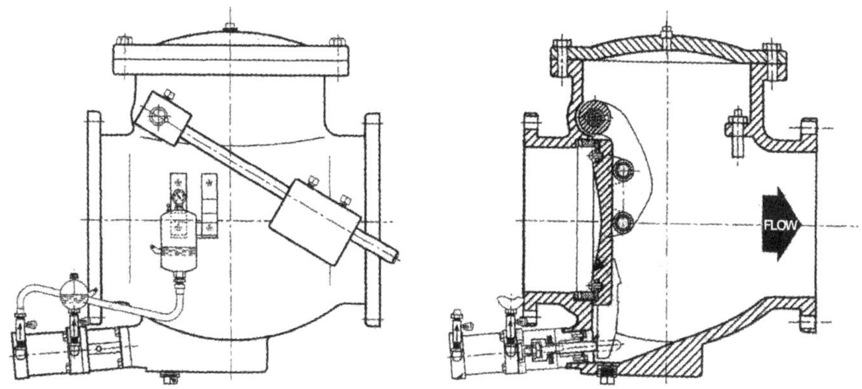

Fig. 3.96 *An example of a damped check valve. The damper allows rapid opening but significantly reduces the rate of closure as the valve disc approaches its seat*

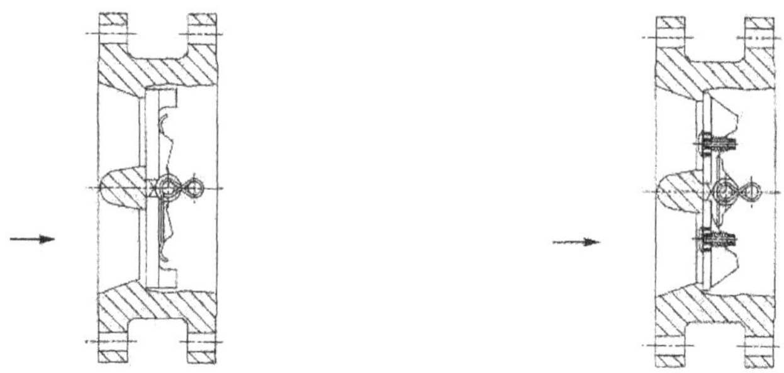

Fig. 3.97 *A 32 in split disc check valve. The one on the left is a conventional check valve, but that on the right has pressure relief valves built into the valve doors. Following closure, these relief valves open to provide a small reverse flow through the valve, thereby relieving the pressure rise on the downstream side (courtesy of Goodwin International plc)*

over several pipeline periods so that beneficial wave reflections from the remote end of the line, or from multiple junctions downstream, will limit the pressure rise on final closure.

The Anti-Pressure Surge (APS) valve, on the other hand, is intended to close in the usual manner, influenced to some extent by the usual springs to initiate movement towards their seats of the valve doors. However (see Fig. 3.97), following final closure, small pressure relief valves in the valve doors will open to deal with any subsequent rise in pressure downstream, and/or any tendency towards cavitation on the upstream side.

SECTION 3.9

Other sources of information

This book is intended to assist engineers new to the field of fluid transients gain an insight into the subject and to be a reference source for data items needed to undertake transient flow analyses. During the life of the previous edition [Thorley (1991)] there have been several more international conferences and, perhaps more significantly, the World Wide Web has evolved.

The material contained within these pages should enable designers to undertake risk assessments of systems intended primarily for conveying liquids, prepare initial assessments of the risks to which they may be exposed, identify the critical design cases, propose strategies for overcoming them, formulate a specification for a computer analysis of the proposed system and be able to discuss the significance of the results produced by it with 'experts'.

For readers who wish to delve more deeply into the theoretical aspects of the subject, numerical modelling and the development of computer codes, the Bibliography that follows, in addition to the various conference proceedings, will provide appropriate sources of further reading.

The later section is a selection of sites currently available (Autumn 2003) on the World Wide Web (www) where information can be found on relevant products and components.

3.9.1 Bibliography

The following texts are all highly recommended for anyone wishing to extend their studies of fluid transients. They all pursue the theoretical development of the partial differential equations governing unsteady and transient flows, leading on to techniques for developing numerical solutions with the aid of digital computers.

To check if any new books have appeared, a good starting point would be to do a web search on a site such as www.amazon.com (or www.amazon.co.uk in the

United Kingdom) using keywords such as Fluid Transients, Pressure Surge, etc. – or an author's name if known.

CHAUDHRY, M.H.
Applied Hydraulic Transients
Second Edition (1987)
Van Nostrand Reinhold Company, New York.
ISBN: 0–442–21514–2

This book describes methods of analysis for developing computer solutions of hydraulic transient problems in hydro-electric and conventional power plants, water supply and the oil industry, including open-channel flows. Several interesting case studies are discussed.

FOX, J.A.
Transient Flow in Pipes, Open Channels and Sewers (1989)
Ellis Horwood Ltd, Chichester, UK.
ISBN: 0–7458–0265–6

This text covers unsteady flow in pipes and channels, and shows how to build mathematical models that can be used as the basis for computer programs for the solution of such problems.

SWAFFIELD, J.A. and BOLDY, A.P.
Pressure Surges in Pipe and Duct Systems (1993)
Avebury Technical Press, Aldershot, UK.
ISBN: 0–291–39796–4

A comprehensive introduction to the unsteady flow of liquids in full-bore and free surface flow is followed by a description of methods of solution based on both 'approximate methods' and the method of characteristics for computer simulations.

WYLIE, E.B. and STREETER, V.L. with Lisheng Suo
Fluid Transients in Systems (1993)
Prentice-Hall, New Jersey, USA.
ISBN: 0–13–322173–3

The latest edition of the books by Professors Wylie and Streeter. A wide range of transient flow situations is covered, from basic principles to methods of analysis and control, embracing liquid and gaseous flows, complex flows, etc., and reflecting work undertaken at the University of Michigan.

3.9.2 World Wide Web

Since the first edition of this book, the World Wide Web started and has evolved into a powerful resource for all kinds of information. Listed below are some of the current websites that may be found useful. However, sites such as these are continually evolving; companies change their names and new companies start up. The examples given below are a very small start, and readers are encouraged to search the web themselves to find the latest sites relevant to their interests, using one of the various 'search engines', such as www.google.com (or try www.google.co.uk if in the United Kingdom) or www.yahoo.com.

When using such sites, a 'keyword' or 'keywords' are needed. This could be a company name or the author of a technical paper or book. Alternatively, a topic such as 'air vessels' or 'check valves' can be used as a keyword. It should be borne in mind that several terms might be used for the same type of component; even the topic of 'water hammer' is known by various names in different industries and countries.

To be sure of capturing all the most relevant links to sites, it may be necessary to use more than one term. For example, as well as 'air vessels', the term 'surge vessels' should be considered. Similarly, 'non-return valves' and 'one-way valves' might be used as well as 'check valves', and so on. Some of the frequently used terms for various components are listed in the table below as an *aide-mémoire*.

Table 3.7 Keywords for Web Searches

Topic/component	Alternatives/synonyms
Air valve	Air admission valve, air relief (or release) valve, vacuum-breaking valve
Air vessel	Air chamber, surge vessel, surge tank, surge chamber, accumulator (when small), air cushion surge chamber (when large)
Bursting disc	Rupture disc
Check valve	Non-return valve, one-way valve
Feed tank	One-way surge tank
Fluid transient(s)	Water hammer, pressure surge, pressure transient, hydraulic transient
Open-channel flow	Free surface flow, two-phase stratified flow, two-component stratified flow
Pressure relief valve	Surge relief valve, prv, surge anticipatory valve

Web address	Topic/component(s)
www.apcovalves.com	Wide range of valves, including check valves, vacuum relief and air release, pressure relief, etc.
www.apv.com	Pumps and valves
www.awwa.org	American Water Works Association publishes a guide on air valve selection
www.beckertech.com	Search engine for gas and liquid pipeline component supplies
www.bpe950.com	Dresser and Grove valves and regulators, including pressure regulating valves
www.cmctechnologies.com	Safety relief valves, bursting discs
www.charlatte.fr	Air vessels and accumulators, including large hybrid and bladder tanks – in French and English
www.cranevalve.com	Wide range of valves and actuators, including check valves and air valves
www.flo-dyne.net	Formerly Fluid Kinetics: surge tanks and air vessels
www.flosafe.com	Pressure relief and safety relief valves
www.flowmaster.com	Software for transient flow analysis
www.fresnovalves.com	Pressure relief and safety relief valves
www.gaindustries.com	Wide range of valves, including check valves, vacuum relief and air release, etc.
www.glenfield.co.uk	Check valves, air and vacuum valves for water and sewage
www.goodwin.co.uk	Dual-plate check and anti-pressure surge valves
www.kypipe.com	Software for transient flow analysis
www.mokveld.com	Nozzle check valves and surge relief for the oil and gas industries
www.noreva.de	Standard nozzle and wafer check valves for oil and water systems
www.pressuresystems.com.au	Safety valves, bursting discs, flame arresters, etc., for petrochemical and other applications
www.saint-gobain-pipelines.co.uk	Air valves and check valves for water and sewage
www.seetru.com	Safety relief valves
www.TechSavvy.com	Technical search engine for identifying suppliers of components, international design codes, etc.
www.ventomat.com	Vacuum relief and air valves for water, sewage and related applications
www.zemarc.com	Accumulators and valves for refuelling and fire protection systems

References

The publications cited in this section are only those to which reference has been made in the text. There are many more in the literature, especially on numerical techniques and computer methods, and for complex flows of liquids, solids and gases – see, for example, the more extensive lists in the texts quoted in the Bibliography, proceedings of the various conferences on pressure surges and the symposia frequently arranged for the Winter Annual Meetings of the American Society of Mechanical Engineers.

AICE (1998) *Guidelines for Pressure Relief and Effluent Handling Systems*, American Institute of Chemical Engineers.

Allievi, L. (1903) Teorìa generale del mòto perturbato dell'acqua néi tubi in pressióne, *Annali Della Societa Degli Ingegneri ed Architetti Italiani*, Milan.

Allievi, L. (1913) Teorìa del cólpo d'arième, *Atti del Collegio Degli Ingegneri ed Architetti Italiani*, Milan.

Allievi, L. (1925) *Theory of Water Hammer* (translated by E. E. Halmos), Riccardo Garoni, Rome.

Ames, W.F. (1979) *Numerical Methods for Partial Differential Equations*, Academic Press.

Anderson, A.A. (Ed.) (2000) Safe design and operation of industrial piping systems, *Proc. 8th Int. Conf. on Pressure Surges*, BHR Group, Cranfield, Beds., UK.

Arastu, A.H., Noble, D.N, LaFramboise, W.L. and Rhoads, J.E. (1999) Diagnostic evaluation of a severe water hammer event in the fire protection system of a nuclear power plant, *3rd ASME/JSME Fluids Engineering Conference*, San Francisco.

Arregui, F.J., Garcia-Serra J., Kruisbrink, A.C.H., Cabrera, E., Fuertes, V.S., Palau, C.V. and Gascón, L. (2003) Air valves dynamic behaviour, *Int. Conf. on Pumps, Electro-mechanical Devices and Systems (PEDS)*, Valencia, Spain.

AWWA (2001) Air-release, air/vacuum and combination valves, Publication M51, American Water Works Association.

Baker, P.J. and Ramos, H. (2000) Selection of transient analysis software for pipeline design – towards a European Standard, *Proc. 8th Int. Conf. on Pressure Surges*, BHR Group, Cranfield, Beds., UK.

Balutto, A. (1998) The application of controlled air transfer technology to new and existing pipeline systems, www.ventomat.com.

Balutto, A. and Muller, M.P. (1995) Points to consider when sizing and positioning air release and vacuum break valves for water pipelines, Dynamic Fluid Control (Pty) Ltd, South Africa.

Balutto, A. and Muller, M.P. (1996) Air valve technology reviewed – introducing controlled air transfer technology, Dynamic Fluid Control (Pty) Ltd, South Africa.

Blum, L.J. (1994) Laboratory tests on air valve performance (2-inch combination air and vacuum relief valves), US Bureau of Reclamation, Hydraulic Branch, Denver, Colorado.

Bonin, C.C. (1960) Water hammer damage to Oigawa Power Station, *Trans. ASME, J. Engng for Power*, 111–119.

Boulos, P.F., Wood, D.J. and Funk, J.E. (1990) A comparison of numerical and exact solutions for pressure surge analysis, *Proc. 6th Int. Conf. on Pressure Surges*, BHRA, Cranfield, Beds., UK, pp. 149–159.

Bournaski, E., Bergant, A. and Kruisbrink, A.C.H. (2003) Air-valve tests at large scale in WL Delft Hydraulics, IAHR Congress, Thessaloniki, Greece, August 2003.

Burrows, R. and Qiu, D.Q. (1996) Effect of air pockets on pipeline surge pressure, *Proc. Inst. Civil Engrs, Water Marit. and Energy*, **112**, 349–361.

Chaudhry, M.H. (1970) Resonance in pressurized piping systems, *Proc. ASCE., J. Hydraul. Div.*, **98**, 325–333.

Chaudhry, M.H. (1987) *Applied Hydraulic Transients*, 2nd edition, Van Nostrand Reinhold, New York.

Cohn, A.R. and Nalley, R.R. (1979) Using regulators for pressure relief, *J. Instrum. Soc. Am.*, **9**.

Collier, S.L. (1983) *Mud Pump Handbook*, Gulf Publishing Co.

Conway, H.D. (1950) *Mechanics of Materials*, Prentice-Hall, New York, pp. 278–280.

Faithfull, E.M. (1989) Surge analysis, Thesis submitted in partial fulfilment of the requirements of the degree of Master of Science in Water Resource System Engineering, The University of Newcastle on Tyne.

Faithfull, E.M. (1990) Private communications with the author.

Fox, J.A. (1989) *Transient Flow in Pipes, Open Channels and Sewers*, Ellis Horwood Ltd, Chichester.

Goodall, D.C., Kjorholt, H., Tekle, T. and Broch, E. (1988) Air cushion surge chambers for underground power plants, *Int. Water Power and Dam Constr.*, **40**(11), 29–34.

Gray, C.A.M. (1953) The analysis of the dissipation of energy in waterhammer, *Proc. ASCE*, **79**, 1176–1194.

Graze, H.R. (1989) Rational design of air chambers to prevent accidents in fluid systems, *Proc. Int. Congress on Cases and Accidents in Fluid Systems*, Sao Paulo, Brazil, Vol. 1, pp. 87–127.

Graze, H.R. and Forrest, J.A. (1974) New design charts for air chambers, *Proc. 5th Australasian Conf. on Hydraulics and Fluid Machinery*, Canterbury, New Zealand, pp. 34–41.

Graze, H.R and Horlacher, H.B. (1986) Design charts for throttled (by-pass) air chambers, *Proc. 5th Int. Conf. on Pressure Surges*, BHRA, Cranfield, Beds., UK, pp. 309–322.

Graze, H.R. and Horlacher, H.B. (1989) Design of optimum sized air cushion surge chambers, *Proc. 6th Int. Conf. on Pressure Surges*, BHRA, Cranfield, Beds., UK, pp. 383–397.

Griffiths, P.T.A. (1972) Surge problems of the Hydro-Electric Commission of Tasmania, *Proc. 1st Int. Conf. on Pressure Surges*, BHRA, Cranfield, Beds., UK, paper E5.

Halliwell, A.R. (1963) Velocity of a water-hammer wave in an elastic pipe, *Proc. ASCE, J. Hydraul. Div.*, **89**(HY4), 1–21.

Hamilton, M. and Taylor, G. (1996) Pressure surge – case studies, *Proc. Int. Conf. on Pressure Surges and Fluid Transients in Pipelines and Open Channels*, BHR Group, Cranfield, Beds., UK, pp. 15–27.

Herforth, H.-E. and Heuser, G. (1986) Pressure surges in a fire water distribution system of an international airport – problems and experiences, *Proc. 5th Int. Conf. on Pressure Surges*, BHRA, Cranfield, Beds., UK, pp. 281–284.

Hills, P.D. (1999) Private communication with author.

IAHR (2000) Hydraulic transients with water column separation, *Report of Working Group 1971–1991*, IAHR, Delft, The Netherlands.

Inst. Pet. (2001) *Guidelines for the Safe and Optimum Design of Hydrocarbon Pressure Relief and Blowdown Systems*, Institute of Petroleum.

Jaeger, C. (1960) A review of surge tank stability criteria, *Trans. ASME, J. Basic Engng*, 765–775.

Jaeger, C. (1963) The theory of resonance in hydro-power systems. Discussion of incidents occurring in pressure systems, *Trans. ASME, J. Basic Engng*, **85**, 631–640.

Jaeger, C. (1977) *Fluid Transients in Hydro-electric Engineering Practice*, Blackie, London.

Joukowsky, N.E. (1900) Uber den hydraulischer Stoss in Wasserleitungsrohren, *Memoires de l'Academie Imperiale des Sciences de St. Petersburg*, Ser. 8, **9**(5); translated by Miss Olga Simin, Water hammer, *Proc. Am. Water Works Ass.*, **24**, 341–424.

Kaye, G.W.C. and Laby, T.H. (1966) *Tables of Physical and Chemical Constants*, 13th edition, Longmans, Green and Co., London.

Kephart, J.T. and Davis, K. (1961) Pressure surges following water column separation, *Trans. ASME, J. Basic Engng*, **83**, 456–460.

Kletz, T.A. (1998) *What Went Wrong? Case Histories of Process Plant Disasters*, Gulf Publishing.

Korteweg, D.J. (1878) Uber die Fortpflanzungsgeschwindigkeit des Schalles in elastisches Rohren, *Annln, der Phys. und Chem.*, Ser. 5, **5**.

Kruisbrink, A.C.H. (1990) Modelling of safety and relief valves in waterhammer computer codes, *Proc. 3rd Int. Conf. on Developments in Valves and Actuators for Fluid Control*, BHRA, Cranfield, Beds., UK.

Kruisbrink, A. C. H. (1996) The dynamic behaviour of check valves, Thesis submitted in partial fulfilment of PhD, City University, London; also published (1997) Delft University Press, ISBN 90–407–1479–7.

Kruisbrink, A. C. H., Arregui, F., Carlos, M. and Bergant, A. (2004) Dynamic performance characteristics of air valves, accepted for presentation at *9th Int. Conf. on Pressure Surges*, BHR Group, Cranfield, Beds., UK.

Lamb, H. (1898) On the velocity of sound in a tube as affected by the elasticity of the walls, *Mem. Manchester Lit. and Phil. Soc.*, **42**(1).

Linton, P. (1972) Note on pressure surge calculations by the graphical method – pump stoppage after power failure, *BHRA Technical Note TN447*, 3rd edition.

Locher, F.A. and Wang, J.S. (1996) Severe water hammer with the majority of a system shut down – a case study, *Proc. 7th Int. Conf. on Pressure Surges and Fluid Transients in Pipelines and Open Channels*, BHR Group, Cranfield, Beds., UK, pp. 211–223.

Locher, F.A., Huntamer, J.B. and O'Sullivan, J.D. (2000) Caution – pressure surges in process and industrial systems may be fatal, *Proc. 8th Int. Conf. on Pressure Surges – Safe Design and Operation of Industrial Pipe Systems*, BHR Group, Cranfield, Beds., UK, pp. 3–18

Main, B.G. (1985) Explosion hazards in off-shore motion compensators, *Proc. Instn Mech. Engrs*, **199**(92).

Marchel, M., Flesch, G. and Suter, P. (1965) The calculation of water hammer problems by means of digital computer. *Proc. ASME Int. Symp. on Waterhammer in Pumped Storage Projects*, pp. 168–188.

Marcos, S. (1998) Air in water pipelines, Dissertation submitted in partial fulfilment of the award of an MSc degree, City University, London.

Merilo, M., Van Duyne, D.A., Safwat, H.H. and Arastu, A.H. (1990) Reducing the frequency of water hammer in nuclear power plants, in *Proc. Symp. on Transient Thermal Hydraulics and Resulting Loads on Vessel and Piping Systems; Trans. ASME, PVP*, **190**, pp. 1–7.

Miller, D.S. (1990) *Internal Flow Systems,* 2nd edition, BHRA (Information Services), Cranfield, Beds., UK.

Niessner, H. (1980) Comparison of different numerical methods for calculating one-dimensional unsteady flows, Lecture 17 in *Unsteady One-dimensional Flows in Complex Networks and Pressurized Vessels*, Von Karman Institute, Belgium.

Parry, C.F. (1991) *Relief Systems Handbook*, Institute of Chemical Engineers, Rugby, UK.

Paz Soldan, G. (1989) Failure of a concrete duct due to hydraulic transient effects, *Proc. Int. Congress on Cases and Accidents in Fluid Systems*, Sao Paulo, Brazil, Vol. 1, pp. 152–155.

Pearsall, I.S. (1965) The velocity of water hammer waves, *Proc. Instn Mech. Engrs, Part E: J. Process Mechanical Engineering*, **180**(3E), 12–20.

Peterson P.F and Kim J.H. (1993) *Transient Phenomena in Nuclear Reactor Systems*, ASME, New York, ISBN 0–79–181115–8–1.

Pickford, J. (1969) *Analysis of Surge*, MacMillan, London.

Provoost, G.A. (1980) The dynamic behaviour of non-return valves, *Proc. 3rd Int. Conf. on Pressure Surges*, BHRA, Cranfield, Beds., UK, paper J1.

Pulling, W.T. (1976) Literature survey of water hammer incidents in operating nuclear power plants, *Report WCAP-8799*, Westinghouse Corp., Pittsburgh, Pennsylvania.

Rich, G.R. (1963) *Hydraulic Transients*, Dover Publications.

Roccard, Y. (1937) *Les Phenomenes d'Auto-oscillation dans les Installations Hydrauliques*, Hermann, Paris.

Rowe, W.D. (1979) Introduction to risk assessment, in *Energy Risk Management*, Eds G. T. Goodman and W. D. Rowe, Academic Press, pp. 7–19.

Serkiz, A.W. (1983) Evaluation of water hammer experience in nuclear power plants, *Report NUREG-0927*, US Nuclear Regulatory Commission, Washington, DC.

Stephenson, D. (1981) *Pipeline Design for Water Engineers*, 2nd edition, Elsevier Sciences, Amsterdam.

Streeter, V.L. and Lai, C. (1962) Waterhammer analysis including fluid friction, *Trans, ASCE, J. Hyd. Div.*, **88**, HY3, 79–122.

Streeter, V.L. and Wylie, E.B. (1967) *Hydraulic Transients*, McGraw-Hill, New York.

Suo, L. (1988) Wavespeeds in cylindrical tunnels bored through unbounded rock mass, Thesis presented in partial fulfilment of a PhD, University of Michigan, Ann Arbor.

Suter, P. (1966) Representation of pump characteristics for calculation of water hammer, *Sulzer Tech. Rev.*, **66**, 45–48.

Swaffield, J.A. and Boldy, A.P. (1993) *Pressure Surges in Pipe and Duct Systems*, Avebury Technical Press.

Sykes, J.M. (1995) Pressure systems risks: the changing face of legislation. *Proc. IMechE Int. Conf. on Pressure Systems: Operation and Risk Management.*

Taylor, G. and Harrison, M. (1990) Resonances in the pipework of reciprocating pumps – a case study, *Proc. 6th Int. Conf. on Pressure Surges*, BHRA, Cranfield, Beds., UK, pp. 325–332

Tennant, R.M. (1979) *Science Data Book,* Oliver and Boyd, Edinburgh.

Thomas, S. (2003) *Air Management in Water Distribution Systems*, Clear Water Legacy, Burlington, Ontario, Canada.

Thorley, A.R.D. (1976) A survey of investigations into pressure surge phenomena, *Research Memorandum ML83*, City University, London.

Thorley, A.R.D. (1986) Dynamic response of flexible hoses for tanker loading and unloading, Technical Note 3, *Proc. 5th Int. Conf. on Pressure Surges*, BHRA, Cranfield, Beds., UK, TN3, pp. 291–294.

Thorley, A.R.D. (1989) Check valve behaviour under transient flow conditions – a state of the art review. *Trans. ASME, J. Fluids Engng*, **111**, pp. 178–183.

Thorley, A.R.D. (1991) *Fluid Transients in Pipeline Systems*, 1st edition, D. and L. George Ltd.

Thorley, A.R.D. and Chaudry, A. (1996) Pump characteristics for transient flow analysis, *Proc. 7th Int. Conf. on Pressure Surges*, BHRA, Cranfield, Beds., UK, pp. 461–475.

Thorley, A.R.D. and Enever, K.J. (1979) *Control and Suppression of Pressure Surges in Pipelines and Tunnels*, Construction Industry Research and Information Association, London.

Thorley, A.R.D. and Kruisbrink, A.C.H. (1994) Dynamic characteristics for damped check valves. *Proc. 2nd Int. Conf. on Water Pipeline Systems*, BHR Group, Cranfield, Beds., UK, pp. 459–476

Thorley, A.R.D. and Main, B.G. (1986) Spontaneous combustion in vapour cavities subjected to fluid transients in pipelines, *Proc. 5th Int. Conf. on Pressure Surges*, BHRA, Cranfield, Beds., UK, paper F1, pp. 139–149.

Thorley, A.R.D. and Spurrett, R.P. (1990) Cavity dynamics and the risk of explosive combustion in pipelines, *Proc. 6th Int. Conf. on Pressure Surges*, BHRA, Cranfield, Beds., UK, pp. 357–370.

Thorley, A.R.D. and Twyman, J.W.R. (1976) Propagation of transient pressure waves in a sodium-cooled fast reactor, *Proc. 2nd Int. Conf. on Pressure Surges*, BHRA, Cranfield, Beds., UK, pp. A1.1–A1.13.

Thorley, A.R.D. and Wiggert, D.C. (1985) The effect of virtual mass on the basic equations for unsteady one-dimensional heterogeneous flows, *Int. J. of Multiphase Flow*, **11**(2), 149–160.

Thornton, R.E. (1983) High gas temperatures associated with pressure surges in a pipeline, *Proc. 4th Int. Conf. on Pressure Surges*, BHRA, Cranfield, Beds., UK, paper B2, pp. 59–75.

Trenke, C.J. (1979) Failure of riveted forge-welded penstock, *Trans. ASCE, J. Energy Div.*, **105**, pp. 93–102.

Tucker, D.M. and Young, G.A.J. (1960) Estimation of the size of air vessels, Presented to 7th Conference on Hydromechanics, BHRA, Cranfield, Beds., UK, Report SP670.

Twyman, J.W.R., Thorley, A.R.D. and Hewavitarne, R. (1980) Wave propagation in plastically deforming ducts, *Proc. 3rd Int. Conf. on Pressure Surges*, BHRA, Cranfield, Beds., UK, pp. 1–15.

Vetter, G. and Schweinfurter, F. (1990) Elimination of disturbing and dangerous pressure oscillations caused by high pressure positive displacement pumps, *Proc. 6th Int. Conf. on Pressure Surges*, BHRA, Cranfield, Beds., UK, pp. 309–324.

White, D.F. (1960) The unintentional ignition of hydraulic fluids inside high pressure pneumatic systems, ASNE J., 405–413.

Wood, D.J., Dorsch, R.G. and Lightner, C. (1966) Wave-plan analysis of unsteady flow in closed conduits, *Proc. ASCE, J. Hydraul. Div.*, **92**(HY2), pp. 83–110.

Wood, D.J. and Jones, S.E. (1972) Waterhammer charts for various types of valve, *Proc. ASCE, J. Hydraul. Div.*, HY1, **99**(HY1), pp. 167–178.

Wylie, E.B. and Streeter, V.L. (1983) *Fluid Transients*, FEB Press, Ann Arbor, Michigan.

Wylie, E.B. and Streeter, V.L. with Suo, L. (1993) *Fluid Transients in Systems*, Prentice-Hall.

Suggested Further Reading

Anderson, A.A. and Robbie, J.F. (1986) The behaviour of surge tanks with horizontal expansion galleries, *Proc. 5th Int. Conf. on Pressure Surges*, BHRA, Cranfield, Beds., UK, pp. 261–271.

Beetes, R. and Watts, J. (Eds) (1992) Unsteady flows and fluid transients, *Proc. Int. Conference on Unsteady Flows and Fluid Transients*, A.A.Balkema, Rotterdam.

Boldy, A.P. (Ed.) (1996) Pressure surges and fluid transients in pipelines and open channels, *Proc. 7th Int. Conf. on Pressure Surges*, BHR Group, Cranfield, Beds., UK.

Hancox, W.T., Ferch, R.L., Liu, W.S. and Nieman, R.E. (1980) One-dimensional models for transient gas–liquid flows in ducts, *Int. J. of Multiphase Flow*, **6**, 25–40.

Martin, C.S. and Padmanabhan, M. (1979) Pressure pulse propagation in two-component slug flow, *Trans. ASME, J. Fluids Engng*, **101**, 44–52.

Safety Assessment Federation (1997) *Potential hazards created by water hammer in steam systems*, Safety Assessment Federation, ISBN 1–901212–15–7.

Sharp, B.B. and Sharp, D.B. (1996) *Water Hammer – Practical Solutions*, Arnold.

SIGTTO (1987) Guidelines for the alleviation of excessive surge pressures on ESD, Society of International Gas Tanker and Terminal Operators, London.

Thorley, A.R.D. and Guymer, C. (1976) Pressure surge propagation in thick-walled conduits of rectangular cross section, *Trans. ASME, J. Fluid Engng*, **89**(3), 455–460.

Wiggert, D.C. and Sundquist, M.J. (1979) The effect of gaseous cavitation in fluid transients, *Trans. ASME, J. Fluid Engng*, **101**, 79–86.

Index

Printed and bound by CPI Group (UK) Ltd, Croydon, CR0 4YY

Printed and bound by CPI Group (UK) Ltd, Croydon, CR0 4YY

13/09/2026

14961034-0003